Statistics for Ecologists Using R and Excel

Data collection, exploration, analysis and presentation

Mark Gardener

DATA IN THE WILD SERIES

Pelagic Publishing | www.pelagicpublishing.com

Published by Pelagic Publishing
www.pelagicpublishing.com
PO Box 725, Exeter, EX1 9QU

Statistics for Ecologists Using R and Excel®
Data collection, exploration, analysis and presentation

ISBN 978-1-907807-12-1 (Pbk)
ISBN 978-1-907807-13-8 (Hbk)

British Library Cataloguing in Publication Data
A catalogue record for this book is available from the British Library.

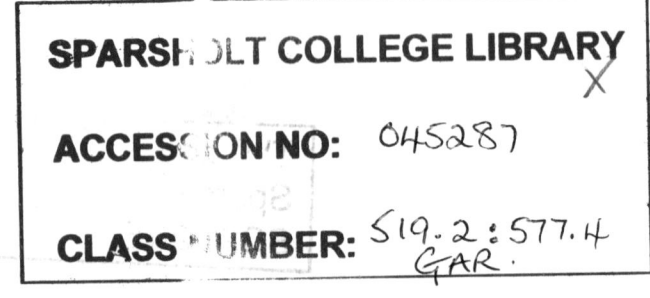

About the author

Mark began his career as an optician but returned to science and trained as an ecologist. His research is in the area of pollination ecology. He has worked extensively in the UK as well as Australia and the United States. Currently he works as an associate lecturer for the Open University and also runs courses in data analysis for ecology and environmental science.

Acknowledgements

I am especially grateful to Nigel Massen at Pelagic Publishing for his help and perseverance throughout the production of this book.

Thanks go to Anne Goodenough for patiently and thoroughly reviewing the manuscript, your comments and views were most helpful.

With a book of this nature data examples are always useful. Some of the data illustrated here were collected by students and I gratefully acknowledge their efforts and send thanks for allowing me to use these data as examples.

Finally my heartfelt thanks go to Christine, for putting up with me throughout the entire process.

Software used

Various versions of Microsoft's Excel® spreadsheet were used in the preparation of this manuscript. Most of the examples presented show version 2007 for Microsoft Windows® although other versions may also be illustrated (including Excel X for Apple Macintosh®).

Several versions of the R program were used and illustrated including 2.8.1. for Windows and 2.11.1 for Macintosh: R Foundation for Statistical Computing, Vienna, Austria. ISBN 3-900051-07-0, URL http://www.R-project.org/.

Downloading free code examples

Free code examples and further information from the author on using R and Excel for statistics can be found at:

http://www.pelagicpublishing.com/statistics-for-ecologists-resources.html

Reader feedback

We welcome feedback from readers – please email us at info@pelagicpublishing.com and tell us what you thought about this book. Please include the book title in the subject line of your email.

Publish with Pelagic Publishing

We publish scientific books to the highest editorial standards in all life science disciplines, with a particular focus on ecology, conservation and environment. Pelagic Publishing produces books that set new benchmarks, share advances in research methods and encourage and inform wildlife investigation for all.

If you are interested in publishing with Pelagic please contact editor@pelagicpublishing.com with a synopsis of your book, a brief history of your previous written work and a statement describing the impact you would like your book to have on readers.

Contents

Introduction

This is not just a statistics textbook! Although there are plenty of statistical analyses here, this book is about the processes involved in looking at data. These processes involve planning what you want to do, writing down what you found and writing up what your analyses showed. The statistics part is also in there of course but this is not a course in statistics. By the end I hope that you will have learnt some statistics but in a practical way, i.e. *what statistics can do for you*. In order to learn about the methods of analysis, we'll use two main tools: a Microsoft Excel spreadsheet (although Open Office will work just as well) and a computer program called R. The spreadsheet will allow you to collect your data in a sensible layout and also do some basic analyses (as well as a few less basic ones). The R program will do much of the detailed statistical work (although we will also use Excel quite a bit). Both programs will be used to produce graphs. This book is not a course in computer programming; we'll learn just enough about the programs to *get the job done*.

It is important to recognise that there is a process involved. This is the scientific process and may be summarised by four main headings:

- Planning
- Data recording
- Data exploration
- Reporting results

The book is arranged into these four broad categories. The sections are rather uneven in size and tend to focus on the analysis. The section on reporting also covers presentation of analyses (e.g. graphs).

Although the emphasis is on ecological work and many of the data examples are of that sort, I hope that other scientists and students of other disciplines will see relevance to what they do.

Mark Gardener 2011

1. Planning

1.1 The scientific method

Science is a way of looking at the natural world. In short, the process goes along the following lines:

- You have an idea about something.
- You come up with a hypothesis.
- You work out a way of testing this hypothesis/idea.
- You collect appropriate data in order to apply a test.
- You test the hypothesis and decide if the original idea is supported or rejected.
- If the hypothesis is rejected, then the original idea is modified to take the new findings into account.
- The process then repeats.

In this way, ideas are continually refined and our knowledge of the natural world is expanded. We can split the scientific process into four parts (more or less): planning, recording, analysing and reporting.

Table 1. Stages in the scientific method

Planning	Recording	Analysing	Reporting
This is the stage where you work out what you are going to do. Formulate your idea(s), undertake background research, decide what your hypothesis will be and determine a method of collecting the appropriate data and a means by which the hypothesis may be tested.	The means of data collection is determined at the planning stage although you may undertake a small pilot study to see if it works out. After the pilot stage you may return to the planning stage and refine the methodology. Data is finally collected and arranged in a manner that allows you to begin the analysis.	The method of analysis should have been determined at the planning stage. Analytical methods (often involving statistics) are used to test the null hypothesis. If the null hypothesis is rejected then this supports the original idea/ hypothesis.	Disseminating your work is vitally important. Your results need to be delivered in an appropriate manner so they can be understood by your peers (and often by the public). Part of the reporting process is to determine what the future direction needs to be.

1.1.1 Planning stage

This is the time to get the ideas. These may be based on previous research (by you or others), by observation or stem from previous data you have obtained. On the other hand, you might have been given a project by your professor, supervisor or teacher. If you are going to collect new data, then you will determine what data, how much data, when it will be collected, how it will be collected and how it will be analysed all at this planning stage. Looking at previous research is a useful start as it can tell you how other researchers went about things. If you already have old data from some historic source then you still need to plan what you are going to do with it. You may have to delve into the data to some extent to see what you have – do you have the appropriate data to answer the questions you want answered? It may be that you have to modify your ideas/questions in light of what you have.

1.1.2 Recording stage

Finally, you get to collect data. The planning step will have determined (possibly with the help of a pilot study) how the data will be collected and what you are going to do with it. The recording stage nevertheless is important because you need to ensure that at the end you have an accurate record of what was done and what data were collected. Furthermore, the data need to be arranged in an appropriate manner that promotes the analysis. It is often the case, especially with old data, that the researcher has to spend a lot of time rearranging numbers/data into a new configuration before anything can be done. Getting the data layout correct right at the start is therefore important.

1.1.3 Analysis stage

The means of undertaking your analysis should have been worked out at the planning stage. This is where you apply the statistics and data handling methods that make sense of the numbers collected. Helping to understand the data is vastly aided by the use of graphs. As part of the analysis, you will determine if your original hypothesis is supported or not.

1.1.4 Reporting stage

Of course there is some personal satisfaction in doing this work, but the bottom line is that you need to tell others what you did and what you found out. The means of reporting are varied and may be informal, as in a simple meeting between colleagues. Often the report is more formal, like a written report or paper or a presentation at a meeting. It is important that your findings are presented in such a way that your target audience understands what you did, what you found and what it means. In the context of conservation for example, your research may determine that the current management is working well and so nothing much needs to be done apart from monitoring. On the other hand, you may determine that the situation is not good and that intervention is needed. Making the results of your work understandable is a key skill and the use of graphs to illustrate your results is usually the best way to achieve this. Your audience is much more likely to dwell on a graph than a page of figures and text.

1.2 Types of experiment/project

As part of the planning process, you need to be aware of what you are trying to achieve. In general, there are three main types of research:

- *Differences*: you look to show that *a* is different to *b* and perhaps that *c* is different again. These kinds of situations are represented graphically using bar charts and box–whisker plots.
- *Correlations*: you are looking to find links between things. This might be that species *a* has increased in range over time or that the abundance of species *a* (or environmental factor *a*) affects the abundance of species *b*. These kinds of situations are represented graphically using scatter plots.
- *Associations*: similar to the above except that the type of data is a bit different, e.g. species *a* is always found growing in the same place as species *b*. These kinds of situations are represented graphically using pie charts and bar charts.

We are not really considering studies that concern whole communities of organisms here. The kinds of approach required for these analyses are dealt with in the companion volume to this work; however, knowledge of the basic statistical approaches dealt with in this volume underpins many community studies.

Once you know what you are aiming at, you can decide what sort of data to collect; this affects the analytical approach as we shall see later.

1.3 Getting data – using a spreadsheet

A spreadsheet is an invaluable tool in science and *Data Analysis*. Learning to use one is a good skill to acquire. With a spreadsheet you are able to manipulate data and summarise it in different ways quite easily. You can also prepare data for further analysis in other computer programs in a spreadsheet. It is important that you formalise the data into a standard format, as we shall see later (in Chapter 2). This will make the analysis run smoothly and allow others to follow what you have done. It also allows you to see what you did later on (it is easy to forget the details).

Your spreadsheet is useful as part of the planning process. You may need to look at old data; these might not be arranged in an appropriate fashion so using the spreadsheet will allow you to organise your data. The spreadsheet will allow you to perform some simple manipulations and run some straightforward analyses, looking at means for example, as well as producing simple summary graphs. This will help you to understand what data you have and what they might show. We will look at a variety of ways of manipulating data as we go along (see Section 3.2).

If you do not have past data and are starting from scratch, then your initial site visits and pilot studies will need to be dealt with. The spreadsheet should be the first thing you look to, as this will help you arrange your data into a format that facilitates further study. Once you have some initial data (be it old records or pilot data) you can continue with the planning process.

1.4 Hypothesis testing

A hypothesis is your idea of what you are trying to determine. Ideally it should relate to a single thing, so "Japanese knotweed and Himalayan balsam have increased their range in the UK over the past 10 years" makes a good overall aim, but is actually two hypotheses. We split up our ideas into parts, each of which can be tested separately:

"Japanese knotweed has increased its range in the UK over the past 10 years."

"Himalayan balsam has increased its range in the UK over the past 10 years."

We can think of hypothesis testing as being like a court of law. In law, you are presumed innocent until proven guilty; you don't have to prove your innocence.

In statistics, the equivalent is the *null hypothesis*. This is often written as H_0 (or H0) and you aim to reject your null hypothesis and therefore, by implication, accept the alternative (usually written as H_1 or H1).

The H_0 is not simply the opposite of what you thought (called the *alternative hypothesis*, H_1) but is written as such to imply that no difference exists, no pattern (I like to think of it as the *dull* hypothesis). For our ideas above we would get:

"There has been no change in the range of Japanese knotweed in the UK over the past 10 years."

"There has been no change in the range of Himalayan balsam in the UK over the past 10 years."

So, we do not say that the range of these species is shrinking, but that there is no change. Getting your hypotheses correct (and also the null hypotheses) is an important step in the planning process as it allows you to decide what data you will need to collect in order to reject the H_0. We will examine hypotheses in more detail later (Section 5.1).

Allied to our hypothesis is the analytical method we will use later to help test and support (or otherwise) our hypothesis. Even at this early stage we should have some idea of the statistical test we are going to apply. Certain statistical tests are suitable for certain kinds of data and we can therefore make some early decisions. We may alter our approach, change the method of analysis and even modify our hypothesis as we move through our planning. This is fine and all part of the scientific process. We will look at ways to choose which statistical test is right for our situation in Section 5.2, where we will see a decision flow-chart (Figure 77) and a key (Table 27) to help us. Before we get to that stage though, we will need to think a little more about the kind of data we may collect.

1.5 Data types

Once you have sorted out more or less what your hypotheses are, your next step in the planning process is to determine what sort of data you can get. It may be that you already have data from previous biological records for example. Knowing what sort of data you have will determine the sorts of analyses you are able to perform.

In general, we have three main types of data:

- *Interval*: these can be thought of as "real" numbers. We know the sizes of them and can do "proper" maths. Examples would be counts of invertebrates, percentage cover, leaf lengths, egg weights, clutch size.
- *Ordinal*: these are values that can be placed in order of size but that is pretty much all you can do. Examples would be abundance scales like DAFOR or Domin (named after a Czech botanist). We know that A is bigger than O but we cannot say that one is twice as big as the other (or be exact about the difference).
- *Categorical* (sometimes called *nominal* data): this is the tricky one because it can be confused with ordinal data. With categorical data you can only say that things are different. Examples would be flower colour, habitat type, sex.

With *interval* data for example, you might count something, keep counting and build up a sample. When you are finished, you can take your list and calculate an average, look to see how much larger the biggest value is from the smallest and so on. Put another way, we have a scale of measurement. This scale might be millimetres or grams or anything else. Whenever we measure something using this scale we can see how it fits into the scheme of things because the *interval* of our scale is fixed (10 mm is bigger than 5 mm, 4 g is less than 12 g). Compare this to the *ordinal* scales described below.

With *ordinal* data you might look at the abundance of a species in quadrats. It may be difficult or time consuming to be exact so you decide to use an abundance scale. The Domin scale shown in Table 2 for example converts percentage cover into a numerical value from 0–10.

Table 2. The Domin scale; an example of an ordinal abundance scale

Domin	% cover
10	91–100
9	75–90
8	51–74
7	34–50
6	26–33
5	11–25
4	4–10
3	<4 many (>10 individuals)
2	<4 some (4–10 individuals)
1	<4 few (1–3 individuals)

The Domin scale is generally used for looking at plant abundance and is used in the British National Vegetation Classification (NVC) classification system (and many other kinds of study). We can see by looking at Table 2 that the different classifications cover different ranges of abundance. For example, a Domin of 8 represents a range of values from about half to three-quarters coverage (51–74%). A value of 6 represents a range from about a quarter to a third coverage (26–33%). The first three divisions of the Domin scale all represent less than 4% coverage but relate to the number of individuals found. The Domin scale is useful because it allows us to collect data efficiently and still permits useful analysis. We know that 10 is a greater percentage coverage than 8 and that 8 is bigger than 6; it is just that the intervals between the divisions are unequal.

There are many other abundance scales, and various researchers have at times worked out useful ways to simplify the abundance of organisms. The DAFOR scale is a general phrase to describe abundance scales that convert abundance into a letter code. There are many examples Table 3 shows a generalised scale for vegetation analysis.

Table 3. An example of a generalised DAFOR scale for vegetation, an example of an ordinal abundance scale

Abundance	Scale	Description
Dominant	D	The dominant vegetation/species highly visible, usually more than 70% cover
Abundant	A	Many individuals or patches visible, usually 30–50% cover
Frequent	F	Several individuals or few patches, cover usually 10–20%
Occasional	O	A small patch or a few individuals, cover usually around 5–8%
Rare	R	Single very small patch or individual, cover usually around 1–3%

There are other letters that might be used to extend your scale. For example C for "common" might be inserted between A and F (ACFOR is a commonly used ordinal scale). You might add E and/or S for "extremely abundant" and "super abundant". You might also add N for "not found". The DAFOR type of scale can be used for any organism, not just for vegetation.

When you are finished, you can convert your DAFOR scale into numbers (ranks) and get an average, which can be converted to a DAFOR letter, but you cannot tell how much larger the biggest is from the smallest – the interval between the values is inexact.

Many of the abundance scales used are derived from the work of Josias Braun-Blanquet, the eminent Swiss botanist. Table 4 shows a basic example of a Braun-Blanquet scale for vegetation cover.

Table 4. The basic Braun-Blanquet scale, an ordinal abundance scale. There are many variations on this scale

Scale	Cover (%)
5	>75
4	51–75
3	26–50
2	5–25
1	1–5
+	<1

With *categorical* data it is useful to think of an example. You might go out and look to see what types of insect are visiting different colours of flower. Every time you spot an insect, you record its type (bee, fly, beetle) and the flower colour. At the end you could make a table with numbers of how many of each type visited each colour. You have numbers but each value is uniquely a combination of two categories.

Table 5. An example of categorical data. This type of table is also called a contingency table. The rows and columns are each sets of categories. Each cell of the table represents a unique combination of categories

	Bee	Butterfly	Beetle	Fly	Moth
Red	87	39	58	46	12
White	55	56	34	64	120
Blue	38	112	11	78	14
Yellow	59	65	23	56	45

Table 5 shows an example of categorical data laid out in what is called a contingency table. The rows are one category (colour) and the columns another category (type of insect).

1.6 Sampling effort

Sampling effort refers to the way you collect data and how much to collect. For example, you have decided that you need to determine the abundance of some plant species in meadows across lowland Britain. How many quadrats will you use? How large will the quadrats need to be? Do you need quadrats at all?

Sample is the term used to describe the set of data that you have. Because you generally cannot measure "everything", you will usually have a subset of stuff that you've measured (or weighed or counted). Let's think about a field of buttercups for a moment. You wish to know how many there are in the field, which is a hectare in size (i.e. 100 metres × 100 metres). You aren't really going to count them all (that would take too long) so you make up a square that has sides of 1 metre and count how many buttercups there are in that. Now you can estimate how many buttercups there are in the whole field. Your sample is 1/10,000th of the area, which is pretty small. The estimate is not likely to be very good (although by random chance it could be). It seems reasonable to count buttercups in a few more 1 m^2 areas. In this way your estimate is likely to get more "on target". Think of it this way: if you carried on and on and on, eventually you would have counted buttercups in every 1 m^2 of the field. Your estimate would now be spot on because you would have counted everything. So as you collect more and more data, your estimate of the true number of buttercups will likely become more and more like the true number.

The problem is, how many times will you have to count the 1 m^2 areas in order to get a good estimate of the true number? We will return to this a little later. Another problem – where do you put your 1 m^2 areas? Will it make a difference? Is a 1 m^2 quadrat the right size? We will look at these themes now.

1.6.1 Quadrat size

If you are doing a British NVC survey, then the size and number of quadrats is predetermined; the NVC methodology is standardised. Similarly if you are making bird species lists for different sites, the methodology already exists for you to follow. Don't re-invent the wheel!

Whenever you collect data, you cannot of course measure everything. You take a sample, essentially a representative subset of the whole. What you are aiming for is to make your sample as representative as possible. If, for example, you were counting the frequency of spider orchids across a site, you would aim to make your quadrat a reasonable size and in-line with the size and distribution of the organism – you would not have the same size quadrat to look at oak trees as you would to look at lichens.

1.6.2 Species area rule

If you are looking at communities, then the wider the area you cover the more species you will find. Imagine you start off with a tiny quadrat, you might find a few species. Make the quadrat double the size and you will find more. Keep doubling the quadrat and you will keep finding more species. If, however, you draw a graph of the cumulative number of species, you will see it start to level off and eventually you won't find any more species. Even well before this, the number of new species is so small that it is not worth the extra effort of the larger quadrat. This idea is called the species area rule.

We can extend the same idea to kick-sampling. We use a standard net for freshwater invertebrate sampling but can vary the time we spend sampling. This is akin to using a bigger quadrat. The longer you sample, the more you get. You can easily see that it is not worth spending 20 minutes to get the 101st species when during 3 minutes you net the first 100.

1.6.3 How many replicates?

When we go out to collect our data, how much work do we have to do? If we are counting the abundance of a plant in a field and are unlikely to count every plant, we take a sample. The idea of sampling is to be representative of the whole without having the bother of counting everything. Indeed attempting to count everything is often difficult, time consuming and expensive.

As we shall see later when we deal with individual statistical tests, there are certain minimum amounts of data that need to be collected. Now, we should not aim to collect just the minimum that will allow a result to be calculated, but aim to be *representative*. If we are sampling a field, we might try to sample 5–10% of the area; however, even that might be a huge undertaking. We should estimate how long it is likely to take us to collect various amounts of data. A short pilot study or personal experience can help with this.

Whenever we sample something from a larger population we are aiming to gain insights into that larger population from our smaller sample. We are going to work out an average of some sort; this might be average abundance, size, weight or something else. We will cover different averages later on (Section 4.1.2). We can use something called a *running mean* to help us determine if we are reaching a good representation of the population

(Section 4.7). In brief, what we do is to take each successive number from a quadrat or net and work out the average. Each time we get a new value we can work out a new average. We can then plot these values on a simple graph. When we have only a few values, the running mean is likely to "wobble" quite a bit. After we collect more data however, the average is likely to settle down. Once our running mean reaches this point, we can see that we've probably collected enough data.

1.6.4 Sampling method

We need to think how we are going to select the things we want to measure. In other words we need a sampling strategy.

Remember the field of buttercups? We can see that it is good to have a lot of data items (a large sample) in terms of getting close to the true mean, but exactly where do we put our sample squares (called quadrats: they do not really need to be square but it is convenient) in order to count the buttercups? Does it even matter? It matters of course because we need our sample to be *representative* of the larger population. We want to eliminate *bias* as far as possible. If we placed our quadrat in the buttercup field we might be tempted to look for patches of buttercups. On the other hand, we may wish to minimise our counting effort and look for areas of few buttercups! Both would introduce bias into our methods.

What we need is a sampling strategy that eliminates bias:

- Random
- Systematic
- Mixed
- Haphazard

Random sampling

In a random sampling method, we use pre-determined locations to carry out our sampling. If we were looking at plants in a field for example, we could measure the field and use random numbers to generate a series of x, y co-ordinates. We then place our quadrats at these co-ordinates. This works nicely if our field is square. If our field is not square we can measure a large rectangle that covers the majority of the field and ignore co-ordinates that fall outside the rectangle. For other situations we can work out a method that provides co-ordinates to place our quadrats. Basically, the locations are pre-determined before we start.

In theory, every point within our area should have an equal chance of being selected and our method of creating random positions should reflect this. What happens if we get the same location twice (or more)? There are two options: either we skip it and make a new pair of co-ordinates or we sample the same area twice. The first option is called *random sampling with replacement* and as long as you determine that that is what you are going to do before you start, then there is probably no problem. If you decide that you are not going to skip duplicates then you use the co-ordinates twice. In practice, this means that you use the same data and record it both times. It is important that you do not ignore it. If you have ten co-ordinates, which includes a duplication, then you will still need to get ten values when you have finished. Obviously you do not need to place the quadrat a second time and count the buttercups again, you simply copy the data.

Random sampling is good for situations where there is no detectable pattern. For example, if we were sampling in mediaeval fields we might have a ridge and furrow system. The old methods of ploughing the field create high and low points at regular intervals. These ridges and furrows may affect the growth of the plants (we assume the ridges are drier and the furrows wetter for instance). If we sampled randomly, we may well get a lot more data from ridges than from furrows. Consequently, we are introducing unwanted bias.

In other cases we may be deliberately looking at a situation where there is an environmental gradient of some sort. For example, this might be a slope where we suspect that the top is drier than the bottom. If we sample randomly then we may once again get bias data because we sampled predominantly in the wetter end of the field (or the drier end). We need to alter our sampling strategy to take into account the situation.

Systematic sampling

In some cases we are deliberately targeting an area where an environmental gradient exists. What we want is to get data from right across this gradient so that we get samples from all parts. Random sampling would not be a good idea so we use a set system to ensure that we cover the entire gradient.

Systematic sampling often involves *transects*. A transect is simply the term used to describe a slice across something. For example, we might wish to look at the abundance of seaweed across a beach. The further up the beach, the drier it gets because of the tide so what we do is to create a transect that goes from the top of the beach (high water) to the bottom of the beach (low water). In this way we cover the full range of the gradient from very dry (only covered by water at high tides) to very wet (in the sea). In the case of our ridge and furrow system, we create a transect that goes across the features and crosses many ridges and furrows (i.e. it is at right angles to the ridge/furrow system). When we sample this transect, we collect data from both ridges and furrows.

With our beach scenario, we have various options regarding how we place our quadrats. We can start at one end and keep placing quadrats end-to-end and cover the entire beach. This would be called a *belt transect* because we have covered the entire transect. It is more likely that we would split the transect into intervals, which we sample at various distances. This is called an *interrupted belt transect*. It is important to determine what you are actually measuring before you start: in the beach case we would probably work out the height above low water rather than an exact distance along the transect; for the ridge and furrow example, we would determine how far apart the ridges and furrows were (assuming this to be a regular interval). We would then set the distance between *sampling stations* so that we had equal numbers of samples from both ridges and furrows.

One transect might not be enough because we may miss a wider pattern (Figure 1). We ought to place several transects and combine the data from them all. In this way we are covering a wider part of the habitat and being more representative of the whole, which is the point.

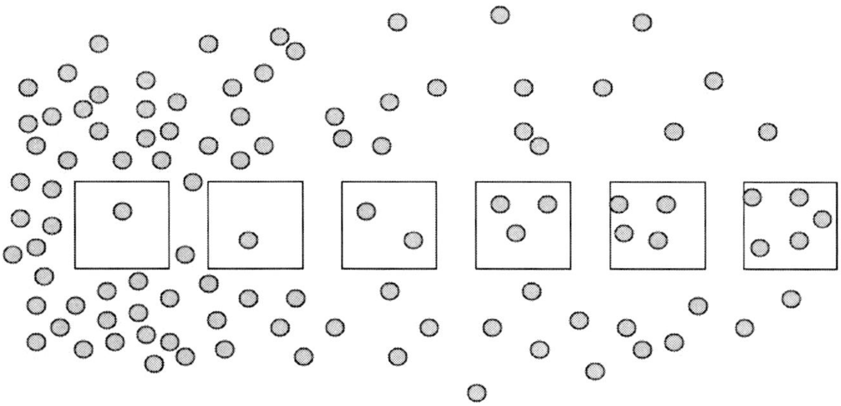

Figure 1. The transect does not show the true pattern here and we should use a wider one (or several side by side) to be more representative

We also need to determine how long the transect should be. We might, for example, be looking at a change in abundance of a plant species along a transect, which may relate to an environmental factor. We need to make sure that we make the transect long enough to cover the change in abundance but not so long that we over-run too far.

Mixed sampling

There are occasions where we may wish to use a combination of systematic and random sampling. In essence, what we do is to set up several transects and sample at random intervals along them. Think for example of a field where we wish to determine the height of some plant species. We could set up random co-ordinates but once we get to each co-ordinate how do we select the plant to measure? One option would be to measure the height of the plant nearest the top left corner. Each quadrat is placed randomly but we have a system to pick which plant to measure. We've eliminated bias because we determined this strategy before we started. Another option would be to place transects (a simple piece of string would do; the transect would then be a *line transect*) at intervals across the field. We then measure plants that touch the string (transect) or the nearest to the string at some random distance along. There are many options of course and you must decide what seems best at the time. The point is that you are trying to eliminate bias and get the most representative sample you can.

Another example might be in sampling for freshwater invertebrates in a stream. We decide that we wish to look for differences between fast-running riffles and slow-moving pools. We need some systematic approach to get a balance between riffles and pools. On the other hand, we do not want to pick the "most likely" locations; we need an element of randomness. We might identify each pool and riffle and assign a number to each one, which we then select at random for sampling. Again the idea is to eliminate any element of bias.

Haphazard sampling

There are times when it is not easy to create an area for co-ordinate sampling, for example, you may be examining leaves on various trees or shrubs that have a "dark" and a "light" side. It might be quite difficult to come up with a quadrat that balances in the foliage and you might attempt to grab leaves at random. Of course we can never be really random. In this case we say that the leaves were collected *haphazardly*. To further eliminate bias, we might grab branches haphazardly and then select the leaf nearest the end.

Whenever we get a situation where we stray from either a set system or truly random, we describe our collection method as haphazard.

1.6.5 Precision and accuracy

Whenever we measure something we use some appropriate device. For example, if we were looking at the size of water beetles in a pond we would use some kind of ruler. When we record our measurement, we are saying something about "how good" our recording device is. We might record beetles sizes as 2 cm, 2.3 cm or perhaps 2.36 cm. In the first instance we are implying that our ruler only measures to the nearest centimetre. In the second case we are saying that we can measure to the nearest millimetre. In the third case we are saying that our ruler can measure to 1/10th of a millimetre. If we were to write the first measurement as 2.0 cm then we would be saying that our beetle was between 1.9 and 2.1 cm.

What we are doing by recording our results in this way is setting the level of *precision*. If we used a different ruler we might get a slightly different result, for example, we measure a beetle with two rulers and get 2.36 and 2.38 cm. The level of precision is the same in both cases (0.01 cm) but they cannot both be correct (the problem may lie with the ruler or the operator). Imagine that the real size of our beetle is 2.35 cm. The first ruler is more accurate than the second ruler.

So *precision* is how fine the divisions on your scale of measurement are. *Accuracy* is how close to the real result your measurement actually is. Ideally you should select a level of precision that matches the equipment you have and in the scale of the thing you are measuring. It seems a little pointless to measure the weight of an elephant to the nearest gram for example.

1.7 Tools of the trade

Learning to use your spreadsheet is time well spent. It is important that you can manipulate data and produce summaries, including graphs. We will cover a variety of aspects of data manipulation as well as the production of graphs later. Many statistical tests can be performed using a spreadsheet but there comes a point when it is better to use a dedicated computer program for the job. There are many on the market, some are cheap (or even free) and others are expensive. Some programs will interface with your spreadsheet and others are totally separate. Some programs are specific to certain types of analysis and others are more general.

Here we will focus on two programs. The spreadsheet we will use is Microsoft Excel. This is common and widely available. There are alternatives and indeed the Open Office spreadsheet uses the same set of formulae and can be regarded as equivalent. The analytical program we will use is called R; this is described first.

1.8 The R program

The program called R is a powerful environment for statistical computing. It is available free at the Comprehensive R Archive Network (CRAN) on the Internet. It is open source and available for all major operating systems.

R was developed from a commercial programming language called S. The original authors were called Robert and Ross so they called their program R as a sort of joke. This is what the R website says about the program:

> *R is an open-source (GPL) statistical environment modeled after S and S-Plus. The S language was developed in the late 1980s at AT&T labs. The R project was started by Robert Gentleman and Ross Ihaka (hence the name R) of the Statistics Department of the University of Auckland in 1995. It has quickly gained a widespread audience. It is currently maintained by the R core-development team, a hard-working, international team of volunteer developers. The R project web page is the main site for information on R. At this site are directions for obtaining the software, accompanying packages and other sources of documentation.*
>
> *R is a powerful statistical program but it is first and foremost a programming language. Many routines have been written for R by people all over the world and made freely available from the R project website as "packages". However, the basic installation (for Linux, Windows or Mac) contains a powerful set of tools for most purposes.*

Because R is a programming language it can seem a bit daunting; you have to type in commands to get it to work; however, it does have a Graphical User Interface (GUI) to make things easier and it is not so different from typing formulae into Excel. You can also copy and paste text from other applications (e.g. word processors). So if you have a library of these commands, it is easy to pop in the ones you need for the task at hand.

R will cope with a huge variety of analyses and someone will have written a routine to perform nearly any type of calculation. R comes with a powerful set of routines built in at the start but there are some useful extra "packages" available on the website. These include routines for more specialised analyses covering many aspects of scientific research as well as other fields (e.g. economics).

There are many advantages in using R:

- It is free, always a consideration.
- It is open source; this means that many bugs are ironed out.
- It is extremely powerful and will handle very complex analyses as easily as simple ones.
- It will handle a wide variety of analyses. This is one of the most important features: you only need to know how to use R and you can do more or less any type of analysis; there is no need to learn several different (and expensive) programs.
- It uses simple text commands. At first this seems hard but it is actually quite easy. The upshot is that you can build up a library of commands and copy/paste them when you need them.

- Documentation. There is a wealth of help for R. The CRAN site itself hosts a lot of material but there are also other websites that provide examples and documentation. Simply adding CRAN to a web search command will bring up plenty of options.

1.8.1 Getting R

Getting R is easy via the Internet. The R Project website is a vast enterprise and has local mirror sites in many countries. The first step is to visit the main R Project webpage (http://www.r-project.org) where you can select the most local site to you (this speeds up the download process a bit).

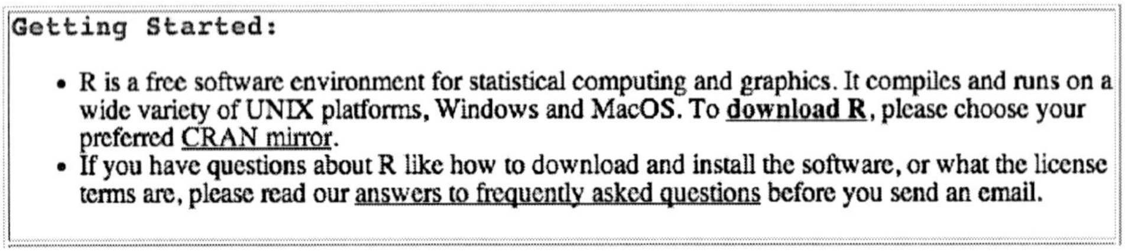

Getting Started:
R is a free software environment for statistical computing and graphics. It compiles and runs on a wide variety of UNIX platforms, Windows and MacOS. To download R, please choose your preferred CRAN mirror.
If you have questions about R like how to download and install the software, or what the license terms are, please read our answers to frequently asked questions before you send an email.

Figure 2. Getting R from the R Project website. Click the download link and select the nearest mirror site

Once you have clicked the download link, you have the chance to select a mirror site. These mirrors sites are hosted in servers across the world and using a local one will generally result in a speedier download.

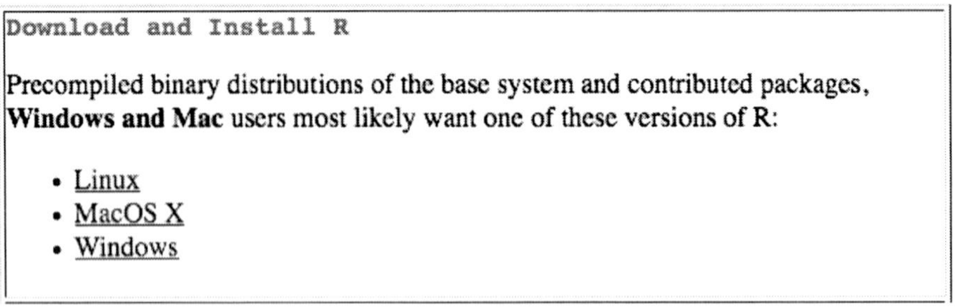

Download and Install R
Precompiled binary distributions of the base system and contributed packages, Windows and Mac users most likely want one of these versions of R:
Linux
MacOS X
Windows

Figure 3. Getting R from the R Project website. Once you have selected the mirror site for your location you can choose the file to download

Once you have selected a mirror site, you can click the link that relates to your operating system. If you use a Mac then you will go to a page where you can select the best option for you (there are versions for various flavours of OSX). If you use Windows then you will go to a Windows-specific page. If you are a Linux user then read the documentation; you generally install R through the terminal rather than via the web page.

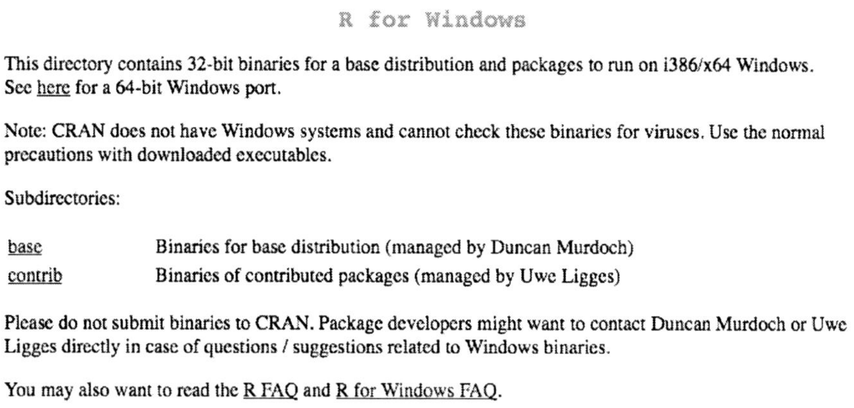

R for Windows
This directory contains 32-bit binaries for a base distribution and packages to run on i386/x64 Windows.
See here for a 64-bit Windows port.
Note: CRAN does not have Windows systems and cannot check these binaries for viruses. Use the normal precautions with downloaded executables.
Subdirectories:
base Binaries for base distribution (managed by Duncan Murdoch)
contrib Binaries of contributed packages (managed by Uwe Ligges)
Please do not submit binaries to CRAN. Package developers might want to contact Duncan Murdoch or Uwe Ligges directly in case of questions / suggestions related to Windows binaries.
You may also want to read the R FAQ and R for Windows FAQ.

Figure 4. Getting R from the R Project website. The Windows-specific page allows you to get the version that is right for your Windows OS

Once you have navigated to the Windows page, you will see something similar to Figure 4. Most users will want to select the base link, which will take you to a page where you can (finally) get the latest version of the installer file.

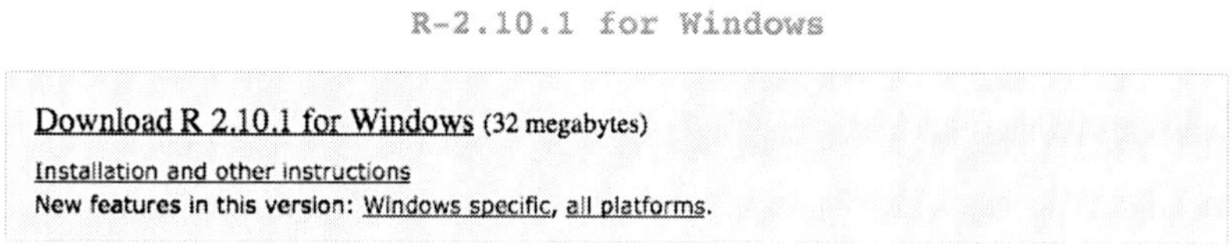

Figure 5. Getting R from the R Project website. The final link will download the latest version (the one shown was current as of March 2010)

Now the final step is to click the link and download the installer file. This is an EXE file and it will download in the usual manner according to the setup of your computer.

1.8.2 Installing R

Once you have downloaded the install file, you need to run it to get R onto your computer. If you use a Mac you need to double-click the disk image file to mount the virtual disk. Then double-click the package file to install. If you use Windows then you need to find the EXE file and run it. If you use Vista or later then it is a good idea to right-click the file and run as administrator.

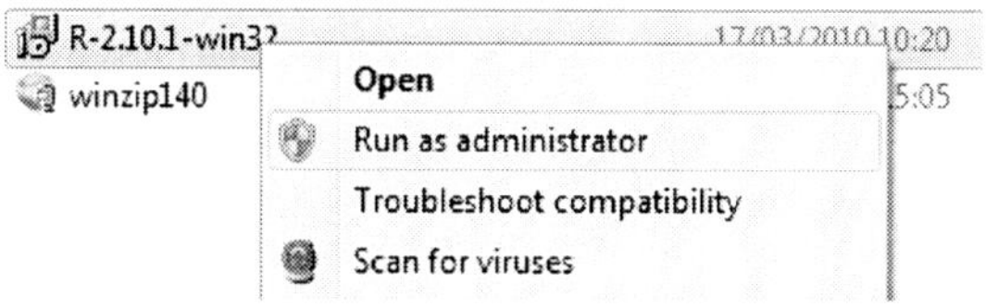

Figure 6. Installing R. If you have Windows Vista or later it is a good idea to right-click the install file and run as administrator

The installation process asks a few basic questions, allowing you to select a language other than English for example. It is usual to accept the default location for the R files (a directory called R). There are a few additional files that you can install. The PDF manuals and the HTML files are both useful. In Figure 7 we have also selected Tcl/Tk files. These are not strictly necessary but can be useful and do not take up too much room.

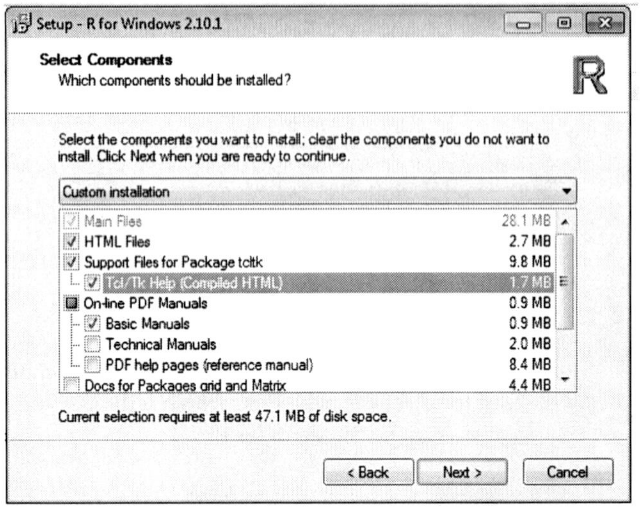

Figure 7. Installing R on Windows. Apart from the basic files, there are a few additional options

The next screen asks if you wish to use customised startup options. In most cases for installing programs you are strongly suggested to say "no" and to accept the defaults; however, I suggest that with R you select "yes" and modify the startup slightly from the default.

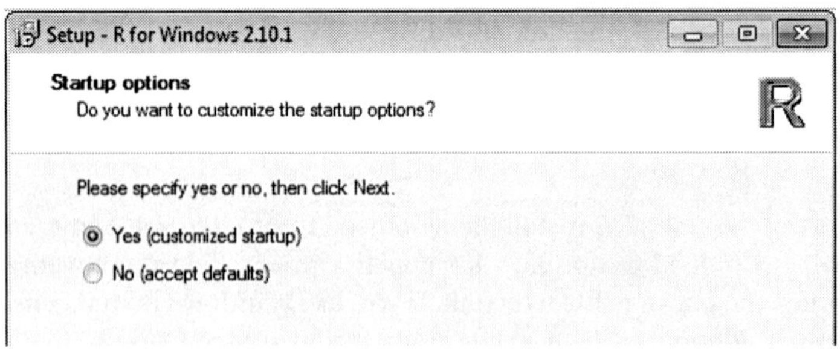

Figure 8. Installing R on Windows. Customised setup is a good idea as we can select better options

R is working on Windows and by default it will take over the whole screen (like a normal program); however, we have the option of setting R to work with separate small windows. An advantage of this is that it makes it easier to have other programs running alongside. This is helpful as we can copy and paste data and commands from other programs (typically a spreadsheet or notepad). Figure 9 shows the options as MDI or SDI.

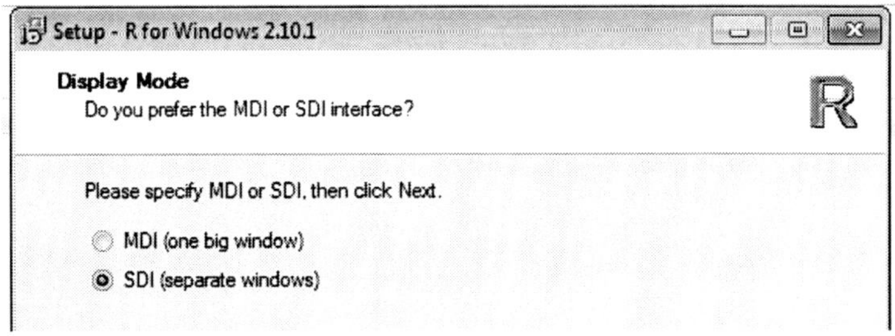

Figure 9. Installing R on Windows. Separate windows are very useful rather than one big one

Setting the option to SDI uses separate windows. This is more space-efficient and allows us to see other program windows whilst running R. Next we are given a few more choices. You may wish to have a quick-launch icon and desktop icon. I personally prefer not to (there are always too many icons on the desktop) and Figure 10 shows the options I would use.

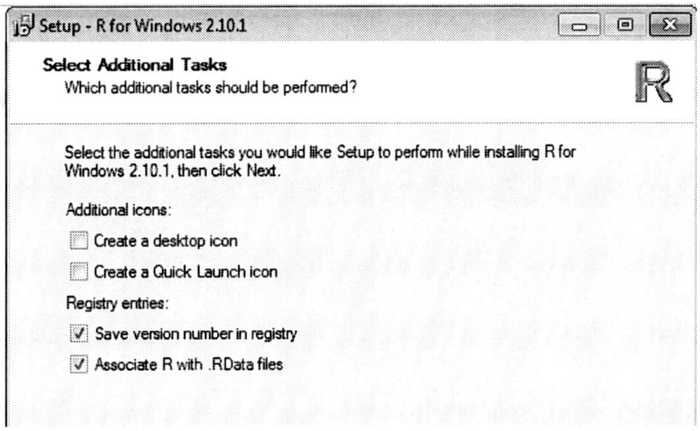

Figure 10. Installing R on Windows. Options for additional icons and registry entries

The two options related to registry entries and it seems sensible to select them both. The installer will copy the relevant files and you will soon be ready to run R.

1.8.3 Running R on Windows Vista or later

We will learn how to use R in Section 3.1 but first it seems an appropriate time to mention a few things specific to later versions of Windows. The R program holds data in memory as you work. Once you are done and exit the program, you are given the option to save the workspace. We will cover this later in Section 3.1.7 but the important point to note is that R will save everything we have done so far and have it waiting for us when we re-open the program. This is very handy but later versions of Windows are more careful about allowing things to be saved to the computer. The R program needs to be given special permission.

From the start button and programs menu you need to find the R program. Figure 11 shows what this looks like on Windows 7.

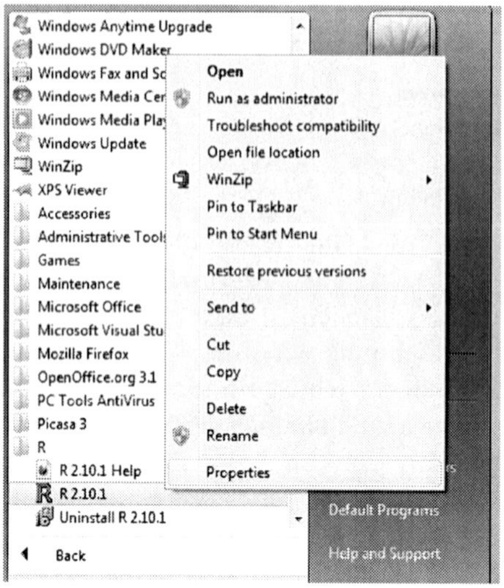

Figure 11. Installing R on Windows. On Windows Vista or Windows 7 (as here) the main program needs to be told to run as administrator. This can be set permanently by selecting the properties menu

Once you have found the R program you need to right-click and a new menu pops up. You could select *Run as administrator* but this would have to be done every time. It is better to select *Properties*, from the bottom of the menu and then we can set this permanently (it could be changed back if you wanted).

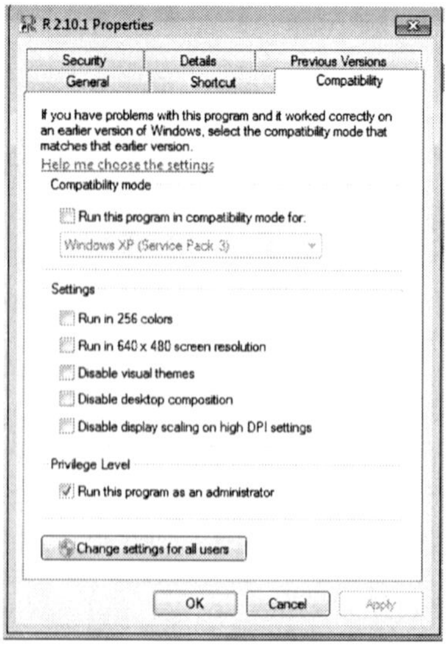

Figure 12. Installing R on Windows Vista/7. The R program needs to be altered to run as administrator

After you hit the *Properties* menu, a new window appears. Click the *Compatibility* tab and you will see something like Figure 12. Check the box that says *Run this program as administrator* and then OK. R will run fine under Vista or Windows 7. You do not need to select compatibility to use R on Windows XP.

Now R is ready to work for you. If you added a desktop icon or a quick launch button then you do not need to do anything else. These are shortcuts that point to the original program file. Since we altered the original program file to run as administrator we should have no trouble.

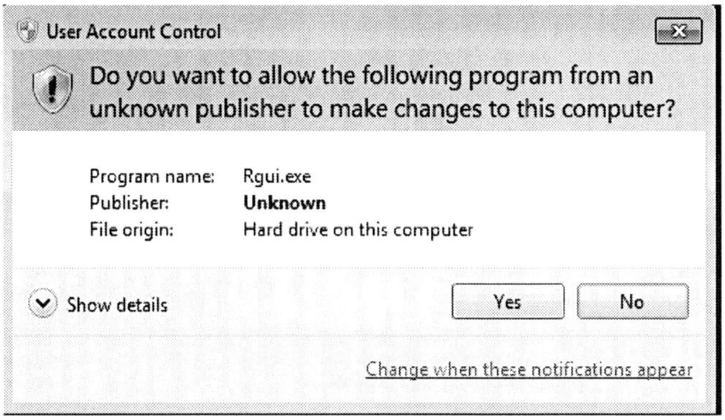

Figure 13. Installing R on Windows. In spite of altering R to run as administrator you may still get the User Account Control window

When you run R in Windows 7 you may still get the User Account Control window appearing (as shown in Figure 13); presumably since you clicked the program to get it started you will want to respond "yes" at this point.

1.9 Excel

There are many versions of Excel and your computer may already have a version installed when you purchased it. The basic functions that Excel uses have not changed for quite some while so even if your version is older than we describe here, you should be able to carry out the same manipulations. We will mainly describe the use of Excel 2007 for Windows. If you have purchased a copy of Excel (possibly as part of the Office suite) then you can install this following the instructions that came with your software. Generally, the defaults that come with the installation are fine although it can be useful to add extra options, especially the *Analysis ToolPak*, which we will describe below.

1.9.1 Installing the Analysis ToolPak

The *Analysis ToolPak* is an add-in for Excel that allows various statistical analyses to be carried out without the need to use complicated formulae. The add-in is not installed as standard and you will need to setup the tool before you can use it. The add-ins are generally ready for installation once Excel is installed and you usually do not require the original disk.

In order to install the *Analysis ToolPak* (or any other add-in) we need to click the Office button (at the top left of the screen) and select *Excel Options*.

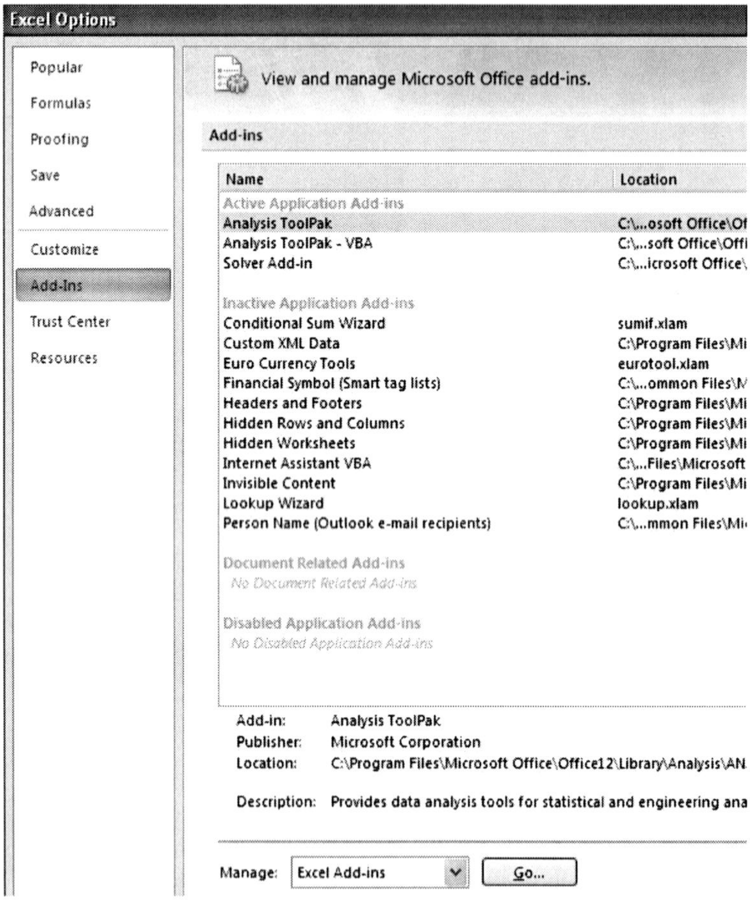

Figure 14. Selecting Excel add-ins from the *Options* menu

In Figure 14 we can see that there are several add-ins already active and some not yet ready. To activate (i.e. install) the add-in, we click the *Go* button at the bottom of the screen. We then select which add-ins we wish to activate (Figure 15).

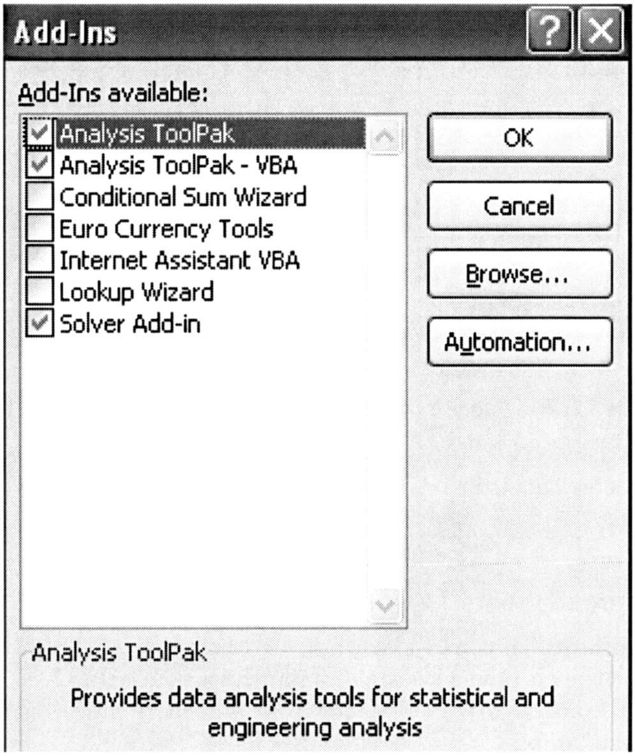

Figure 15. Selecting the add-ins for Excel

Once we have selected the add-ins to activate, we click the *OK* button to proceed. The add-ins are usually available to use immediately after this process.

To use the *Analysis ToolPak* we use the *Data* button on the ribbon and select the *Data Analysis* button (Figure 16).

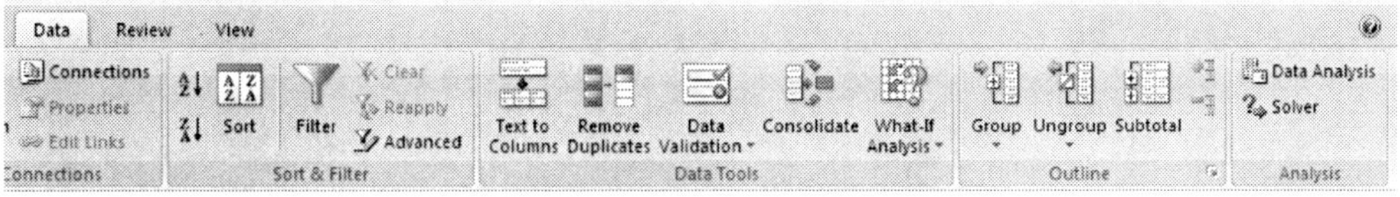

Figure 16. The *Analysis ToolPak* is available from the *Data Analysis* button on the Excel *Data* ribbon

Once we have selected this, we are presented with various analysis tools (Figure 17).

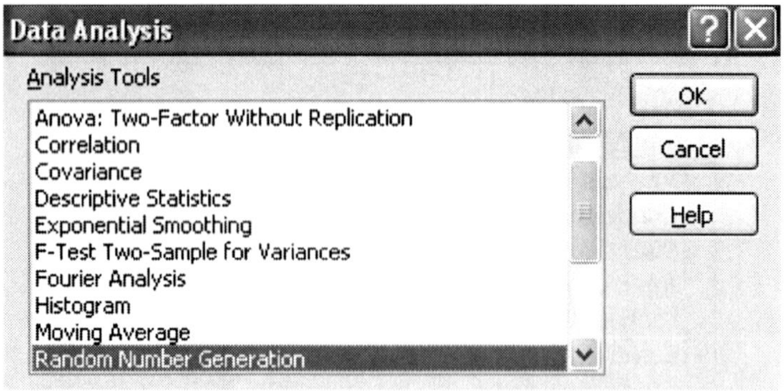

Figure 17. The *Analysis ToolPak* provides a range of analytical tools

Each tool requires the data to be set out in a particular manner; help is available using the *Help* button.

1.9.2 Other spreadsheets

The Excel spreadsheet that comes as part of the Microsoft Office suite is not the only spreadsheet and there are others available; of particular note is the Open Office program. This is available from http://www.openoffice.org and there are versions available for Windows, Mac and Linux.

Other spreadsheets generally use the same functions as Excel, so it is possible to use another program to produce the same result. Graphics will almost certainly be produced in a different manner and we will demonstrate graphics with Excel 2007 for Windows in Section 12.4.3.

2. Data recording

The data you write down is of fundamental importance to your ability to make sense of it at a later stage. If you are collecting new data then you are able to work out the recording of the data as part of your initial planning. If you have past data then you may have to spend some time re-arranging before you can do anything useful.

2.1 Collecting data – who, what, where, when

It is easy to write down a string of numbers in a notebook. You might even be able to do a variety of analyses on the spot; however, if you simply record a string of numbers and nothing else you will soon forget what the numbers represented. Worse still, nobody else will have a clue what the numbers mean and your carefully collected data will become useless.

All recorded data needs to conform to certain standards in order to be useful at a later stage. The minimum you ought to record is:

- *Who*: the name of the person that recorded the data.
- *What*: the species you are dealing with.
- *Where*: the location that the data were collected from.
- *When*: the date that the data were recorded.

There are other items that may be added, depending upon your purpose, as we shall see later.

2.1.1 Biological data and science

Your data are important. In fact they are the most important part of your research. It is therefore essential that you record and store your data in a format that can be used in the future. There are some elements of your data that may not seem immediately important but which nevertheless are essential if future researchers need to make sense of them.

We need to write down our data in a way that makes sense to us at the time and also will make sense to future scientists looking to repeat or verify our work. Table 6 shows some biological data in an appropriate format. Not all the data are shown here (the table would be too big).

Table 6. An example of biological data: bat species numbers at various sites around Milton Keynes (only part of the data are shown)

Species	Recorder	Date	Site	GR	Quantity
Pipistrellus pipistrellus	Atherton, M	5-Aug-06	Ouzel Valley	SP880375	21
Myotis daubentonii	Atherton, M	5-Aug-06	Ouzel Valley	SP880375	23
Nyctalus noctula	Atherton, M	5-Aug-06	Ouzel Valley	SP880375	26
Plecotus auritus	Atherton, M	5-Aug-06	Ouzel Valley	SP880375	54
Myotis natteri	Atherton, M	5-Aug-06	Ouzel Valley	SP880375	54
Pipistrellus pipistrellus	Atherton, M	5-Aug-06	Newport	SP874445	43
Myotis daubentonii	Atherton, M	5-Aug-06	Newport	SP874445	11
Nyctalus noctula	Atherton, M	5-Aug-06	Newport	SP874445	9

Every record (a row in our table) always has who, what, where and when. This is really important for several reasons:

- It allows the data to be used for multiple purposes.
- It ensures that the data you collect can be checked for accuracy.
- It means that you won't forget some important aspect of the data.
- It allows someone else to repeat the exercise exactly.

In the example above, you can see that someone (M Atherton) is trying to ascertain the abundance of various species of bat at sites around Milton Keynes in the UK. It would be easy for him to forget the date because it doesn't seem to matter that much. But if someone tries to repeat his experiment, they need to know what time of year he was surveying at. Alternatively, if environmental conditions change, it will be essential to know what year he did the work.

If you fail to collect complete biological data, or fail to retain and communicate all the details in full, then your work may be rendered unrepeatable and therefore useless as a contribution to science.

Once your biological data are compiled in this format, you can sort them by the various columns, export the grid references to mapping programs, and convert the data into tables for further calculations using a spreadsheet. They can also be imported into databases and other computer programs for statistical analysis.

Data collection in the field

When you are in the field and using your field notebook, you may well use shortcuts to record the information required. There seems little point in writing the site name and grid reference more than once for example. You may decide to use separate recording sheets to write down the information. These can be prepared in advance and printed as required. Once again there will be items that do not need to be repeated, a single date at the top of every sheet would be sufficient for example, however, when you transfer the data onto a computer it is a simple matter to copy the date or your name in a column.

In general, we aim to create a column for each item of data that we collect. If we were looking at species abundance at several sites for example, then we would need at least two columns, one for the abundance data and one for the site. In our field notebook or recording sheet we may keep separate pages for each site and end up with a column of figures for each site. When we return to base and transfer the data to the spreadsheet, we should write our data in the "standard format", i.e. one column for each thing (as in Table 6).

Supporting information

As part of our planning process (including maybe a pilot study), we should decide what data we are going to collect. Just because we *can* collect information on 25 different environmental variables does not mean that we should. The date, location and the name of the person collecting the data are basic items that we always need but there may also be additional information that will help us to understand the biological situation as we process the data later. These things include field sketches and site photographs.

A field sketch can be very helpful because you can record details that may be hard to represent in any other manner. A sketch can also help you to remember where you placed your quadrats; a grid reference is fine but meaningless without a map! Photographs may also be helpful and digital photography enables lots of images to be captured with minimum fuss; however, it is also easy to get carried away and forget what you were there for in the first place. Any supporting information should be just that – support for the main event: your data.

2.2 How to arrange data

As in the example in Table 6, it is important to have data arranged in appropriate format. When we enter data into our spreadsheet we ought to start with a few basics these correspond to the: *who, what, where, when*. There are extra items that *may* be entered depending on the level of study. These will pretty much correspond to your needs and the level of detail required. If you are collecting data for analysis then it is also important to set out your data in a similar fashion. This makes manipulating the data more straightforward and also maintains the multi-purpose nature of your work. You need to move from planning to recording and on to analysis in a seamless fashion. Having your data organised is really important!

When you collect biological data, enter each record on a separate line and set out your spreadsheet so that each column represents a factor.

Table 7. Data table layout (only part of the data are shown). Showing the most important elements in biological records

Who	Where	When	What	How many	Other (site)
MG	SJ4314	14-Aug-06	*U. dioica*	12	Pond
MG	SJ4314	14-Aug-06	*U. dioica*	8	Pond
MG	SJ4314	14-Aug-06	*U. dioica*	7	Pond
MG	SJ4314	14-Aug-06	*U. dioica*	32	Wood

Table 7 shows some data. Since the contents of the first four columns are identical, you might consider leaving them out and making an entry at the top (you would certainly do this in your field notebook or recording sheet). This is certainly one option but it is just as easy to fill out the top row and copy the entries down the remaining cells. In this case we are examining the abundance of nettles (*Urtica dioica*) at two sites near Preston Montford in Shropshire. We want to see if there are differences between two sites. It would seem easier to simply make two columns, one for the quadrats at the pond site and one for the wood. This is certainly an option but if we stick to this layout we can easily reproduce the two columns anytime we wish (using various tools in the spreadsheet). In fact many analytical programs prefer the layout as it is with the data in one column and the explanatory variable/factor in another. Here is another example.

When we begin to collect our data, we have a variety of ways of writing the results down. For example, imagine that we are looking at the abundance of a plant species in quadrats at two different sites. Our natural instinct would be to write down the abundance of the plant in two columns, one of reach site. If we examined other sites we would create extra columns. Table 8 shows the data written in this manner.

However, we might also write all the abundance data down in a single column, and then use a second column to tell us which site the data were recorded from. Table 9 shows this layout.

Table 8. Data table layout. Abundance of *Ranunculus acris* at two sites in Buckinghamshire. Here the abundance is shown in two columns, one for each site

Upper	Lower
2	4
3	6
4	7
3	8
4	7
6	6
5	5
3	4
4	2

Table 9. Data table layout. Abundance of *R. acris* at two sites in Buckinghamshire. Here the data are shown in two columns, one for the site and one for the abundance

Site	Count
Upper	2
Upper	3
Upper	4
Upper	3
Upper	4
Upper	6
Upper	5
Upper	3
Upper	4
Lower	4
Lower	6
Lower	7
Lower	8
Lower	7
Lower	6
Lower	5
Lower	4
Lower	2

The first layout (Table 8) is perhaps the manner in which we would naturally want to write down our results in our field notebook. For simple projects this is perfectly acceptable. However, when we have a lot of data or more complex situations the second format (Table 9) is a better choice. Many computer programs prefer the data to be in this layout and for complex statistical analyses it is essential.

For example, Table 10 shows a small part of a complex dataset. Here we have recorded the abundance of several butterfly species. We could have recorded the species in several columns, one for each; however, we also have different locations. These locations are themselves further subdivided by management. If we wrote down the information separately we would end up with several smaller tables of data and it would be difficult to carry out any actual analyses. By recording the information in separate columns we can carry out analyses more easily.

Table 10. Data table layout. Complex data are best set out in separate columns. Here butterfly abundance is recorded for four different factors

Transect	Year	ssp	Man	Count
N	1996	pbf	no	1.15
N	1997	pbf	no	1.54
N	1998	pbf	no	0
N	1996	pbf	yes	1.54
N	1997	pbf	yes	4.62
N	1998	pbf	yes	0
N	1996	pbf	yes	2.78
N	1997	pbf	yes	1.67
N	1998	pbf	yes	0
S	1996	swf	yes	7.11
S	1997	swf	yes	25.53
S	1998	swf	yes	2.37

The data in Table 10 can be split into various subsections using our spreadsheet and the filter command (Section 3.2.2). We can also use the pivot table function to review the data (Section 3.2.7).

Now we have gone through the planning process. Ideally, we would have worked out a hypothesis and know what data we need to collect to support our hypothesis (or to reject it). We will look at hypothesis testing in more detail in Chapter 5. We ought to know at this stage what type of analysis we are going to run on our data (Table 27 in Chapter 5).

Once we have collected data and written it in our spreadsheet in an appropriate format, we are ready to begin to explore the data. This is the subject of the following chapters. Ideally, we should begin by sketching our data graphically and we will introduce this in Chapters 4 and 6. We will cover graphs in Section 12.4, although there will be some examples throughout the text. After we have a graphical overview of our data, we should summarise it numerically (Chapter 4). Subsequent chapters deal with choosing and carrying out the various analytical methods we might use to support or reject our initial hypothesis. Before we get to that, we need to become a bit more familiar with some of the tools that we are going to use, namely Excel and R. Familiarising ourselves with these tools is the subject of Chapter 3.

3. Beginning data exploration – using software tools

In order to make sense of your data you will need to use some of the tools we have come across already, namely your spreadsheet and the R program. We will cover the various specifics as we go along but before we go any further we need to become a bit more familiar with the R program.

3.1 Beginning to use R

Once you have installed R, run it using the regular methods: you may have a shortcut on the desktop or use the *start* button. If you are using later versions of Windows then you need to alter the program permissions (see Section 1.8.3 about running as administrator). Once you have run the program, you will see the main input window and a welcome text message. This will look something like Figure 18 if you are using Windows. There is a > and cursor | to show that we can type at that point. In the examples, we shall show the > to indicate where we have typed a command, and lines beginning with anything else are the results of our typing.

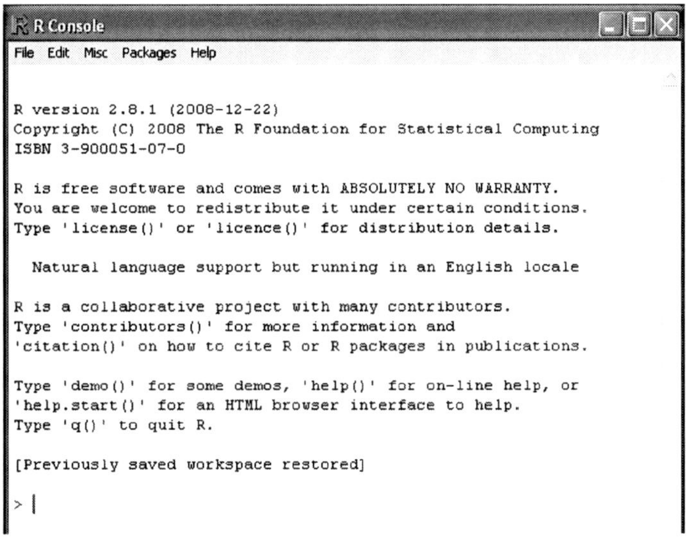

Figure 18. The R program interface is a bit sparse compared to most Windows programs

The program appearance (GUI) is somewhat sparse compared to most Windows programs. You are expected to type commands in the window. This sounds a bit daunting but is actually not that hard. Once you know a few basics, you can start to explore more and more powerful commands because R has an extensive help system. There are many resources available on the R website and it is worth looking at some of the recommended documents (most are PDF) and working through those. Of course this book itself will provide a good starting point!

After a while you can start to build up a library of commands in a basic text editor; it is easy to copy and paste commands into R. It is also easy to save a snapshot of the work you have been doing for someone else to look over.

3.1.1 Getting help

R has extensive help. If you know the name of a command and want to find out more (there are often additional options), then type one of the following:

```
> help(topic)
> ?topic
```

You replace the word *topic* with the name of the command you want to find out about. You can also bring up a Windows-type help window:

```
> help()
```

There is an HTML version of the help files too and this will open in your default browser if you type:

```
> help.start()
```

These commands are listed in the opening welcome message (Figure 18).

Online help

There are loads of people using R and many of them write about it on the Internet. The R website is a good start but just adding CRAN to a web search may well get you what you need. CRAN is the acronym for the "Comprehensive R Archive Network".

3.1.2 Starting with R

R can function like a regular calculator. Start by typing some maths:

```
> 23 + 7
> 14 + 11 + (23*2)
> sqrt(17-1)
> pi*2
```

Notice that R does not mind if you have spaces or not. Your answers look something like the following:

```
[1] 30
[1] 70
[1] 4
[1] 6.283185
```

The [1] indicates that the value displayed is the first in a series (and in this case the only value). If you have several lines of answers, then each line begins with a [*n*] label where *n* is a number relating to "how far along" the list the first in that line is. This maths is very useful but you really want to store some of the results so that you can get them back later and use them in longer calculations. To do this you create variable names. For example:

```
> answer1 = 23 + 21 + 17 + (19/2)
> answer2 = sqrt(23)
> answer3 = answer1 + (answer1 * answer2)
```

This time you will notice that you do not see the result of your calculations. R does not assume that just because you created these objects you necessarily want to see them. It is easy enough to view the results, we need to type in the name of the thing we created:

```
> answer1
[1] 70.5
> answer2
[1] 4.795832
> answer3
[1] 408.6061
```

In older versions of R, the = sign was not used and instead a sort of arrow was used:

```
> answer4 <- 93 - 21 + sqrt(41)
> 77/5 + 9 -> answer5
```

This is actually a bit more flexible than the = sign (try replacing the arrows above with = and see what happens) but for most practical purposes, = is quicker to type.

Data names

All stuff in R needs a name. In a spreadsheet, each cell has a row and column reference, e.g. B12, which allows us to call up data as we like. R does not set out data like this so we must assign labels to everything so we can keep track of it. We need to give our data a name; R expects everything to have a name and this allows great flexibility. Think of these names as like the memory on a calculator; with R you can have loads of different things stored at the same time.

Names can be more or less any combination of letters and numbers. Any name we create must start with a letter, other than that we can use numbers and letters freely. The only other character allowed is a full stop. R is case sensitive so the following are all different:

```
data1
Data1
DATA1
data.1
```

It is a good idea to make your names meaningful but short!

3.1.3 Inputting data

The first thing to do is to get some data into R so we can start doing some basic things. There are three main ways to do this.

Combine *c()* command

The *c()* command (short for combine, or concatenate) reads data from the keyboard so;

```
> data1 = c(24, 16, 23, 17)
```

Here we make an object called *data1* and combine the four numbers in the brackets to make it. Each value has to be followed by a comma. The spaces are optional, R ignores them, but they can be useful to aid clarity and avoid mistakes when typing.

You may easily add more items to your data using the *c()* command for example:

```
> data2 = c(data1, 23, 25, 19, 20, 18, 21, 20)
```

Now we have a new item called *data2*. We could also have overwritten our original item and effectively added to it without creating a new object:

```
> data1 = c(data1, 23, 25, 19, 20, 18, 21, 20)
```

Here we put the old data at the beginning because we wanted to add the new values after, but we could easily place the old data at the other end or even in the middle.

Scan() command

The *scan()* command reads data values from the keyboard but unlike *c()* it does not need to have commas. This makes it easier to enter data:

```
> data3 = scan()
```

R now waits for you to enter data. You can enter after a few items and carry on with new lines as you wish but once you are finished, enter a blank line. This signifies to R that you are finished.

```
> data3 = scan()
```

```
1: 2 5 76 23 14.4 7 19
8: 21 16.3 112 3
12: 6
13:
Read 12 items
```

In the example above, we entered data over three lines. Each value is separated with only a space rather than a comma, so it is a little easier to type in.

You can also use the *scan()* function in conjunction with your spreadsheet (or other programs) and the copy and paste functions. If you copy Excel data to the clipboard you can then switch to R and run the *scan()* function. When you paste you will find the data transferred to R. The last thing you need to do is to press ENTER and tell R you are done. Alternatively you might switch back to Excel and copy another set of values, but we always enter the blank line at the end to tell R to stop scanning. The following examples illustrate the process. We start by typing a name and the *scan()* command.

```
> data.item = scan()
1:
```

The program now waits and a 1: is shown on the screen to indicate that something is expected. We switch to our spreadsheet and copy a column of cells containing data. Next we switch back and paste (using the keyboard shortcuts or the *Edit* menu).

```
> data.item = scan()
1: 23
17
9
8
```

We see the data enter R but the program is not finished. We could paste more data or finish by pressing ENTER to insert a blank line.

```
> data.item = scan()
1: 23
2: 17
3: 9
4: 8
5:
Read 4 items
```

The final display shows us that the last item entered was blank (we see a 5:) and a message tells us how many data items were received (four in this case – item five was the blank). We could easily copy a row of cells rather than a column; the process is the same.

We often have quite extensive datasets and the *scan()* command along with copy and paste can be fairly tedious in these cases. We do have a more efficient way to move data from our spreadsheet into R but we will return to this later (Section 3.3). For the time being, we have a couple of methods of getting numbers into R to work with.

3.1.4 Seeing what data items you have

Once you have a few objects in the memory, you will naturally want to see what you have. Do this by simply typing the name of a variable. For small datasets this is not a problem but if you have lots of data then this might be a bit tedious (R does its best to wrap the text to fit the screen).

```
[1] 24 16 23 17
[1] 24 16 23 17 23 25 19 20 18 21 20
```

The examples above show our *data1* and *data2* items.

List items *ls()* command

The *ls()* command will show you what objects reside in the memory of R. This will usually comprise of the samples that you entered but you may also see other objects that arise from calculations that R performs.

```
> ls()
[1] "beetle.cca"  "biol"    "biol.cca"    "env"     "op"
```

The example above shows five objects, some are data typed in directly and others are results from various calculations.

Remove items *rm()* command

You might wish to remove some clutter and using the *rm()* command allows you to delete objects from R.

```
> rm(data1)
```

This will now get rid of the *data1* object that we created earlier.

3.1.5 Summary statistics in R

Now that you have a few objects in R, you will want to explore them. R provides a lot of basic commands that will do this.

Summary() command

The *summary()* command will provide basic summary statistics about your chosen object. Some statistical commands create many sub-objects that do not display by default, so typing *summary()* will generally provide a lot of extra information. The basic command displays something like this:

```
> summary(data1)
   Min. 1st Qu.  Median    Mean 3rd Qu.    Max.
  16.00   16.75   20.00   20.00   23.25   24.00
```

Here we see that the basic stats are the middle values (mean and median) as well as the ends (max and min) and something in between, the inter-quartiles. If you perform a *summary()* on other types of data you may get something different.

Statistical summary functions

There are a number of useful statistical functions that we may select; here are some examples:

```
> length(data1)
> mean(data1)
> median(data1)
> max(data1)
> min(data1)
> sd(data1)
```

The first shows us how many items there are in our data (the number of observations, also called replicates). The next determines the mean (a kind of average). The *median()* command determines another kind of average. We can determine the largest and smallest values using the *max()* and *min()* commands. The final *sd()* command works out the standard deviation (a measure of variability). Try them out on some data and become familiar with them; we will meet some of them again shortly.

3.1.6 Previous commands and R history

Now you have a few basic commands at your disposal. R stores your commands in an internal list, which you can access using the up arrow. The up and down arrows cycle through recent commands, allowing you to edit a previous command. The left and right arrows move through the command line. Alternatively, you can click the command you want to edit. When you exit R, the previous commands will usually be stored automatically.

Saving history of commands

You can save (or load) a list of previous commands using the GUI. In Windows, the *File* menu gives the option to load or save a history file (Figure 19). This may be useful if you have been working on a specific project and wish to recall the commands later.

Saving the workspace

You can also save the workspace; this includes all the objects currently in the memory. Imagine you are working on several projects, one might be a bird survey, another a plant database and the third an invertebrate study. It might be useful to keep them all separate. R allows you to save the workspace as a file that you can call up later. You can also send the file to a colleague (or tutor) who can access the objects you were working on (Figure 19).

Saving a snapshot of your R session

If you have been working on a series of analyses it may be useful to save an overview of what you have been doing. You can do this using the *File* > *Save To File* menu option (Figure 19). R will save the input window as a plain text file, which you can open in a text editor. You can use this to keep a note of what you did or send it to a colleague (or tutor). You can also use the text file as the basis for your own library of commands; you can copy and paste into R and tweak them as required, R is very flexible.

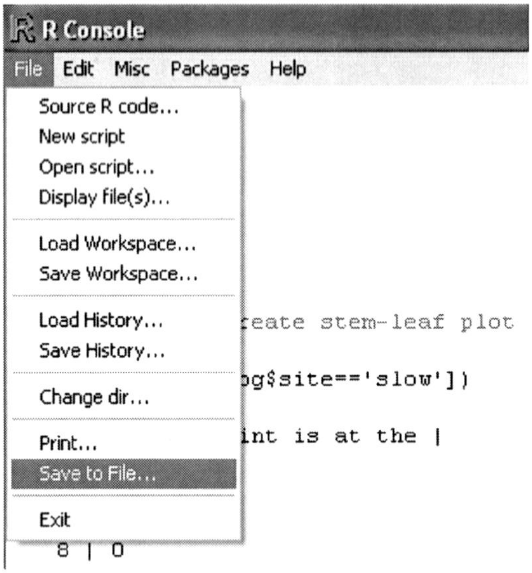

Figure 19. The *File* menu in R allows you to save a snapshot of the R working area to a plain text file

Annotating work

R accepts input as plain text. This means you can paste text from other applications. It is useful to have a library of commands at your disposal. You can annotate these so that you know what they are. R uses the hash character # to indicate that the following text is an annotation and is therefore not executed as a command.

```
> vegemite(vegdata, use= "Domin") # from the vegan
package
```

Here the note reminds the user that the vegan library (or package) needs to be loaded because this command is not part of the original R program (see Section 10.1.2 for notes about additional libraries of R commands).

3.1.7 Exiting R *q()* command

When you are ready to end your session in R, you can quit using several methods. You can use the buttons in the GUI like a regular program or use the *quit()* or *q()* command.

```
> q()
```

R will now ask if you want to save the workspace (Figure 20). It is a good idea to say *yes*. This will save the history of commands, which is useful. It will also save any items in memory. When you run R later, you can type *ls()* to list the objects that R has saved.

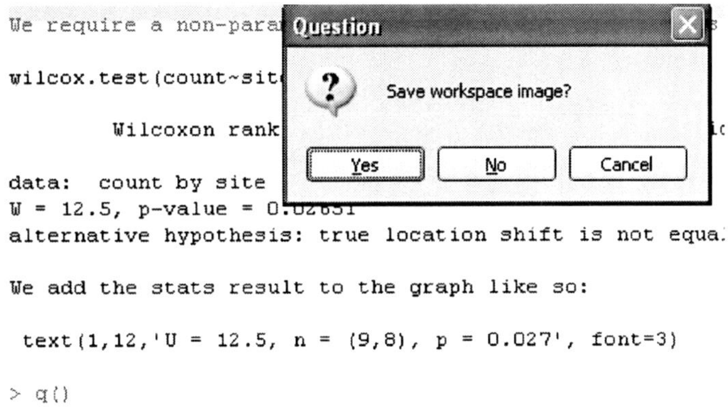

Figure 20. When quitting R you are prompted to save the workspace, thus preserving the data in memory

Now we have been introduced to the R program, it is time to look at things we can do in Excel; this is the subject of the following section.

3.2 Manipulating data in a spreadsheet

Usually we collect data in a field notebook and then transfer them to a spreadsheet. This allows us to tidy up and order the data in a way that will allow us to explore it. It is also a convenient way to store and transfer data. In Section 2.2 we looked at how to arrange the data to begin with; however, this is only the start – there are a few things that we can do once we have our data.

3.2.1 Sorting

It is often useful to re-order the data and this is possible by sorting using the *Data > Sort* menu. Figure 21 shows the layout for Excel 2007 but the options are similar for all versions.

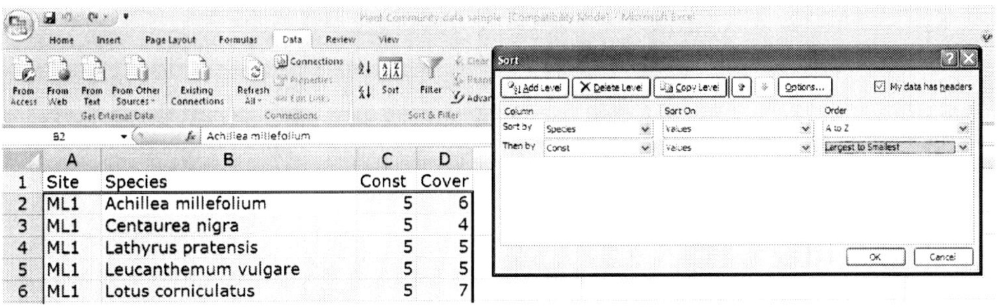

Figure 21. Sorting data in Excel, which provides a range of sorting options

The menu box now allows you to select the columns you wish to sort (and whether you want an ascending or descending order). It is useful to have column names so you recall what the data represent. Ensure the button is ticked where it says *My data has headers.*

3.2.2 Data filtering

Often we need to examine part of our data. For example we may have collected data on the abundance of plant species at several sites. If we wish to look at a single site or a single species (or indeed both) then we can use the filtering ability of Excel. The exact command varies a little according to the version of the program you are using; Figure 22 shows Excel 2007 and the filtering option using the *Home* menu (in other versions this may be in the *Data* menu).

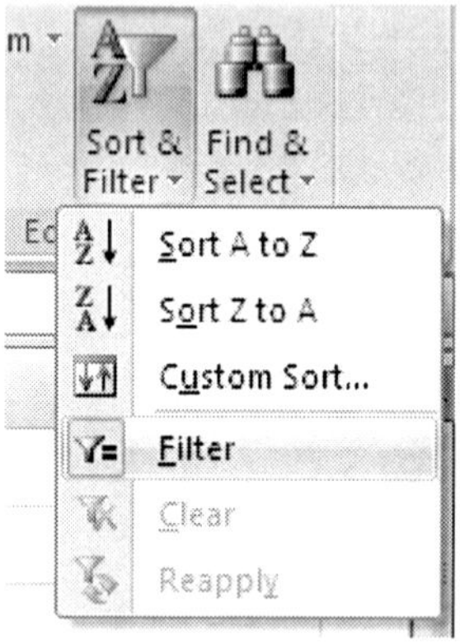

Figure 22. Using the filter option (in the *Home* menu) to select parts of the data

We can highlight the columns of data we wish to filter. Excel will select all the columns by default so we do not make an active selection. Once the filter is applied, we see a little arrow in each column heading. It is important to give each column a heading. We can then click the arrow to bring up a list.

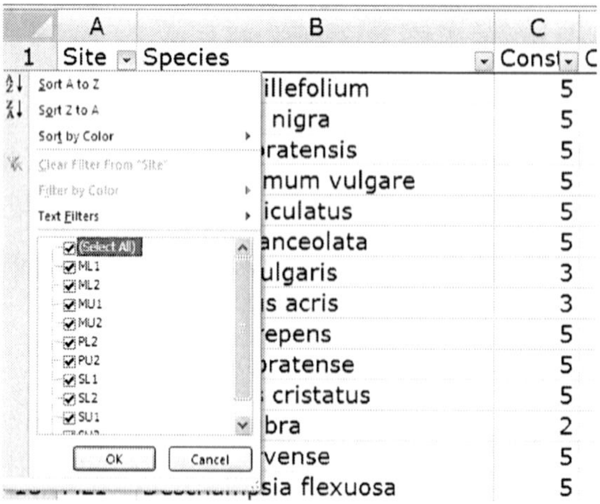

Figure 23. Filtering in Excel. Once a filter is applied, it can be invoked using the drop-down menu

Figure 23 shows the filter in action. We can select to display a variety of options from the list available. Here we have site names so could select all, several or just one site to display. By adding a filter to subsequent columns we can narrow the focus.

3.2.3 Paste special

When copying data it is possible to rotate the rows and columns. Some programs require the species to be the rows and the samples to be the columns but other programs need the samples to be the rows. It is simple enough to rotate the data. First of all we highlight the data we want and copy to the clipboard. Next we place the cursor where we want the data to appear (make a new file or worksheet before copying if you want) and click *Edit > Paste Special*. If you are using a later version of Excel then you can select a variety of paste options in the *Home* menu (right-clicking also works).

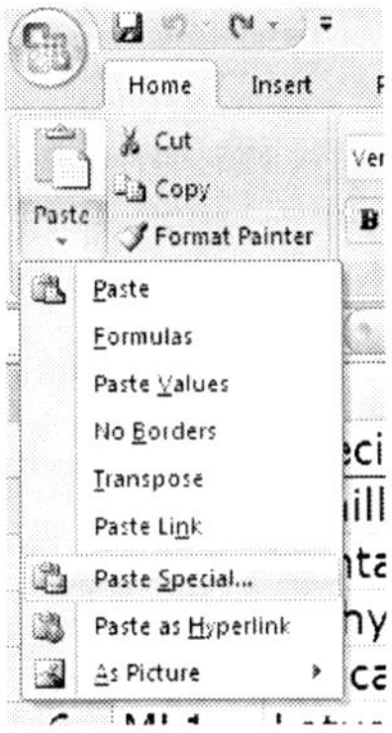

Figure 24. The paste options are found in the *Home* menu of later versions of Excel

You now have a variety of options. If your data are created using formulae then it is advisable to select *Values* from the options at the top. This will not keep the formulae but because it would not be recognised correctly (since you are moving the items) then this is desirable. If you want to preserve formatting (e.g. italic species names) then you can select this option. In this case we also wish to rotate the data so that rows become columns and columns become rows so we select the Transpose button.

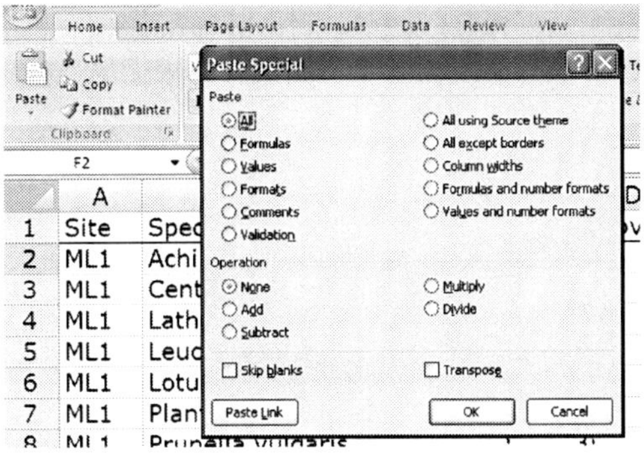

Figure 25. Using paste special in Excel. There are a variety of options for moving blocks of data from one place to another

The samples (sites) are now represented as rows and the species are columns.

3.2.4 File formats

You may need to save your data in a format that is not an Excel workbook. Some programs require plain text files, whilst others will accept text where the columns are separated by commas or tab characters. To do this you need to select the worksheet you want to save and click *File* > *Save As*. In Office 2007 you access this via the *Office* button as shown in Figure 26. The comma separated variables (CSV) format in particular is useful as many programs (including R) can read this.

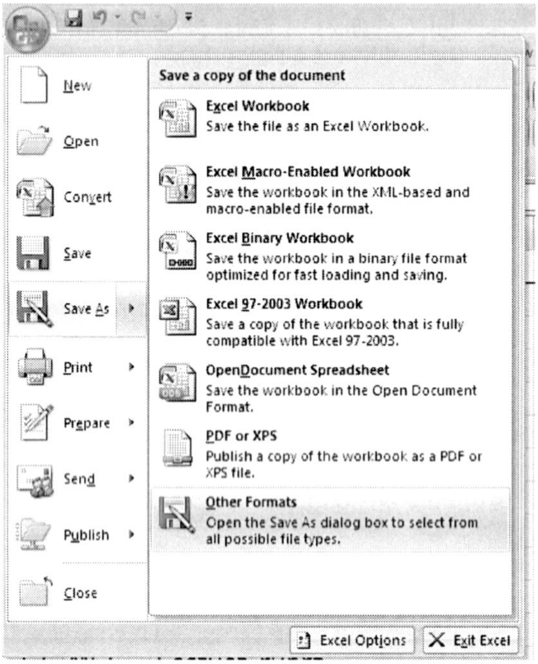

Figure 26. The *Save As* menu in Excel 2007

At the bottom of the menu box there will be a drop-down menu that allows you to select the file type required. CSV is a common format but formatted text and plain text are also used at times.

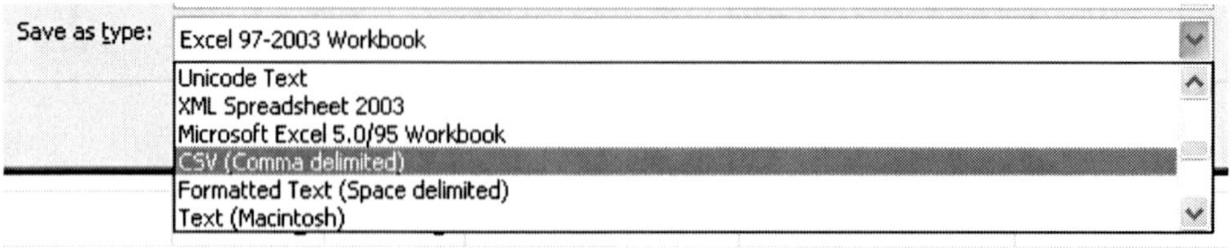

Figure 27. The CSV or comma delimited format is used by many other programs

The CSV format is plain text resulting in the loss of some formatting and you will only be able to save the active tab; multiple worksheets are not allowed. Because of these reasons, Excel will give you a warning message to ensure you know what you are doing. The resulting file is plain text and, when you view it in a word processor, the data is separated by commas. It is a good idea to keep a master XLS file; you can then save the bits you want as CSV to separate files as required.

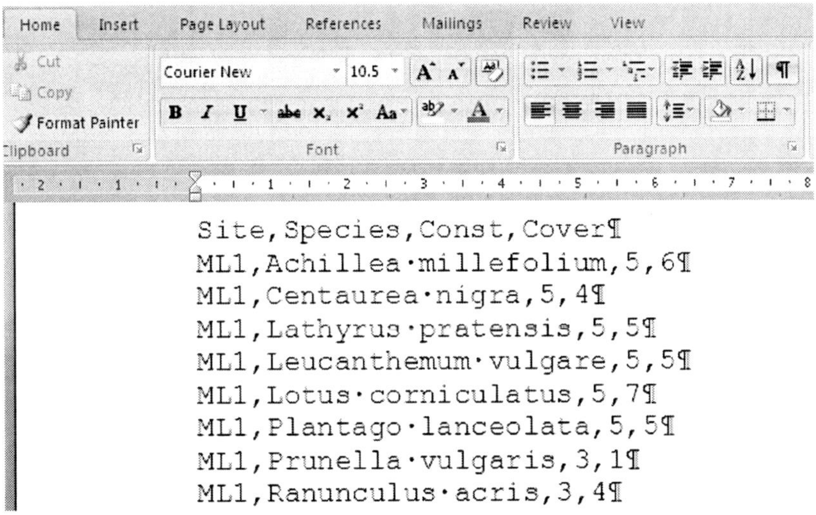


```
Site,Species,Const,Cover¶
ML1,Achillea·millefolium,5,6¶
ML1,Centaurea·nigra,5,4¶
ML1,Lathyrus·pratensis,5,5¶
ML1,Leucanthemum·vulgare,5,5¶
ML1,Lotus·corniculatus,5,7¶
ML1,Plantago·lanceolata,5,5¶
ML1,Prunella·vulgaris,3,1¶
ML1,Ranunculus·acris,3,4¶
```

Figure 28. CSV format. In this format, commas separate data and when Excel reads the file it will place each item into a separate column when it comes across a comma. Many analytical programs use the CSV format

In Figure 28 we see what a CSV file looks like in a word processor (Word 2007). CSV stands for "comma separated variables" and as this name suggests, we see the data separated by commas. When the CSV data are read by a spreadsheet, a comma causes a jump to a new column.

3.2.5 Opening a CSV file in Excel

When we open a CSV in our spreadsheet, the program will convert the commas into column breaks and we see what we expect from a regular spreadsheet. We can do all the things we might wish in a spreadsheet, such as creating graphs; however, if we want the spreadsheet elements to remain, we must save the file as an Excel file. When we hit the *Save* button we will get a warning message to remind us that the file we opened was not a native Excel file; we then have the choice of choosing the format.

3.2.6 Lookup tables

In Section 1.5 we looked at different data types. We may have recorded our results in a Domin scale or ACFOR scale, both examples of ordinal data. When we write our results in our field notebook, we record this scale but later we may wish to present the data in some other way.

In the following example (Figure 29) we have *constancy* values in column C as regular numbers (1–5); these simply refer to how many of the five quadrats each species was found in. For a report, we want to replace the number with roman numerals as this makes a clear distinction between the cover and constancy and makes for clearer reading. We could do a search and replace, although if we did this then we would lose the original data and also we would need to do it five times. In other situations we may have more than five things to replace so a more efficient solution is to use a replacement table.

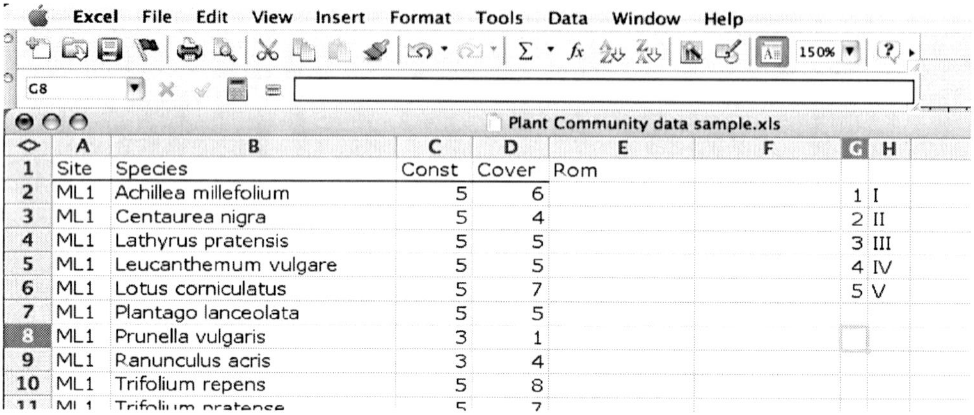

Excel File Edit View Insert Format Tools Data Window Help
C8
Plant Community data sample.xls
A B C D E F G H
1 Site Species Const Cover Rom
2 ML1 Achillea millefolium 5 6 1 I
3 ML1 Centaurea nigra 5 4 2 II
4 ML1 Lathyrus pratensis 5 5 3 III
5 ML1 Leucanthemum vulgare 5 5 4 IV
6 ML1 Lotus corniculatus 5 7 5 V
7 ML1 Plantago lanceolata 5 5
8 ML1 Prunella vulgaris 3 1
9 ML1 Ranunculus acris 3 4
10 ML1 Trifolium repens 5 8

Figure 29. Lookup tables are a useful way to create new fields in your data. Here we want to replace our constancy values with roman numerals. Note that the lookup table must be sorted in ascending order

Here we have made a table that contains the items we want to look for (1–5) and the items that we want to replace them with (I–V). **It is important that the lookup table is sorted in ascending order** by the first column, which contains the items you are searching for. After creating your table, it is worth checking this and doing a sort. We create a new column to hold the new data.

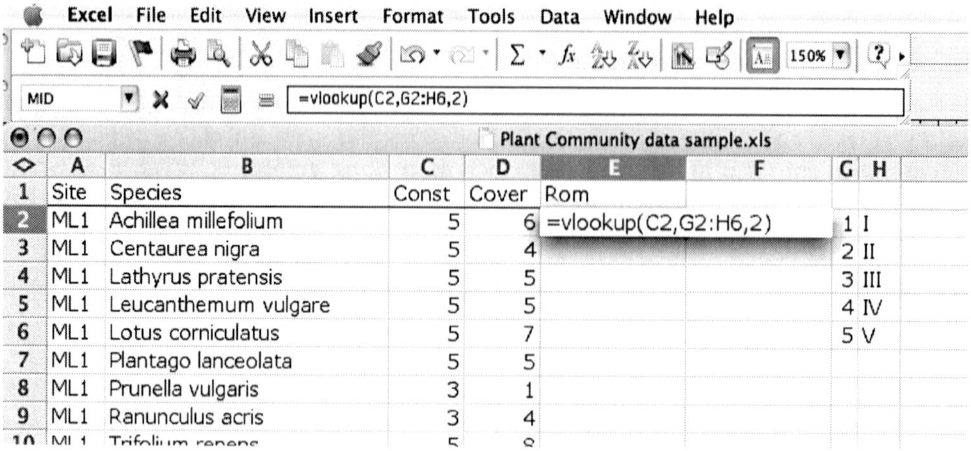

Figure 30. The *vlookup* command is used to replace one value with another

To replace the values we use the VLOOKUP function in Excel. All functions start by typing the = sign. Our formula has three parts: the first is the item we are looking for. In this case it is C2; that is 5 in the *Const* column C. The second is the reference location of the table. In this case it starts at G2 and carries over to H6. The best way to enter the location is to highlight it with the mouse. The table could be in a different sheet (this is in fact often preferable) in which case the name of the sheet appears before the reference, e.g. Sheet3!G2:H6 (this will happen automatically if you select the range of cells with your mouse). The third part of our formula tells Excel to look in the second column of our table (where we have the roman numerals) and that is what will end up in the cell.

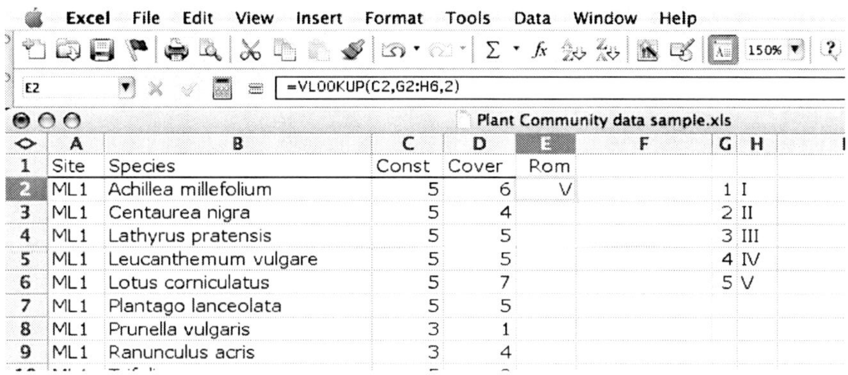

Figure 31. The lookup table has been used to create the first entry in a new field

Now we have the roman numeral V in the cell as a replacement for the regular 5. The advantage over search and replace is that we now have both values and have not lost the original. The next step is to copy the formula down the rest of the column; however, if we do that there will be a problem: Excel will replace the C2 on the next line with C3. That is fine because that is what we want; however, it will also replace the location of our table with G3:H7 and that it **not** what we want. We want the row and column references to remain fixed. We need to edit our formula to tell Excel that the G2:H6 reference is not to change. We do this using the dollar sign.

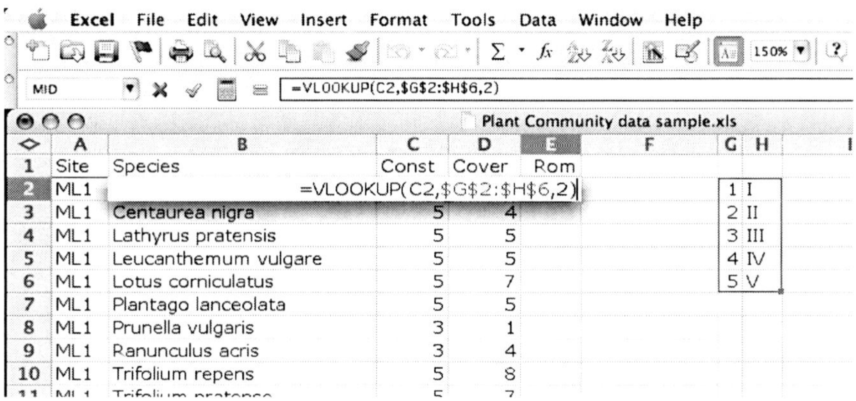

Figure 32. The lookup formula needs some tweaking before it can be copied down the rest of the cells

We click in the formula bar and add dollar signs in front to give $G$2:$H$6, everything else can stay the same. Press ENTER and the result looks the same as before. Now we can copy the cell to the clipboard and paste into the rest of the column.

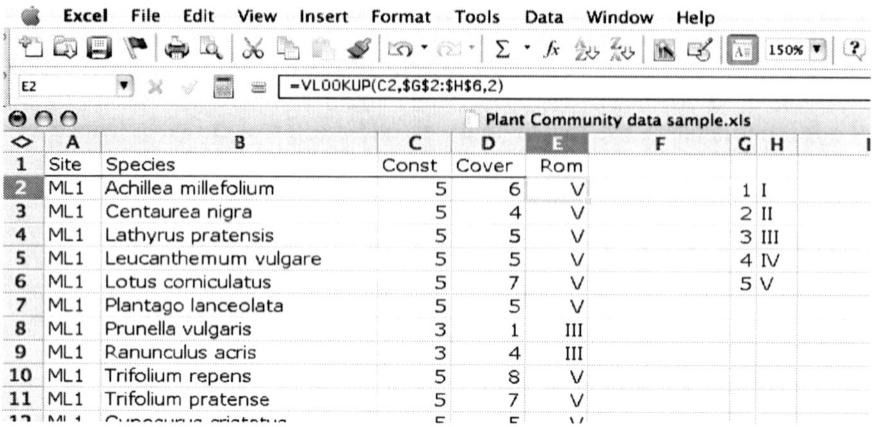

Excel File Edit View Insert Format Tools Data Window Help
150%
E2
=VLOOKUP(C2,$G$2:$H$6,2)
Plant Community data sample.xls
A B C D E F G H I
1 Site Species Const Cover Rom
2 ML1 Achillea millefolium 5 6 V 1 I
3 ML1 Centaurea nigra 5 4 V 2 II
4 ML1 Lathyrus pratensis 5 5 V 3 III
5 ML1 Leucanthemum vulgare 5 5 V 4 IV
6 ML1 Lotus corniculatus 5 7 V 5 V
7 ML1 Plantago lanceolata 5 5 V
8 ML1 Prunella vulgaris 3 1 III
9 ML1 Ranunculus acris 3 4 III
10 ML1 Trifolium repens 5 8 V
11 ML1 Trifolium pratense 5 7 V

Figure 33. The completed table. The lookup table used to change arabic to roman numerals

The formula has been copied down the column correctly and we have successfully replaced all the arabic numerals with roman numerals. The similar HLOOKUP formula reads across the row of a table and returns the value in the rows below.

In Figure 35 we have some data on seaweed abundance on a rocky shore. We have collected data at various points from low tide up towards the high tide mark. For convenience we have used an ACFOR scale (A for abundant, C for common and so on in decreasing fashion). This is perfectly fine and sensible but if we wish to draw a graph or undertake some statistics we will probably want to change the scale from an alphabetic one to a numeric one. In other words we want to replace the letters with numbers.

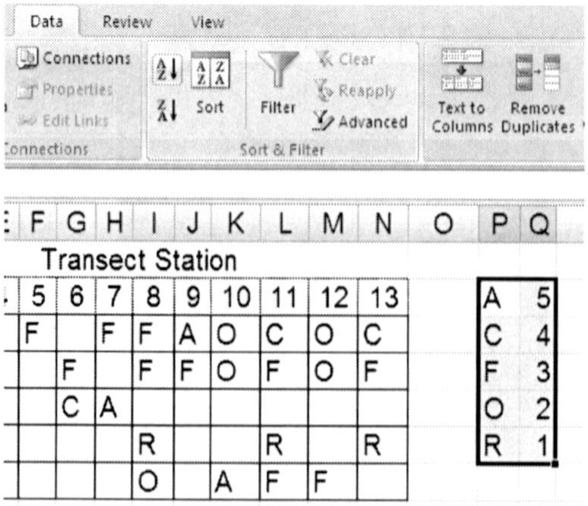

Figure 34. When using lookup tables it is important that the table of reference values is sorted in ascending order

Once again we need to use a lookup table and here we set out the values we wish in cells P2–Q7. We have our original values in the first column and the values we want to replace them with in the second column.

SUM =IF(I3="","",VLOOKUP(I3,$P$2:$Q$6,2))

	A	B	C	D	E	F	G	H	I	J	K	L	M	N	O	P	Q
1		Transect Station															
2	**Species**	1	2	3	4	5	6	7	8	9	10	11	12	13		A	5
3	*Fucus vesiculosus*					F		F	F	A	O	C	O	C		C	4
4	*Ulva lactuca*						F		F	F	O	F	O	F		F	3
5	*Enteromorpha spp.*		F	C			C	A								O	2
6	*Chondrus crispus*								R			R		R		R	1
7	*Corallina officinalis*								O		A	F	F				
8																	
9		Transect Station															
10	**Species**	1	2	3	4	5	6	7	8	9	10	11	12	13			
11	*Fucus vesiculosus*					3		3	?))	5	2	4	2	4			
12	*Ulva lactuca*						3		3	3	2	3	2	3			
13	*Enteromorpha spp.*		3	4			4	5									
14	*Chondrus crispus*								1			1		1			
15	*Corallina officinalis*								2		5	3	3				

Figure 35. Using a lookup table to convert ordinal data as an alphabetic character to a numerical value. Here we have to consider empty cells as well and use an If statement

We have chosen a simple 1–5 scale here but we could have replaced the letters with anything, including different letters. It is still important to get the lookup table sorted in ascending order so before we start we should select the appropriate cells and use the sort button (Figure 34 shows how to do this in Excel 2007).

As before we use a VLOOKUP function to read each cell, look down the first column of our replacement table and replace whatever it finds with the appropriate value in the second column. The problem now is that we have blank cells. These of course correspond to quadrats where no seaweed was found. It would be tedious to insert a zero for everything you did not find. What we must do is to take into account blank cells in the spreadsheet.

Using the IF() function to ignore blank cells

The IF function allows us to take some condition into account; we can do one thing or another according to the result. This allows us to take into account blank cells for example (and many things besides). The basic form of the function is:

IF(*compare with something, what to do if it is TRUE, otherwise do this*)

There are three parts: the first part allows you to decide something, then we use a comma and insert what to do if this decision was true, finally we add another comma and write down what to do if the result of the decision was not true (i.e. false).

To look for a blank cell we use a pair of double quotes. This is the first part of our formula IF(*cell* = ""...) meaning that we look to see if the cell is blank. If it is blank then we want to keep it that way so we add a pair of double quotes, which will force the cell to remain blank. Finally we insert our VLOOKUP part. In practice we would select a non-empty cell and create our VLOOKUP formula. Then we could add the IF part at the beginning. Once we are happy the formula works on one cell, we can copy and paste over all the data. Doing it this way means we are less likely to get confused and make a mistake.

3.2.7 Pivot tables

Pivot tables are a useful method of re-arranging and summarising data. They are especially useful for extracting data from spreadsheets where you have data in "recording" format, i.e. each column contains a single variable/factor. In Table 11 we see some data in this format.

Table 11. Data in biological recording format may be rearranged using a pivot table

Count	Habitat	Obs
3	grass	1
4	grass	2
3	grass	3
5	grass	4
6	grass	5
12	grass	6
21	grass	7
4	grass	8
5	grass	9
4	grass	10
7	grass	11
8	grass	12
6	heath	1
7	heath	2
8	heath	3
8	heath	4
9	heath	5
11	heath	6
12	heath	7
11	heath	8
19	arable	1
3	arable	2
8	arable	3
8	arable	4
9	arable	5
11	arable	6
12	arable	7
11	arable	8
9	arable	9

Our data comprise of three columns: the first is the abundance of butterflies observed on a transect walk, the second is the site (as a habitat type) and the final column is the transect number or observation (replicate is another term used). We can see that we have 12 observations in the *grass* habitat, 8 in the *heath* and 9 in the *arable*. This column is helpful because it will allow us to index the data, as we shall see now. With the data in recording format, we can carry out statistical tests using R (or any other program) but it may also be useful to "unstack" the two main columns and create three samples, one for each habitat. We can use a pivot table to do this.

We can find the option to create a pivot table in the *Insert* menu in Excel 2007. In earlier versions the pivot table option was found in the *Data* menu. Once we select the *PivotTable* option we get the chance to select our data. If the active cell is adjacent to or within a block of data, then the block of data will be selected automatically (Figure 36).

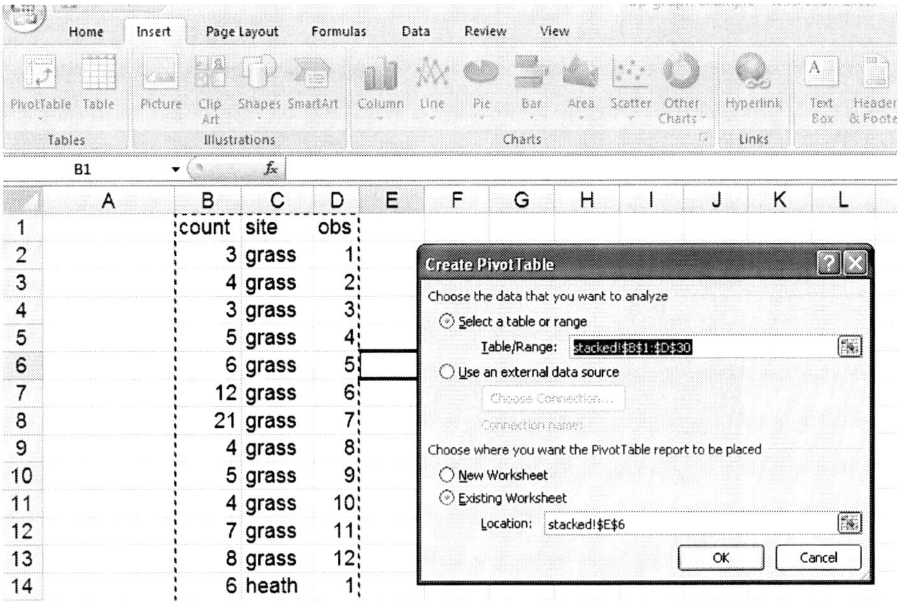

Figure 36. The *Insert* menu is used to start a pivot table. On older versions of Excel, the *Data* menu contained the option

The menu that pops up is largely self-explanatory. In most instances we will want a new table to be placed in a new worksheet, this makes it easier to save as a CSV and is generally a tidier approach. In this instance we have selected to place the table in the existing worksheet to help the reader (you) to visualise the process. We select the cell where the top left part of the table will go.

Once we click OK, we are presented with a blank table and a dialog box allowing us to build our pivot table report (Figure 37).

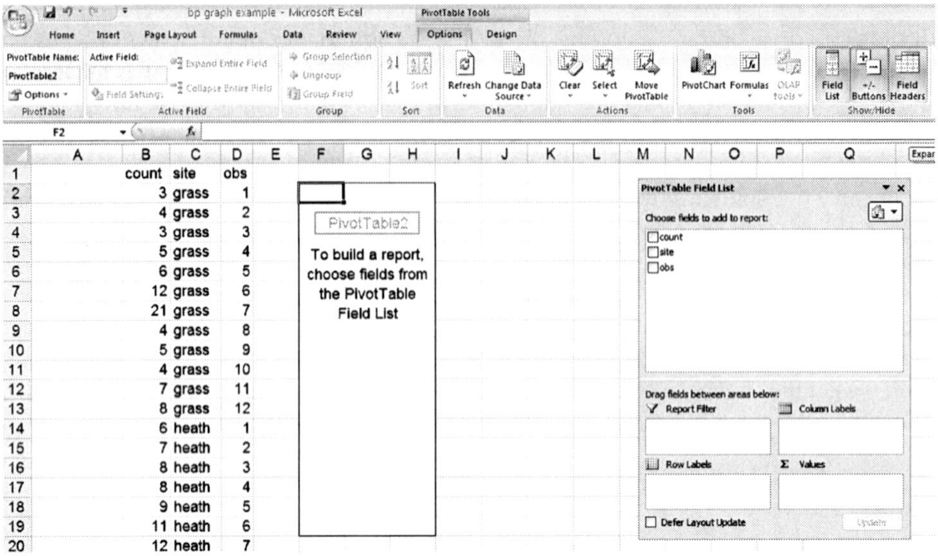

Figure 37. Once the data for the pivot table is selected, the report can be generated

The default layout in Excel 2007 is slightly different to previous versions. The older version is perhaps to be preferred as it allows you to visualise the table better than the new version. On the *PivotTable Tools* menu on the ribbon, we can see an *Options* button (far left, Figure 37). If we click this we can set a variety of options including the old style layout (Figure 38).

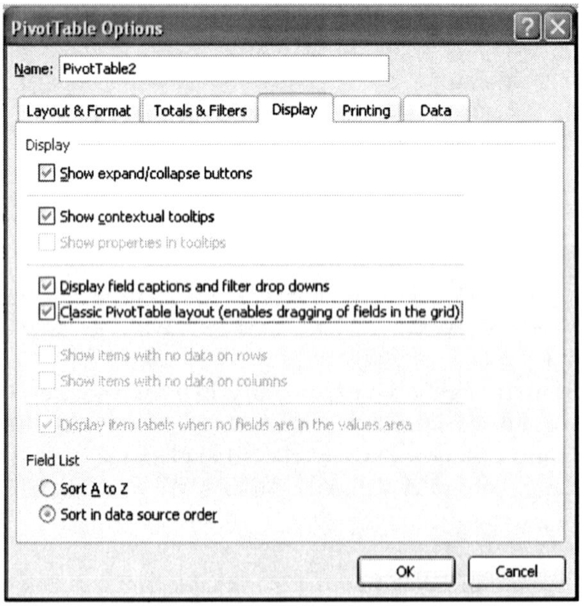

Figure 38. Pivot table options include old style layout, which enables drag and drop

We can select the *Classic* style from the *Display* tab in the Options panel. Whilst we are here we can set other options too. We do not really want row and columns sums so we should turn them off on the *Tools and Filters* tab (Figure 39).

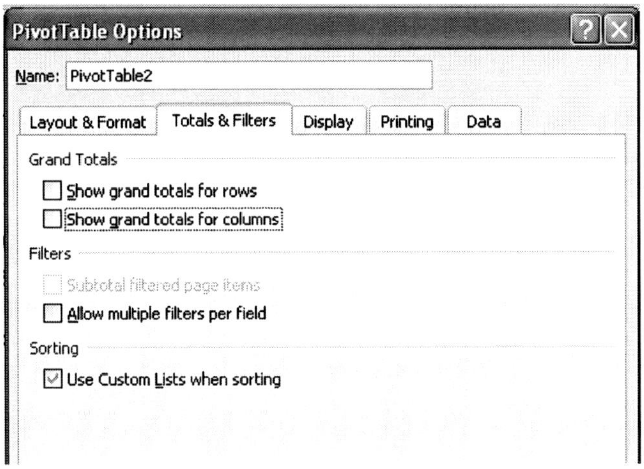

Figure 39. Pivot table options include setting of totals for rows and columns. Usually we do not want these

Once we have set the options, we return to the pivot table itself (Figure 40).

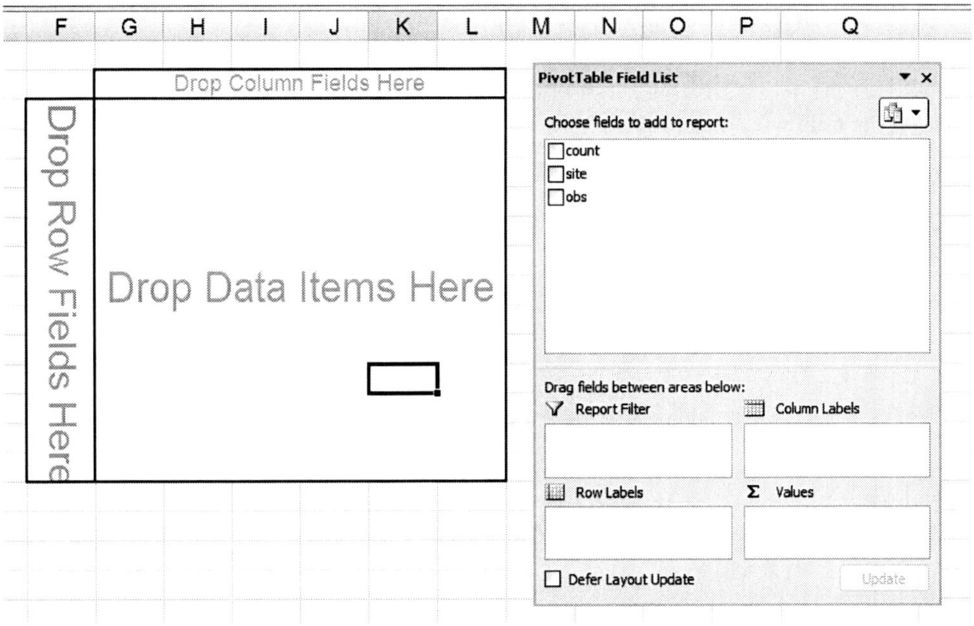

Figure 40. Blank pivot table ready to be created using drag and drop

There are four main areas in our table. The main body of the table has areas for row and column fields as well as the main data. For the time being we will focus on those. We want to create a table that has a column for each of our sites, so we can start by selecting the *site* item in the *Field list* box and dragging it into the top of the pivot table (Figure 41).

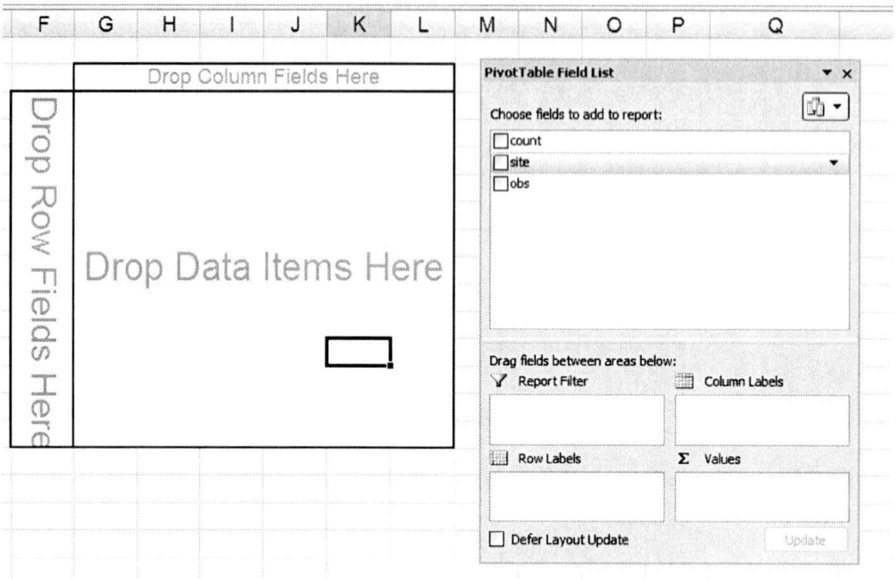

Figure 41. Items that will form the pivot table can be dragged from the field list into the table. Here we select *site* from the list on the right and drag to the pivot table on the left

As we drag the item across we will see our table begin to be populated (Figure 42). We also see that the field we dragged is also placed in the *Column Labels* box.

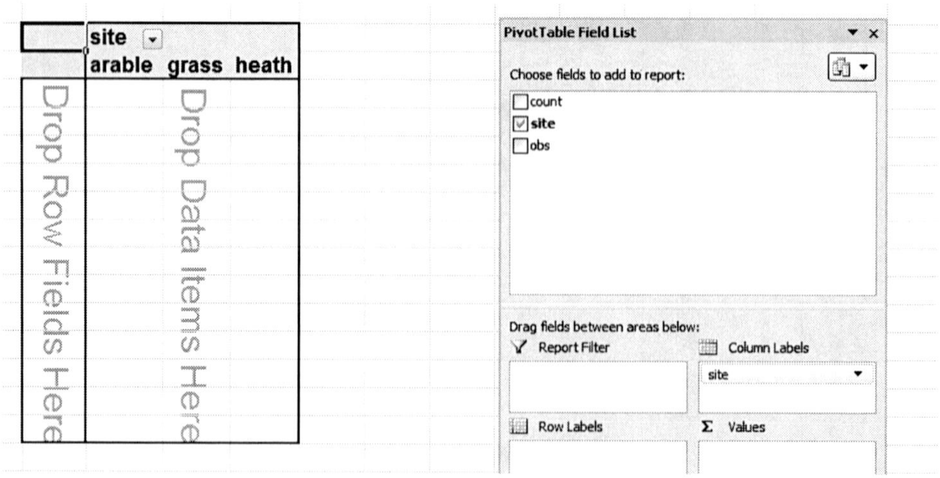

Figure 42. As items are dragged into the table it begins to be built up

This provides us with another way of creating or modifying the table as we can drag items between these four boxes; however, as we said earlier, it is easier to visualise the building of your table using the *classic* layout. Now we have the columns in place we should think about the rows. We want the *obs* field to be our rows so we drag this item into the *Row field* part of the table (Figure 43).

	site		
obs	arable	grass	heath
1		Drop Data Items Here	
2			
3			
4			
5			
6			
7			
8			
9			
10			
11			
12			

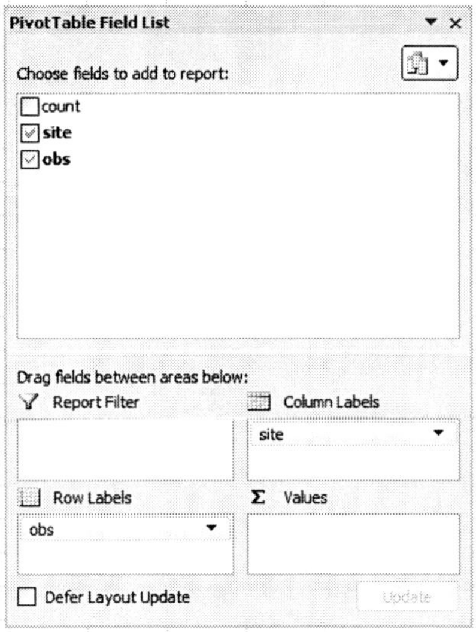

Figure 43. The row items are placed. Now the pivot table has rows and columns but lacks data

We can now see that only the *data items* part of the table is empty. We fill this in by dragging in the *count* field item (Figure 44).

Sum of count	site		
obs	arable	grass	heath
1	19	3	6
2	3	4	7
3	8	3	8
4	8	5	8
5	9	6	9
6	11	12	11
7	12	21	12
8	11	4	11
9	9	5	
10		4	
11		7	
12		8	

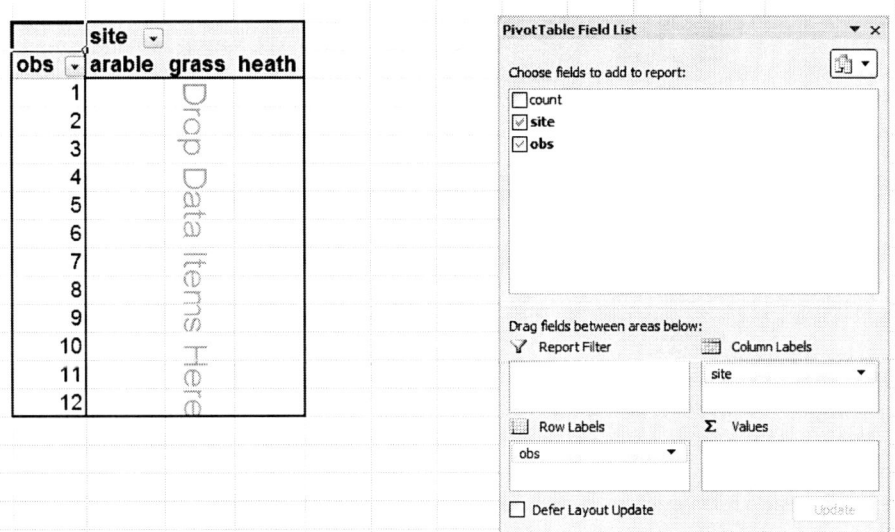

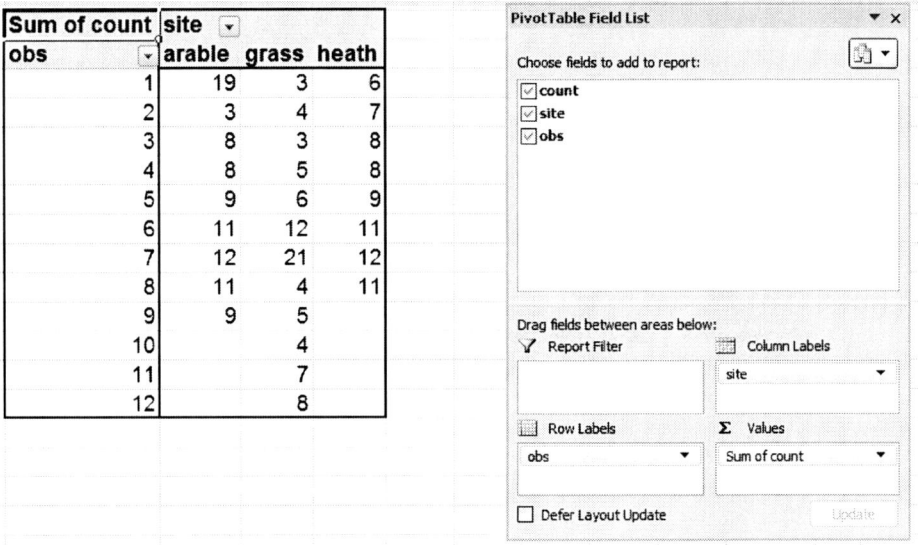

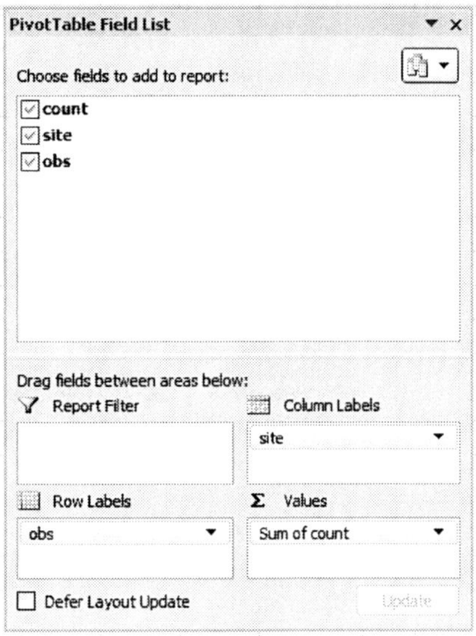

Figure 44. The pivot table is now completed apart from final formatting

Our table is now nearly complete. We have all the data we require and all we need to do now is to finish the formatting. The *PivotTable Tools* menu on the ribbon gives us a variety of options (Figure 45) and also we can right-click in the table itself to bring up these options.

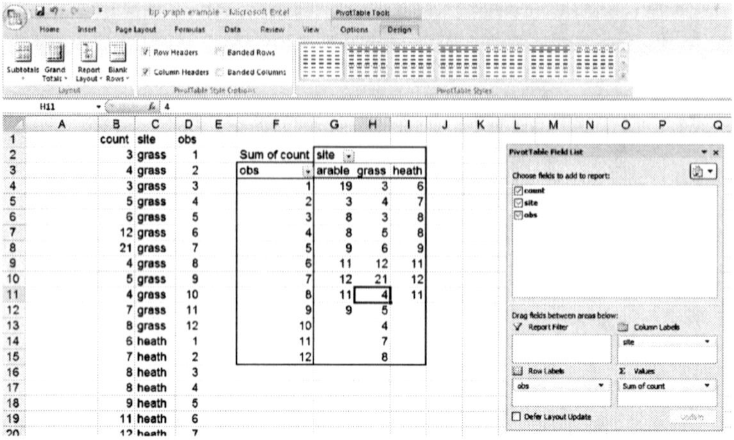

Figure 45. The *PivotTable Tools* menu allows formatting and display options to be modified

Usually the main data is summarised as the *sum*. In this case that is exactly what we want. Each cell in our table represents a unique combination of observation and site; in other words the original data. If we omitted the *obs* field item we would arrange our data solely by site. Our *sum of count* items would then represent the sum of all the observations for each site. One way we can alter the way the data are presented is by clicking the *Σ values* box where we can change to *average* for example.

In Figure 46 we see another example of data. Here we have a survey of some common bird species. Each species has been recorded in several locations, each of which is a different habitat type.

	A	B	C	D	E	F	G
1	Species	Site	GR	Date	Recorder	Quantity	Habitat
2	Blackbird	Springfield	SP873385	04-Jun-07	Starling, C	47	Garden
3	Chaffinch	Springfield	SP873385	04-Jun-07	Starling, C	19	Garden
4	Great Tit	Springfield	SP873385	04-Jun-07	Starling, C	50	Garden
5	House Sparrow	Springfield	SP873385	04-Jun-07	Starling, C	46	Garden
6	Robin	Springfield	SP873385	04-Jun-07	Starling, C	9	Garden
7	Song Thrush	Springfield	SP873385	04-Jun-07	Starling, C	4	Garden
8	Blackbird	Campbell	SP865395	04-Jun-07	Starling, C	40	Parkland
9	Chaffinch	Campbell	SP865395	04-Jun-07	Starling, C	5	Parkland
10	Great Tit	Campbell	SP865395	04-Jun-07	Starling, C	10	Parkland
11	House Sparrow	Campbell	SP865395	04-Jun-07	Starling, C	8	Parkland
12	Song Thrush	Campbell	SP865395	04-Jun-07	Starling, C	6	Parkland
13	Blackbird	Ouzel Valley	SP880375	04-Jun-07	Starling, C	10	Hedgerow
14	Chaffinch	Ouzel Valley	SP880375	04-Jun-07	Starling, C	3	Hedgerow
15	House Sparrow	Ouzel Valley	SP880375	04-Jun-07	Starling, C	16	Hedgerow
16	Robin	Ouzel Valley	SP880375	04-Jun-07	Starling, C	3	Hedgerow
17	Blackbird	Linford	SP847403	04-Jun-07	Starling, C	2	Woodland
18	Chaffinch	Linford	SP847403	04-Jun-07	Starling, C	2	Woodland
19	Robin	Linford	SP847403	04-Jun-07	Starling, C	2	Woodland
20	Blackbird	Kingston Br	SP915385	04-Jun-07	Starling, C	2	Pasture
21	Great Tit	Kingston Br	SP915385	04-Jun-07	Starling, C	7	Pasture
22	House Sparrow	Kingston Br	SP915385	04-Jun-07	Starling, C	4	Pasture

Figure 46. Common bird species and habitat type. These data are in full recording format and can be summarised with a pivot table

We can use a pivot table to re-arrange these data into a summary table. The first step is to use the *Insert* menu (older versions use the *Data* menu) to bring up the *PivotTable* menu (Figure 47).

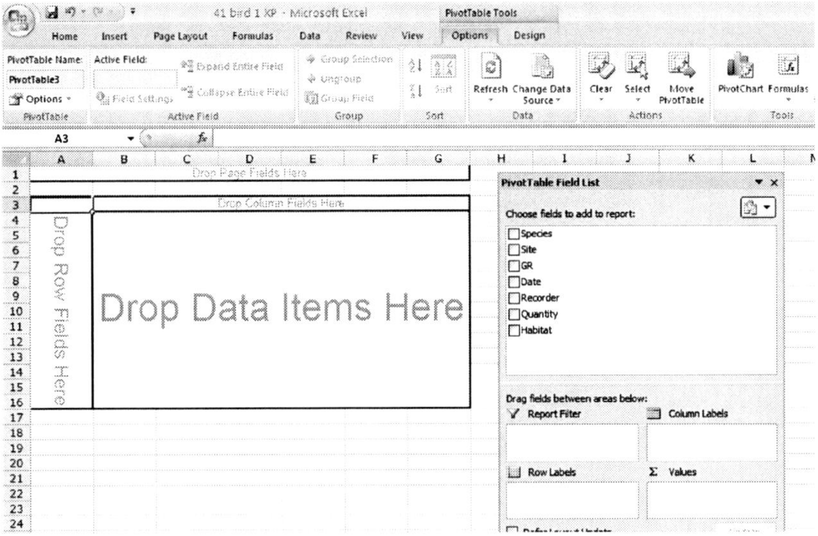

Figure 47. Bird data ready to be compiled into a pivot table

We can use the *Options* button to give us the *classic* look as we did before (Figure 38). This allows us to drag and drop directly into the table and is a little easier to visualise; however, it is okay to use the new style and the boxes provided in the *Field List* box.

We start by dragging the *Species* item into the *Row Field* section (Figure 48).

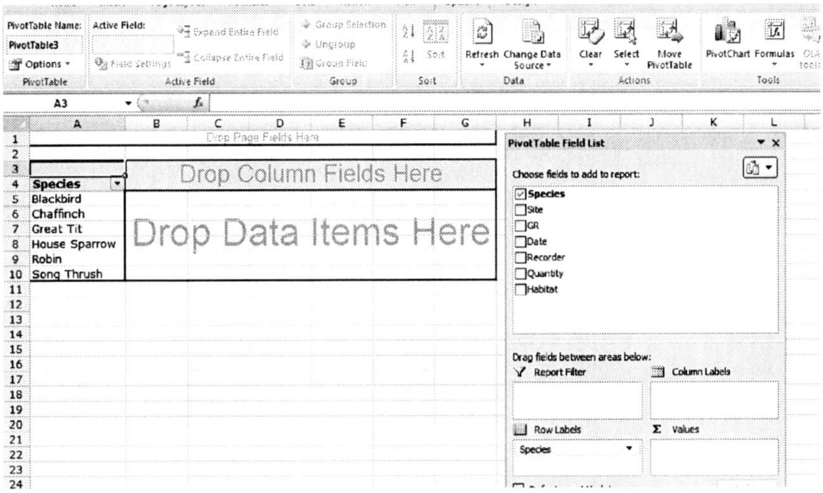

Figure 48. The row items will be the bird species names in our pivot table

As before we see these items reflected in the table immediately. Our column data will be the *Habitat* field so we drag that across into the table. Finally we drag the *Quantity* field into the main data part of the table, which is now more or less completed (Figure 49).

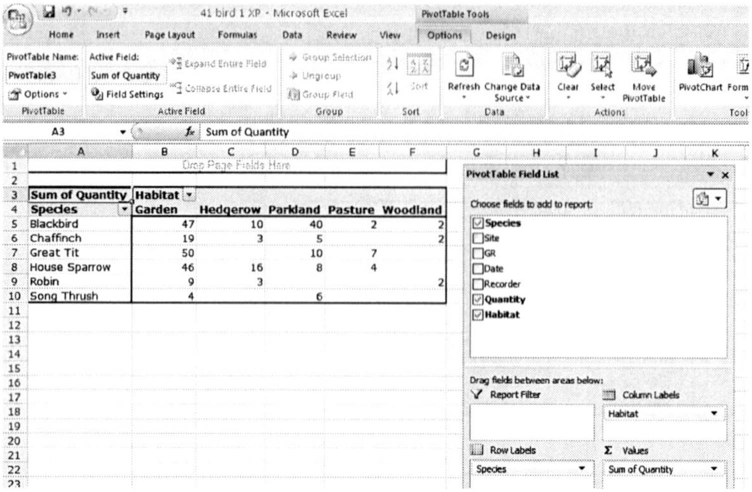

Figure 49. The bird data table is mostly complete but we have missing values in the data that correspond to zero

Our table appears complete but we see missing cells. In the previous example we had differing numbers of observations for each sample so the missing values are fine. In our bird example the missing values really ought to be replaced by zeroes to indicate that we did not find any *Robin* in *Pasture* for example. On the *PivotTable Tools* menu we can set table options (Figure 50).

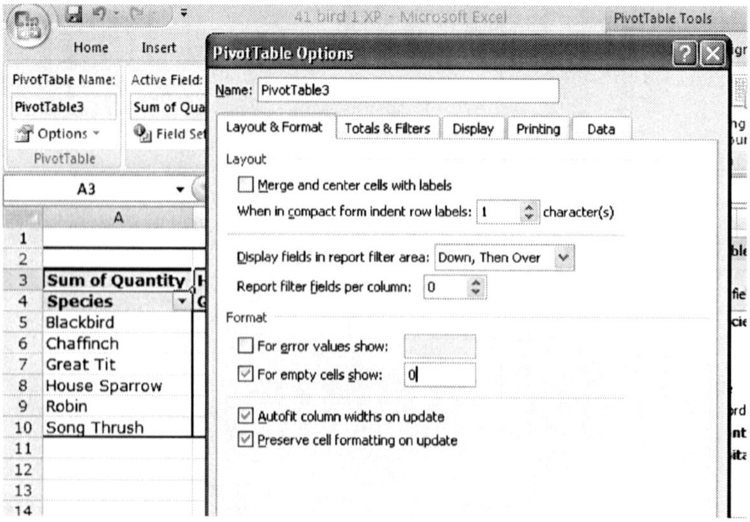

Figure 50. Empty cells can be replaced with any character or value. Here we replace empty cells with 0 (zero)

On the *Layout* tab we can choose to display empty cells with any character or value. Here we will enter a zero. Our final table now looks more like Figure 51.

Sum of Quantity					
	Garden	Hedgerow	Parkland	Pasture	Woodland
Blackbird	47	10	40	2	2
Chaffinch	19	3	5	0	2
Great Tit	50	0	10	7	0
House Sparrow	46	16	8	4	0
Robin	9	3	0	0	2
Song Thrush	4	0	6	0	0

Figure 51. Completed pivot table for bird/habitat data. Empty cells have been replaced with 0 (zero) values

Now that we have our summary table we can copy the data to a new location. To copy the data we could save the pivot table worksheet to a CSV file or copy and paste to a new file. You will need to use *Paste Special* (Section 3.2.3) if you use copy and paste otherwise Excel will try to link back to the original data.

There are many other things that you might do with a pivot table; you can have more than one data field in a row or column for example. Whenever you are inserting data into a spreadsheet you should think about how the data will be used and how they may be re-arranged using a pivot table. This might lead you to add new columns that can be used for indexing. It is a lot easier to do this at an early stage.

3.3 Getting data from Excel into R

At some point we will wish to transfer data into R. The spreadsheet is a useful tool to set out and manipulate data. We can also carry out a range of analyses and produce graphs (Section 12.4.3); however, sooner or later we will want to conduct more in-depth analyses and the R program is invaluable for this. In order to import data we will need to learn a few more commands.

The *read.csv()* command

The combine and scan commands are useful for entering small samples of data but you will usually have a lot more. The *read.csv()* command allows you to read CSV files. These may be prepared easily from spreadsheets (Section 3.2.4). By default R expects the first row to contain variable names.

```
> data4 = read.csv(file.choose())
```

As usual when using R we assign a name to store the data. In the above example we call this *data4*. The *read.csv* part tells R to look for a CSV file. The *file.choose()* part tells R to open an explorer-type window so you can choose the file you want from your computer

(Figure 52), otherwise you must specify the exact filename yourself (this is what you have to do if you are running R in Linux). There are a variety of variations on this command but since the CSV format is so readily produced it seems unnecessary to learn anything other than the basic form of the command (try looking at the help for this command in R).

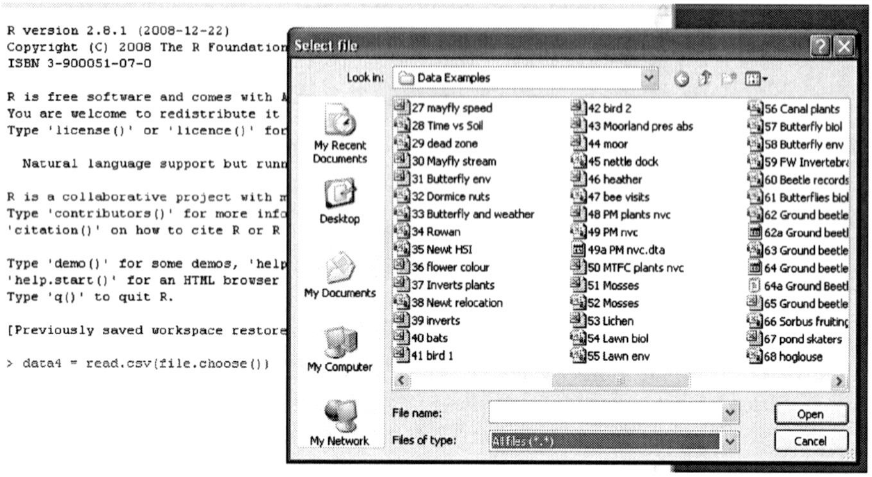

Figure 52. Importing data to R. The *file.choose()* command opens an explorer-type window (here shown in Windows XP)

Now that we have become a bit more familiar with our computer tools, it is time to look at the data itself and what we can do with it.

4. Exploring data – looking at numbers

Statistics is a subject concerned with numbers but what exactly are statistics? The term has a variety of connotations and not all of them give a favourable impression. In biological research you often collect a lot of data. Statistics allow you to make sense of these data. Essentially the purpose of statistics is to take a large quantity of information and present it in a simpler and more meaningful manner. A good start would be to produce a graph to summarise our data (Section 4.2). In Chapter 6, we will discover how graphs can help us explore our data and we will examine graphs in more detail in Section 12.4. Representing our data graphically is a really important step as it helps us to visualise the situation and inform our approach. We may decide that we need to collect more data or even to revise our statistical method. We are leaving much of the details of the graphical work until Section 12.4 because we want to focus on demonstrating the range of possibilities in terms of analysis. We should always bear in mind that graphs are useful as part of the general analytical process and not just something to leave until the final report.

Imagine you were interested in the size of water beetles in a local pond. You visit the pond, capture a beetle and measure it. You now know how big the beetles are in the pond right? Unfortunately this is not the case – as with many other biological data, the size of the beetles is variable. A single measurement is not enough and so we collect more beetles and measure them all. The data are shown in Table 12.

Table 12. Beetle carapace lengths in millimetres

Beetle sizes in mm				
17	26	28	27	29
28	25	26	34	32
23	29	24	21	26
31	31	22	26	19
36	23	21	16	30

If we were recording these data into our notebook or transferring the data into a spreadsheet then we would normally put all the values in a single column; however, for space and readability we have presented the data in Table 12 as five columns.

4.1 Summarising data

In Table 12 we see our beetle sizes in millimetres. Obviously we are excited to have collected all these data and wish to tell our colleagues. How would we go about reporting our data? We could show the table but that might not be impressive and could be quite tedious for the reader. What we need is a way of *summarising* our data. One of the most useful ways we can do this is by using an *average*.

4.1.1 Exploring averages

An average is a way of representing the middle of a set of numbers; statisticians talk about the *central tendency* of a sample. There are several types of average; the most common one used is the *mean*. To work this out we add up all our numbers and divide by the number of items we have. Written in mathematical notation this is:

$$\bar{x} = \frac{\sum x}{n}$$

Figure 53. Formula to calculate the arithmetic mean

The bar over the x signifies a *mean*, the n represents the number of data items and the capital Greek sigma tells us to add together all the x values (the observations).

This is not the only average. We could also work out the *median* value. To do this we would arrange the data into numerical order and then select the middle number. In this case we have 25 items so the middle item would be the 13th largest. It is not easy to write this mathematically, the best we can do is:

$$median = Rank_{(\frac{n}{2})+0.5}$$

Figure 54. Formula representing the median

Here we have the *rank* and n, the number of data items as before. If we had 24 items (an even number) then the median would lie at the 12.5th rank, i.e. between the 12th and 13th values (you take the halfway point).

Finally we could select the most common value in our set of numbers; this is called the *mode*.

Table 13. Summary of beetle carapace data

Summary of data	
Sum	650
n	25
Mean	26
Median	26
Mode	26

Table 13 shows the three averages for our set of data. Also shown are the sum and the number of data items. We can see that in this case the three averages are identical (a lucky coincidence!) but this is generally not the case.

Averages in Excel

Calculating averages in the spreadsheet is easy enough if we know how to create simple formulae. We recall that the formula to calculate the mean is the sum of all the values divided by how many items there are (Figure 53).

To determine the mean in Excel we need to create a formula for the sum, a formula for the count (how many there are) and a formula to work out the mean.

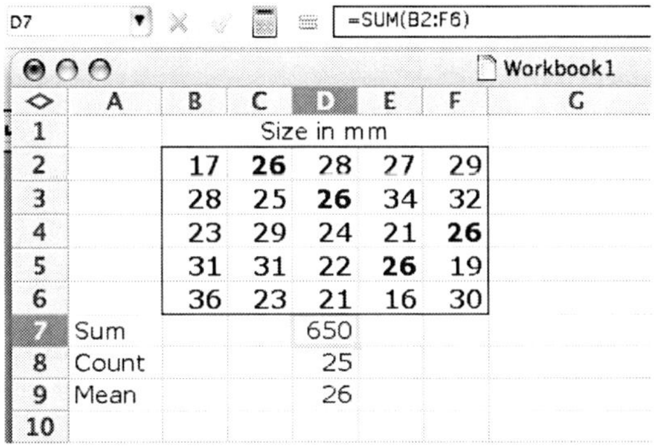

Figure 55. Calculating the mean using Excel formula

In Figure 55 we can see our beetle size data again. To determine the sum of the values we use the SUM function. The values in the brackets simply refer to the rows and columns where the data are to be found. In this case we start at B2 and continue to F6, hence we define the top left and bottom right of the area where the data are found. It would have been better to set out the data in a single column but here the table has been spread over five columns to make it fit on the page easier.

To determine the count we use the COUNT function. In this instance we would use the same range of cells. To calculate the mean from the example above we need to divide D7 by D8. Finally we create a formula in cell D9. This would simply be D7/D8.

This is fine but it would be nice if we could work out the mean in one step; Excel has built-in formulae for working out all three averages. The functions are AVERAGE, MEDIAN and MODE for the mean, median and mode respectively. In our example above we simply replace the name of the function for the one we want and keep the cell range (the B2:F6 part) the same. It is unfortunate the name of the mean function is "average".

Averages in R

We can use the R program to determine averages as well. First of all we need our values. In the example below we have used the *scan()* function to get the data in from the spreadsheet. The data have been named *bd*, which effectively stores them in the memory and allows us to perform various operations on the numbers.

```
> bd
[1] 17 26 28 27 29 28 25 26 34 32 23 29 24 21 26 31 31 22 26 19 36 23 21 16 30
> mean(bd)
[1] 26
> median(bd)
[1] 26
> mode(bd)
[1] "numeric"
> as.numeric(names(which.max(table(bd))))
[1] 26
```

The first command is simply to type the name of the data item. It shows us the data we have. It is not really necessary but it is sometimes helpful just to be sure that you are working on what you thought.

To get the mean of our beetle data (*bd*) we use the *mean()* command. We see the result of 26. To see the median we use the *median()* command. The result is also 26.

We might think that the mode would be *mode(bd)* but as you can see from the example above we get an unexpected result. We actually have to use the rather daunting looking formula instead. The good news is that we almost never use the mode! More good news is that once you know what the command is, you can save it in a text file and paste it in when you need it; simply replace the *bd* with the name of your data item.

If we side-track briefly and look at the last command, we'll see what R did. The command was:

```
> as.numeric(names(which.max(table(bd))))
```

Let's take the last part and type that in by itself:

```
> table(bd)
bd
16 17 19 21 22 23 24 25 26 27 28 29 30 31 32 34 36
 1  1  1  2  1  2  1  1  4  1  2  2  1  2  1  1  1
```

This table command has split the data and shows us how many items there are of each value. This is potentially useful, as we shall see soon. We can readily see that the value 26 occurs most (four times) but we need to be a bit more explicit to tell R what is required.

We might have wanted to work out the mean from the sum and the number of items. Try this for yourself:

```
> sum(bd) / length(bd)
```

The *length()* command is like the COUNT function in Excel. Here we join the *sum()* and *length()* commands together. Try each one separately to convince yourself that each gives the expected result.

4.1.2 Which average to use

The best average to use is the one that best describes the data. In general the mode is the least useful of the averages and is useful only when the dataset contains a large quantity of numbers (maybe $n > 1000$). The mean and median are much more useful. The mean would be the average to use when the data are fairly symmetrical, with more or less equal values above and below the mean value. The median is a more useful average when there is a preponderance of low values (or high ones).

4.2 Distribution

This brings us to our next point. We wish to use the most appropriate average but how do we determine whether we have a preponderance of low (or high) values or if the data are evenly spread? We need to look at the distribution of the data. One easy way to do this is to make a tally plot.

Table 14. Tally plot to show frequency distribution for beetle carapace lengths. The data appear normally distributed

Tally	Bin
X	16
X	18
X	20
XXX	22
XXX	24
XXXXX	26
XXX	28
XXX	30
XXX	32
X	34
X	36

In Table 14 we see the data separated into categories (often called bins); here the lowest bin is for values up to 16, the next covers values > 16 but not above 18 and so on. For each category we put a cross (or a tick or tally mark) against the appropriate bin. When we are done we see a "picture" of our data. In this instance we can see that the tallies are pretty evenly spread around the mid-point, the plot looks fairly symmetrical about the middle (this is called normal distribution or parametric). This would be a good situation to use the mean as our average. If the plot had appeared like the one below (Table 15) then we might make a different choice.

Table 15. Tally plot of data that is not normally distributed. There are more values at the low end and the distribution is not symmetrical

Tally	Bin
x	16
xxxx	18
xxxxx	20
xxxxxx	22
xxx	24
xx	26
xx	28
x	30
x	32
x	34
x	36

In Table 15 we see a different situation. We have a skewed distribution here and the plot is not symmetrical, we have a long tail at the higher end. This indicates that we have more low than high numbers. In this case we would select the median as our average. This skewed distribution is also known as non-parametric.

You can produce a tally plot from your data simply enough using a notebook and pencil; it is the sort of thing you can do as you go along.

Another quick way of looking at the distribution of the data is to use a stem and leaf plot. In Table 16 the data are written out in two columns with the left column being the 10s and the right column the units.

Table 16. Stem and leaf plot of beetle carapace data as an alternative to a tally plot. The advantage over tally plots is that the original data values can be seen

Stem–leaf plot of data	
1 |	679
2 |	112334
2 |	5666678899
3 |	1124
3 |	6

Table 16 shows our data as a stem and leaf plot. From this we can see quite easily that our data are fairly evenly distributed. In this instance each block of 10s has been split into two with 0–4 being in one half and 5–9 in the other. This is done to make the plot more readable and to give us more bins. For other sets of numbers it may be okay to use a single block for each division of 10.

You can create a stem and leaf plot simply in a notebook using a pencil, which makes it a useful tool for when you are out in the field. In the examples shown here the values are in order, which makes it easier to read but is not essential. We can generate a stem plot using R quite simply using the *stem()* command (see Section 12.5.5).

```
> stem(our.data)
```

You simply replace the *our.data* part with the name of the data you are using. There is no easy method of making a stem and leaf plot in Excel.

4.2.1 Histograms

The tally plot (e.g. Table 14) and stem and leaf plots (Table 16) are both useful ways of looking at the distribution of the data. Both can be done in the field as you are collecting data. Another way of presenting the distribution is to create a histogram or bar plot of the data. In this case you have a series of bars, each relating to a bin category, and their height is related to the frequency of the data in that bin.

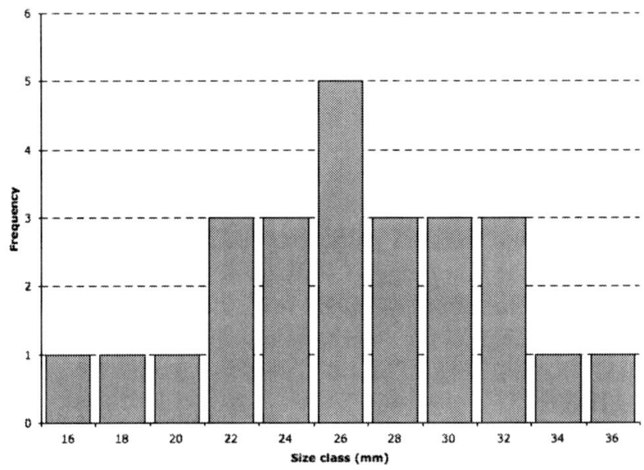

Figure 56. Histogram of beetle carapace data. This plot was produced in Excel

Figure 56 is a histogram of the beetle data in Table 12. The x-axis shows the size classes (i.e. the bin categories) and the y-axis shows the frequency. Here the horizontal gridlines have been preserved (as dashed lines) but often these will be omitted. In a true histogram the x-axis represents a continuous variable (although divided into parts) and the bars should be touching; here they have been separated by a short interval for clarity.

Histograms in Excel

In Excel the graph made is a simple bar chart (see Section 12.4.3 for more details). The frequency data are entered using the FREQUENCY formula.

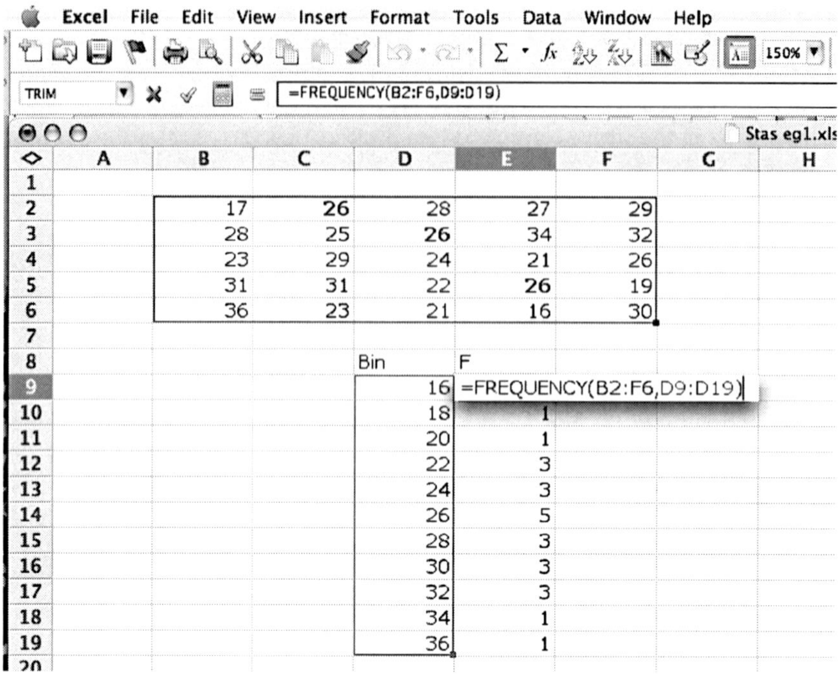

Figure 57. Using the FREQUENCY formula in Excel to produce data for a histogram. The data shown are beetle carapace lengths

In the FREQUENCY formula the first parameter (B2:F6) points to the data, the second parameter (D9:D19) points to the range of bins. This type of formula is called an array formula in Excel and is entered simultaneously in all the cells (here it would be in E9:E19). We can do this by highlighting all the cells and entering the formula. Instead of pressing ENTER, you press CTRL+SHIFT+ENTER (in Windows), see the Excel help for details (on a Mac press cmd+ENTER).

Each bin is written in the spreadsheet as a single numerical value. In reality each bin represents a range of values. In Figure 57 for example the first bin (16) represents values ranging up to 16. The next bin (18) represents values greater than the previous bin but only up to 18.

We can also use the *Analysis ToolPak* (Section 1.9.1) to generate the frequencies that we will need. We still need to create a list of bins but the *ToolPak* takes out the need for us to grapple with the FREQUENCY formula.

We run the histogram routine by clicking *Data Analysis* on the *Data* ribbon menu. We then select *Histogram* from the options and select our data. In the following example (Figure 58) we see the results of the routine; we have a simple table with our bins and the frequencies.

B	C	D	E	F
Beetle Sizes in mm				
17	**26**	28	27	29
28	25	**26**	34	32
23	29	24	21	**26**
31	31	22	**26**	19
36	23	21	16	30

Bin			*Bin*	*Frequency*
16			16	1
18			18	1
20			20	1
22			22	3
24			24	3
26			26	5
28			28	3
30			30	3
32			32	3
34			34	1
36			36	1
			More	0

Figure 58. The *Analysis ToolPak* can produce frequencies to be used in a histogram

Once we have our frequencies we can create our histogram as a bar chart (see Section 12.4.3 for more details).

Histograms using R

In R we can create histograms using the *hist()* function.

```
> hist(bd)
```

Here we create a histogram of our beetle data (which were saved as a named object called *bd*). The resulting output looks like Figure 59.

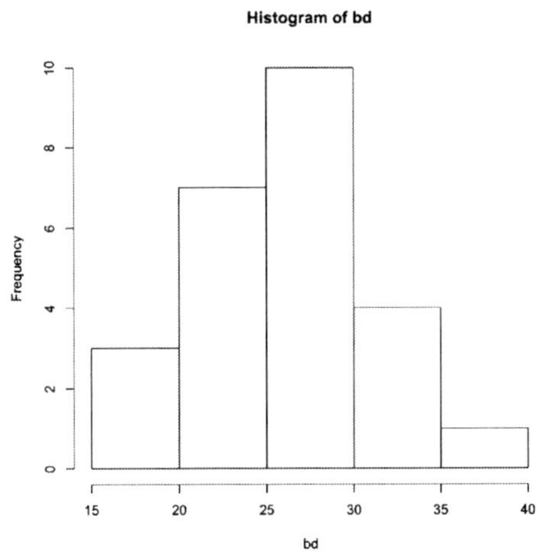

Figure 59. Histogram of beetle size data created using R

It is possible to specify the exact bins in R but in practice it is generally best to let the program work out the best arrangement for itself. Notice how R arranges the values on the category axis (the *x*-axis). The values are placed at the edges of the bars, highlighting the fact that the axis represents a continuous scale rather than discrete categories.

It is also possible to use points (joined with smooth lines) instead of bars but this generally does not look good unless you have a lot of data. In the R program there is a way to produce a density plot, which is a similar chart. Our beetle data have been graphed as a density plot in Figure 60.

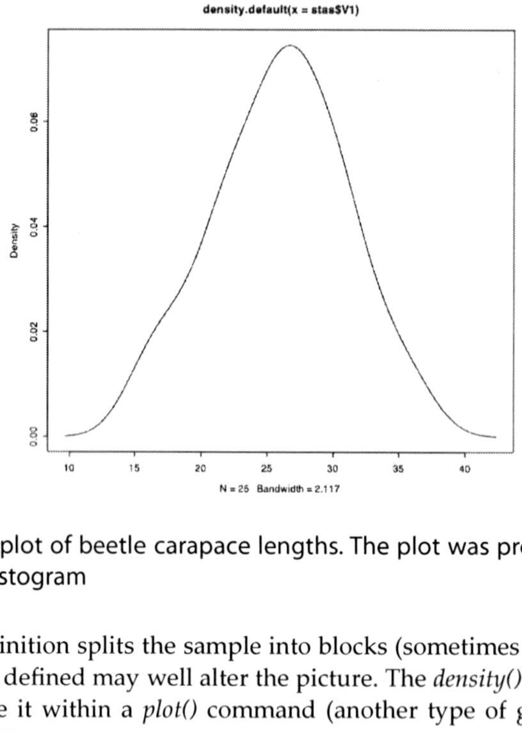

Figure 60. Density plot of beetle carapace lengths. The plot was produced in R and is an alternative to the histogram

A histogram by definition splits the sample into blocks (sometimes called bins) and how the breakpoints are defined may well alter the picture. The *density()* function provides an alternative. You use it within a *plot()* command (another type of graph, which we will cover in Section 12.4):

```
> plot(density(bd))
```

In this case, we used the beetle data we already saved as an item named *bd*. It is possible to create a regular histogram and add the density lines afterwards. This is sometimes useful because it can help you determine if the data are normally distributed or not.

```
> hist(bd, freq = F)
> lines(density(bd))
```

First we create a histogram and add a bit on the end (*freq* = *F*), which tells R to make everything add up to 1 rather than use the actual frequencies. The next command, *lines()*, adds lines to the graph. The lines we want are the density function of our data.

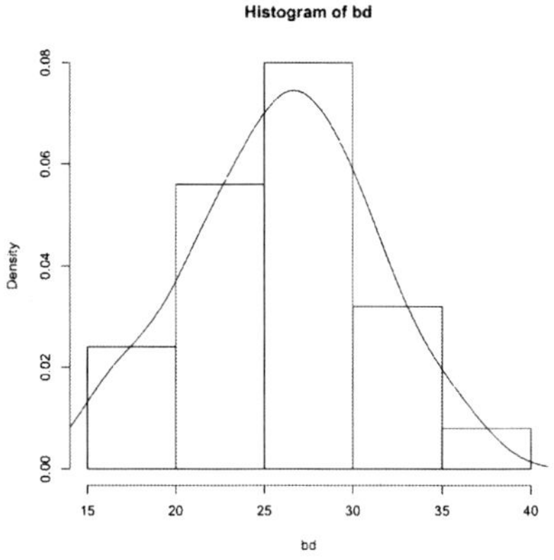

Figure 61. Density plot of beetle carapace lengths superimposed on a histogram of the same data

4.3 A numerical value for the distribution

So far we have seen how to put a value to the middle, or central tendency, of our data (the average) and how to describe the distribution graphically. It would be nice if we could use a number to describe our distribution to go with the number that we have for the average.

4.3.1 Range

The simplest thing would be to present the range of values, i.e. the smallest and largest. For our beetle data (Table 12) we would have a range of 36 – 16 = 20. This is all very well but not all that informative. We could have two sets of data with the same range but very different distributions (see the previous tally plots Table 14 and Table 15 for example); however, we can gain some insight into the distribution if we use the median as well. In the first example, the median will lie more or less half-way between the ends of the range. In the second example, the median will lie closer to the low end.

In Excel we can use several functions to give us the range. There are built-in functions to determine the largest and smallest values for example. Let's return to our beetle data example from Table 12.

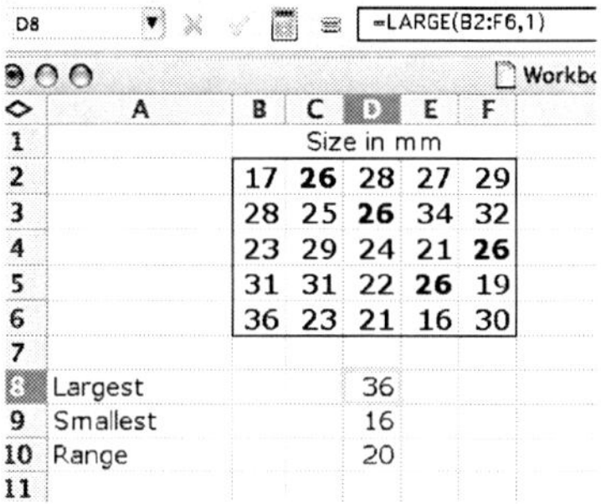
D8 =LARGE(B2:F6,1)

Workbo

	A	B	C	D	E	F
1				Size in mm		
2		17	**26**	28	27	29
3		28	25	**26**	34	32
4		23	29	24	21	**26**
5		31	31	22	**26**	19
6		36	23	21	16	30
7						
8	Largest			36		
9	Smallest			16		
10	Range			20		
11						

Figure 62. Calculating the range using Excel. We use the built-in functions for determining the largest and smallest values and then subtract one from the other

The Excel formula for working out the largest value from a range of values is LARGE. We insert the range of cells into the formula (B2:F6 in this case) and also the number 1 after a comma. This tells Excel to look for the first largest value in this range of cells. We can see that we get 36. The SMALL command does something similar except that this time it looks for the first smallest. We could of course alter the 1 to a 2 and get the second largest or the second smallest but we don't want to do that here.

To determine the range we simply subtract the smallest value from the largest (D8–D9) and in cell D10 in Figure 62 we see that this is done (36 – 16 = 20).

We can do something similar using R, although the largest value (the maximum) is found using the *max()* function whilst the minimum is found using *min()*. In the following example we use the beetle data once again:

```
> bd
[1] 17 26 28 27 29 28 25 26 34 32 23 29 24 21 26 31 31 22 26 19 36 23 21 16 30
> max(bd)
[1] 36
> min(bd)
[1] 16
> max(bd) - min(bd)
[1] 20
```

To get the range we obviously take the smaller value away from the larger one. Here we use the formula together. This also illustrates something about R: the ability to link things together. There is no need for us to calculate each term separately; we do the whole calculation in one step.

4.3.2 Quartiles

Now we have the end points of our distribution (the range) and the middle (median or mean). We can also divide the data into further halves (so we end up with four chunks) and the results are the quartiles. If we take the middle value between the lowest value and the median, we get the lower quartile. If we take the middle value between the highest point and the median, we get the upper quartile. These two values are also known as the first and third quartiles respectively. The median is the second quartile and the endpoints are the zero and fourth (lowest end and highest end). For our beetle data the values are shown in Table 17:

Table 17. Quartile values for beetle (*Haliplus lineatocollis*) carapace lengths. The median is the middle value and the inter-quartiles (Q1, Q3) are half-way between the median and each end

Min	Q1	Median	Q3	Max
16	23	26	29	36

We can get these quartile values using Excel using the QUARTILE function. In Figure 63 we see our beetle size data once more. This time we have used the quartile function to produce all five of the values we require. The quartile function has two parts: the first is the range of cells containing the data you want to summarise. The second part is a number from 0–4. Zero would represent the lowest quartile (i.e. the minimum value), 1 is the lower quartile, 2 is the second quartile (i.e. the median), 3 is the upper or third quartile and 4 gives the fourth quartile or maximum value.

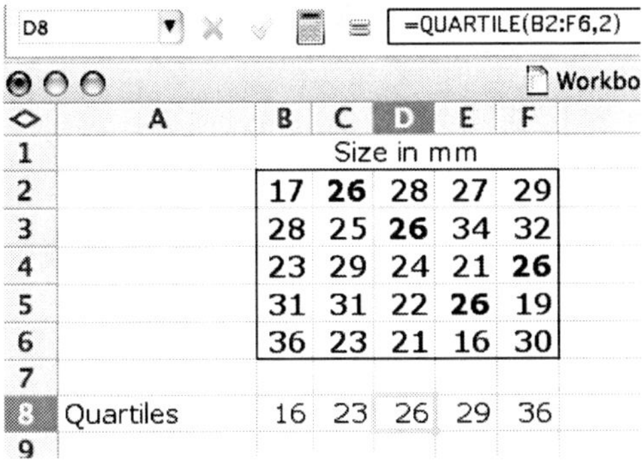

Figure 63. Using the quartile function in Excel. The median value (highlighted in D8) is produced using the second quartile

In R we can produce the quartiles using similar commands; there are two to choose from:

```
> summary(bd)
   Min. 1st Qu.  Median    Mean 3rd Qu.    Max.
     16      23      26      26      29      36
> quantile(bd)
  0%  25%  50%  75% 100%
  16   23   26   29   36
```

The first command, *summary()*, is a general command and will produce a different report depending on what your data are (as we shall see later). Here we get all the quartiles and the mean as well. There is a more specialised command, *quantile()* for producing quartiles, and we see this presented after the *summary()* command.

As well as producing all the quartiles, the command can produce a single value for us. We simply need to tell R which quartile we want. We can see that we have five to choose from. To get the 25% quartile we type:

```
> quantile(bd, 0.25)
```

To get the 75% we type:

```
> quantile(bd, 0.75)
```

Of course if our data are called something other than *bd* we need to change the command to reflect this. The command is called *quantile* and not quartile, like in Excel. Strictly speaking quartiles split things into quarters but a quantile can be any proportion (try using 0.2 in the command above for example).

Now we have a better description of our data using fewer numbers than the original dataset. We can produce a plot of these five values to give a visual representation of our data. The graph is called a box–whisker plot.

4.3.3 Box plot

It is always a good idea to represent our data graphically before we do anything else and the box–whisker plot is a good place to start. We will look at different types of graph briefly in Chapter 6, but we will cover graphs in a lot more detail in Section 12.4.

Figure 64 shows a box plot (also known as a box–whisker plot) of our beetle data from Table 13.

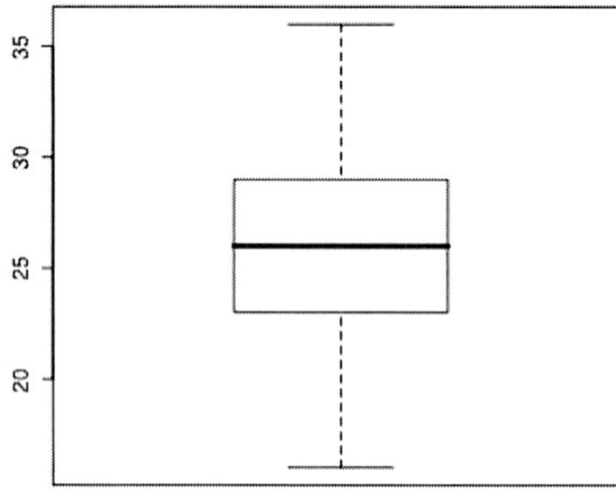

Figure 64. Box–whisker plot of beetle (*Haliplus lineatocollis*) carapace lengths. The stripe shows the median, the whiskers show the maximum and minimum values and the box shows the inter-quartiles

The box plot (Figure 64) shows the five descriptive values: the median is shown as a stripe (sometimes a dot). The box shows the inter-quartile range, i.e. the distance between the first and third quartiles. The whiskers extend out to the full range, i.e. the maximum and minimum. This sort of plot is possible to do in Excel but it is not straightforward and the program needs a lot of coercing (you need to use one of the stock graphs, see Section 12.4.3). It is, however, simple to do in R using the *boxplot()* command.

To create the box plot we put the name of our data (we'll use the *bd* beetle data from before) in the brackets:

```
> boxplot(bd)
```

This is a bit sparse but we'll see later how to add better axis titles and other items to make our plot look better. For the time being we can gain an impression of our data just from looking at the plot. If the data are normally distributed, then the middle stripe will be more or less in the middle and the box–whiskers will be symmetrically arranged either side. If the data are skewed, then the stripe and the box–whiskers will not be symmetrical. It might be helpful to think of the box plot as a kind of contour map showing our sample of data as a kind of hill. Just like reading a map we can visualise the shape of the ground by looking at the contours.

4.3.4 Standard deviation

The range and quartiles provide a useful measure of the distribution of the data. However, there are other measures that may be used, especially if the data appear to be symmetrically arranged (as in the case of our beetle data). One such measure is called standard deviation. To calculate it we use the formula shown below (Figure 65).

$$s=\sqrt{\frac{\sum(x-\bar{x})^2}{n-1}}$$

Figure 65. Formula to calculate standard deviation of a sample

This is simple enough to calculate using Excel (and many calculators); in fact there is even a function to do it: STDEV. The steps required to determine standard deviation are shown in Table 18.

First of all we calculate the mean; in our example this is shown with the sum and the count of items. We could use the AVERAGE command but since we need the count anyway we'll stick with it for now. Next we subtract the mean from each item of data (this is the $x - \bar{x}$ column). Of course we get positive and negative values because some of the original data are larger than the mean and some are smaller. If we simply add these together we get zero. We need to remove the negative signs and to accomplish this we will square each value (a common thing in statistics). The squares are shown in the third column, headed $(x - \bar{x})^2$. We now add these up (getting 612 in this case) and divide by $(n - 1)$. Finally we take the square root (since we squared earlier this brings the value back to the same order of magnitude).

What we have done essentially is to take the average deviation from the mid-point. As our example shows, we could do the calculation in a spreadsheet easily enough but there is also a function that allows us to determine standard deviation directly without all the intermediate steps.

The command is simply STDEV. In the brackets we put the cell range where the data we want to summarise can be found (in Figure 66 the range is B2:F6).

Table 18. Calculating the standard deviation of a sample

	x	$x - \bar{x}$	$(x - \bar{x})^2$
	17	−9	81
	28	2	4
	23	−3	9
	31	5	25
	36	10	100
	26	0	0
	25	−1	1
	29	3	9
	31	5	25
	23	−3	9
	28	2	4
	26	0	0
	24	−2	4
	22	−4	16
	21	−5	25
	27	1	1
	34	8	64
	21	−5	25
	26	0	0
	16	−10	100
	29	3	9
	32	6	36
	26	0	0
	19	−7	49
	30	4	16
Sum	650	0	612
n	25		
mean	26.0		
s	5.05		

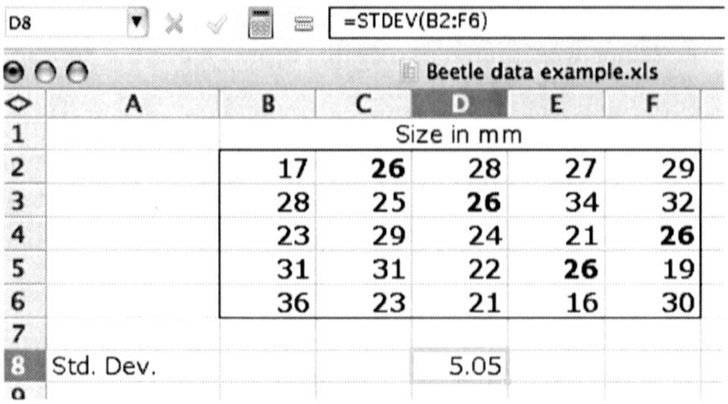

D8
=STDEV(B2:F6)
Beetle data example.xls
A B C D E F
Size in mm
17 26 28 27 29
28 25 26 34 32
23 29 24 21 26
31 31 22 26 19
36 23 21 16 30
Std. Dev.
5.05

Figure 66. Calculating standard deviation in Excel

In R, the command to work out standard deviation is *sd()*. Below we see the command in action on our beetle data, which we saved to an item we called *bd*.

```
> sd(bd)
[1] 5.049752
```

Figure 67 shows two sets of data. They both have the same mean (40) and number of items ($n = 33$) but they have quite different standard deviation.

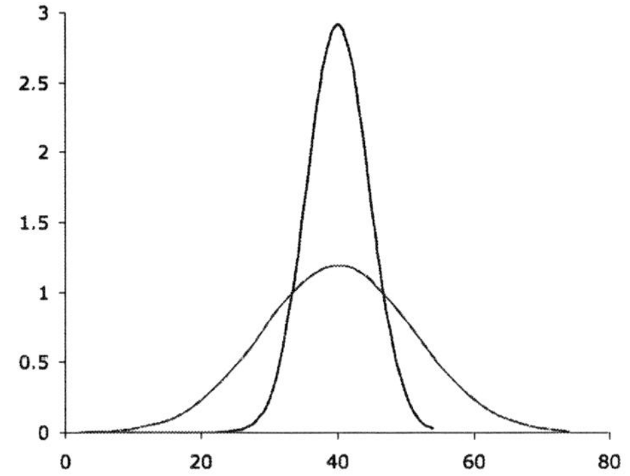

Figure 67. Frequency plots of two samples. Both have the same mean but have differing dispersion (standard deviation for the narrow curve = 4.5, standard deviation for the wider curve = 11)

In Figure 67 we see how the data with the largest standard deviation has the wider, flatter spread on our frequency graph. The thinner plot has a standard deviation of 4.5 whilst the fatter plot has a standard deviation of 11.

Standard deviation has an important property. If you plot your frequency graph and draw vertical lines at 1 standard deviation above and below the mean you will encompass (between the lines) 67% of all the data. If you move your lines out to 2 standard deviations (plus and minus) you encompass 95% of all the data. This property becomes important in statistical testing (Chapter 5).

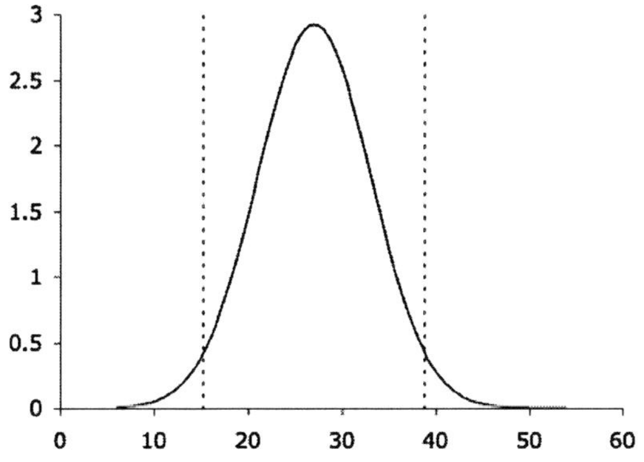

Figure 68. Properties of the normal distribution curve. Dashed lines are shown at +2 and −2 standard deviations from the mean. Only 5% of the data lies outside this range

In Figure 68 the dotted lines are at +2 standard deviations and –2 standard deviations and the space between them is occupied by 95% of the data. Put another way, only 5% of the data lie further than 2 standard deviations from the mean.

Why use (*n* – 1)?

You may have wondered why we divided our formula by ($n - 1$) rather than n. The reason is that we are working on a sample. In most cases you will be collecting data from only some of the available items. For example, our beetles from Table 12 represent some of the individuals that were in the pond. We might spend a long time and get every single beetle out but it seems unnecessary. What we do instead is to collect a representative sample of the population. We are therefore working out estimates of the mean and standard deviation (if we worked out median and range, they would also be estimates). This is taken into consideration by using $n - 1$. Think of it as a kind of correction factor. If n is quite small then subtracting 1 might make quite a large difference. If n is pretty large then subtracting 1 makes a smaller difference. In this case as n approaches the true size of the population (beetles in this instance) then the correction factor gets smaller and smaller. Let's illustrate using a new example.

You are asked to determine how large the leaves on a shrub actually are. This seems simple enough so you rip off a bunch and get out a ruler. It turns out that your handful was ten leaves. Because you didn't measure all the leaves you, divide by $n - 1$ instead of n when you do your calculation of standard deviation. What you have done is to use nine instead of ten as the divisor in your calculation, a 10% difference. In a fit of zeal you go out and get nine more handfuls and measure a further 90 leaves, making 100 in total. Now you divide by 99 (which is $n - 1 = 100 - 1$) and get a new estimate. Now your divisor is only 1% different. As you collect more and more leaves, your correction factor (the –1 part) is getting smaller and smaller.

Table 19. Effects of using *n* – 1 in calculation of standard deviation

n	*n* – 1	% diff.
10	9	10
100	99	1
1000	999	0.1
10000	9999	0.01

Of course the final value of the standard deviation is not altered by the same degree because the formula is a little more complicated. So, how much difference does it actually make? Here is an example that shows the correction factor in action. The readings are for the leaves from the shrub we've been discussing. All the numbers are the maximum length in millimetres. Each handful contains leaves of exactly the same size.

Table 20. Leaf size data (max length in mm) to illustrate the effect of $n-1$ in standard deviation calculation

	31	31	31	31	31	31	31	31	31	31
	33	33	33	33	33	33	33	33	33	33
	31	31	31	31	31	31	31	31	31	31
	36	36	36	36	36	36	36	36	36	36
	29	29	29	29	29	29	29	29	29	29
	32	32	32	32	32	32	32	32	32	32
	35	35	35	35	35	35	35	35	35	35
	34	34	34	34	34	34	34	34	34	34
	38	38	38	38	38	38	38	38	38	38
	27	27	27	27	27	27	27	27	27	27
n	10				50					100
Mean	32.6				32.6					32.6
σ_n	3.14				3.14					3.14
σ_{n-1}	3.31				3.17					3.15

The first column contains the readings from the first ten leaves. At the bottom are some of the summary statistics. We can see that the population standard deviation is 3.14 whilst the sample standard deviation is 3.31. Our correction factor has made quite a difference. In the middle we have presented summary statistics for the first 50 leaves. The mean is of course the same because each handful has leaves of identical size. The population standard deviation is also the same because it's based on n, which we haven't corrected. The sample standard deviation is now much closer. We have used a divisor of ($n - 1 = 49$) which is only a 2% "correction" (compared to the 10% we used for the first 10 leaves). The final column shows the summary for all 100 leaves. The two measures of standard deviation are now very close (the correction factor is only 1% of the divisor).

4.4 Statistical tests for normal distribution

In many cases it is sufficient to look at our histograms or stem plots in order to determine if our data are normally distributed; however, there are ways to examine the distribution of the data mathematically. The properties of the normal distribution are known quite explicitly. If we know the mean and standard deviation, we can plot a graph. The formula to work out the normal distribution is shown in Figure 69.

$$y=\frac{1}{\sigma\sqrt{(2\pi)}}e^{-[(x-\mu)^2/2\sigma^2]}$$

Figure 69. Formula to determine normal distribution

The formula looks quite daunting and you might be glad to know that we shall not delve into any great detail. There are only two quantities that need to be known: standard deviation (σ) and mean (μ). All we really need to know is that it is possible to use this to determine if a sample of data follows the normal distribution. The simplest way to do this is to use the Shapiro–Wilk test in the R program.

4.4.1 Shapiro–Wilk test for normality in R

Once we have some data, we can perform a test for normality quite simply using the *shapiro.test()* command.

```
> shapiro.test(our.data)
```

Here we replace the *our.data* part with the name of our sample of values. In the example below we see the Shapiro–Wilk test carried out on the beetle data we met earlier (Table 12).

```
> Jun
[1] 17 26 28 27 29 28 25 26 34 32 23 29 24 21 26 31 31 22 26 19 36 23 21 16 30
> shapiro.test(Jun)

  Shapiro-Wilk normality test

data:  Jun
W = 0.9882, p-value = 0.9885
```

We can see that the result is not significant ($p = 0.9885$). In other words, the data (*Jun*) are not significantly different from a normal distribution. We also have some data from beetles collected at a different time of year; these data are called *Mar* and the result is shown below:

```
> Mar
[1] 18 21 18 51 46 47 50 16 19 15 17 16 52 17 15 16 17 49 48 16 18 16 17 15 20
> shapiro.test(Mar)

  Shapiro-Wilk normality test

data:  Mar
W = 0.6662, p-value = 2.644e-06
```

In this case we see that the *p*-value is significant (here *p* is a very small number), which indicates that the sample (*Mar*) is skewed, i.e. is significantly different from a normal distribution.

4.5 Distribution type

When your data are symmetrical, as in the beetle data (see the tally plot, Table 14), the distribution is called normal distribution or parametric. In this case the mean makes the best average and the standard deviation the best measure of the spread (also called dispersion). When your data are not normally distributed, i.e. non-parametric or skewed (see the tally plot, Table 15), the median makes the best average and the dispersion is best modelled using the range and quartiles.

Table 21. Which measure of centrality and dispersion (spread) to use according to distribution of data

	Parametric	Skewed
Average:	Mean	Median
Dispersion:	Standard deviation	Range and Quartiles

4.5.1 Standard error

When we collect data we are usually sampling from a larger population. If we think back to the beetle data in Table 12, we collected individuals from a pond and measured them. We did not get every beetle from the pond. What we have is a sub-set of the pond, a sample. When we work out the mean (or the median) we are really estimating the population mean. As we collect more individuals, our estimate should get closer to the real true mean of the population.

Imagine that we took a sample from a population. We can determine the mean in the usual way. Now let us return to the pond and collect another sample. We can work out the mean again. The two means are unlikely to be exactly the same. This situation is represented in Figure 70. The larger curve represents the complete population and shows the frequency distribution (normally distributed in this case). We also see three smaller curves. These represent three samples taken from the larger population (we went to the pond three times).

If we look at the means of the samples, we see that some are be smaller than the overall population mean and some are larger than the population mean. We could carry on taking samples and calculating the means and find out that the means would be normally distributed, just like the population data. If we took these sample means, we could work out how far away they were from the original population mean. We can also work out the standard deviation of these sample means from the population mean.

This deviation of samples from the population mean is called standard error.

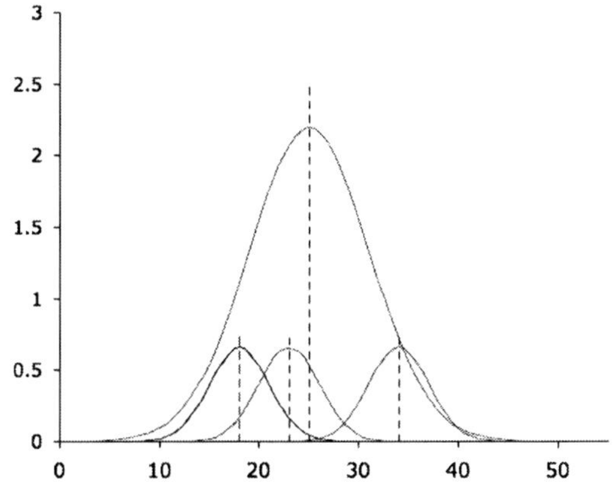

Figure 70. Standard error. The large curve represents a complete population. The small curves represent separate smaller samples taken from this population. Standard error is the standard deviation of these sample means from the population mean

If we look at Figure 70, we can see that if the overall population had a very large standard deviation, the large curve would be wide and fat. We would expect our samples to be a bit more widespread and this would make our standard error larger. If the original population had a small standard deviation, the large curve would be tall and thin. Our samples would probably be closer together and result in a smaller standard error.

We can illustrate using an example. Previously we looked at some leaf sizes and the data were shown in Table 20. In this instance it was remarkable that each handful of ten leaves was identical in size. It would of course be a lot more realistic if each handful of leaves were different sizes. In Table 22 we see more leaf data. This time each handful (our sample) contains ten leaves but they are all different sizes.

Table 22. Leaf sizes in millimetres. Each sample of ten leaves has an associated mean value calculated

	34	35	33	26	35	35	35	32	38	29
	32	34	33	29	37	37	31	37	31	28
	30	38	34	33	33	32	29	35	29	30
	36	32	31	33	35	35	31	34	29	29
	36	34	37	28	27	32	28	32	36	32
	36	31	33	38	33	32	36	33	34	31
	41	34	30	27	34	34	29	27	37	31
	35	34	27	35	28	31	31	36	32	32
	30	33	36	34	29	33	28	34	37	34
	34	33	31	30	36	26	29	35	34	35
μ	34.2	33.7	32.5	31.3	32.8	32.5	30.8	33.7	33.7	31.1

Each sample of ten leaves has a mean calculated. If we take a mean of all 100 leaves we find that it is 32.6. Here we see that the mean of each sample of ten leaves is slightly different. The mean of each sample varies; in some cases it is lower than the true value and in others it is higher. We can think about it like this. We have a total population; in this case we have 100 leaves and their size is normally distributed around a mean value (32.6 mm). Each time we take a sample we have a smaller set of numbers, also distributed normally around a mean. This new mean is an estimate of the true mean of the bigger population.

If we look at the means of the samples we can demonstrate that they are normally distributed using a stem and leaf plot (Table 23).

Table 23. Stem and leaf plot of means of ten samples of leaves (sizes in millimetres)

30	8
31	13
32	558
33	777
34	2

We can determine the standard deviation of these ten means. If we do this we get 1.22, which is a lot smaller than the standard deviation of the 100 leaves (which is 3.11). This new measure of standard deviation is of course the standard error.

We have seen how we could work out standard error in theory but this requires us to know the actual population mean and the whole point of sampling is to avoid having to collect absolutely everything. What we need is a way to estimate it:

$$S.E = \frac{s}{\sqrt{n}}$$

Figure 71. Formula to estimate standard error

The formula (Figure 71) shows how we estimate standard error. We use the standard deviation and divide by the square root of the sample size.

We can do a similar thing even if the data are not normally distributed. We could take medians of samples from a larger population and determine the differences from the overall median; however, the medians of the samples are not likely to be normally distributed, so it is not very useful! What we do instead is to use the quartiles in lieu of the standard error. When we summarise some non-parametric data, we give the median as our central value and the inter-quartile range, i.e. the difference between the third and first quartile (we looked at quartiles in Section 4.3.2).

The effect of both standard error and the inter-quartile range is to smooth out the effects of extreme values in our sample.

Standard error is a useful measure and is used quite a bit in statistical tests. These tests rely on the properties of the normal distribution. Recall from Figure 68 how most of the data lie between the mean and 2 standard deviations. What standard error means in practice is that the more data items you have, the more tightly clustered around the true mean the estimated samples means will be. You can be more confident that you are near the real average when you have more data items. We will come on to the idea of confidence interval next.

4.5.2 Confidence intervals

When we report the results of a simple sampling exercise we need to be succinct. The three important things we need to report are: average, spread and sample size.

If our data appear to be normally distributed then we use the mean and standard deviation as our average and spread (see Table 21). The median and range (or quartiles) would be the way to summarise our data if the sample was not normally distributed.

If we examine the beetle data in Table 12 we can work out the mean and standard deviation. We get $\mu = 26$ and $s = 5.05$. We already know $n = 25$. To report this we could write:

"The mean beetle size was 26 mm (s = 5.05, n = 25)."

The reader can see the situation instantly. There were 25 beetles captured and measured. The average size was 26 mm and the standard deviation was 5.05. You may decide to use the standard error instead of the standard deviation. We work that out to be 1.01 and would write:

"The mean beetle size was 26 mm (SE = 1.01, n = 25)."

Either is acceptable. Similarly when we draw graphs we should represent the spread of data as well as the average and use error bars (we will show how to do this in Section 12.4). We could use standard deviation but the bars would be bigger than if we plotted standard error instead.

If this sounds a little bit like cheating then hold on – the standard error has an important property in statistical testing and this can give us a better impression of the data than standard deviation, as we shall see now.

Calculating confidence intervals

It would be useful to know the general range of sizes that our beetles are likely to be. We can use something called the confidence interval to tell us. It is simple enough to calculate. We need the mean and the standard error. From Figure 68 we know that 95% of the data will be within 2 standard deviations either side of the mean in a normally distributed population. We are dealing with a sample rather than the complete population so we use the standard error to do something similar. We can estimate the confidence interval as 2 × standard error. For our beetles we get $CI = 2 \times 1.01 = 2.02$. This means that 95% of the beetles in our sample ought to be between 23.98 and 28.02 mm in size. We can write this as: 26 ± 2.02 mm ($n = 25$, $CI_{0.95}$) indicating to our reader mean, spread and sample size like before.

Different levels of confidence interval

Our confidence interval calculated above was an estimate. It so happens that we can determine confidence intervals for 95%, 99% and 99.9%. The properties of the normally distributed curve have been studied intensively and statisticians have determined that 95% of the data will lie between ± 1.96 standard deviations (rather than the 2 we mentioned previously). Table 24 shows how many standard deviations we expect for different levels of probability. We can use these values to determine confidence intervals. These values are part of a family of numbers relating to the normal distribution and we'll come across the related *t* statistic later (Section 7.1) when we look at differences between two samples.

Table 24. Values used for calculating confidence intervals

Number of standard deviations	*p* level
1.96	0.05
2.58	0.01
3.29	0.001

The p level is the level of probability. We have been using the percentage of the data that lie within the confidence interval (e.g. 95% of the data will lie within the interval) but we can also think of this in another way. We can turn this around and say that 5% of the data will lie outside of the confidence interval. In other words there is a probability of 0.05 that the data will lie outside the confidence interval. If we use 1.96 instead of 2 to work out our beetle confidence interval (at 95%, $p = 0.05$) we get 26 ± 1.98 mm meaning that 95% of all the beetles ought to be between 24.02 and 27.98 mm in size. Alternatively only 5% will be smaller than 24.02 mm or bigger than 27.98 mm in size.

Confidence intervals in statistical testing

We can use these confidence intervals to say something meaningful about two (or more) samples. For example, let us return to our beetle example. We have a sample from a population and have determined the mean and 95% confidence interval. If we took beetles from another pond we could do the same for the new sample.

If the two confidence intervals do not overlap we can be pretty sure that there is a real difference between the sizes of beetles from the two ponds. In fact we can be 95% certain because our confidence interval tells us that only 5% of the data are likely to lie outside this range.

This idea is behind statistical tests that look at differences between samples and where the data are normally distributed. This is why the properties of the normal distribution are so important. The idea of the confidence interval also leads us to think about things in terms of probability or likelihood. There are few occasions when we can be totally certain of our results. This is simply because biological data is variable. Rather than look for absolute certainty we strive to put a value to the probability that our result is not down to chance. When we set up our hypothesis (see Sections 1.4 and 5.1) we are looking to put a value on how likely it is that our hypothesis is correct.

4.6 Transforming data

The properties of the normal distribution are so well understood that it underpins quite a few statistical analyses. It is so important that wherever possible we strive to collect normally distributed data. Some data we would not expect to be normally distributed. For example, collecting of invertebrates often results in skewed data with a lot of low counts followed suddenly by a large one! At other times we may expect to get normal distribution, e.g. measurements of size or weight, but do not. We can sometimes get over this by collecting more data and filling in the gaps.

If we end up with data that is not normally distributed, there are things that we might do to the numbers to make the distribution more like our ideal. There is no reason to suppose that a regular scale of numbers is inevitably the best to use for our data. For example, we may collect data that varies in orders of magnitude. This may well be a logarithmic scale. Log scales are reasonably common in the natural world (pH being a prime example).

In order to improve the fit to a normal distribution, we simply perform a mathematical operation on our data. Then we examine the new data and see if the distribution is more like the parametric ideal. There are several commonly used mathematical operations:

- Logarithm
- Square root
- Arcsine (also called angular)
- Reciprocal

Essentially what we do is to take our original data and apply the mathematical operation to all the values. Thus we create a new set of values based on the mathematical operation; hopefully this produces a more normally distributed sample.

4.6.1 Logarithmic transformation

The log transformation is potentially useful when we have data that range across one or more orders of magnitude. These data may arise when dealing with organisms of very different sizes for example. We apply the transformation by using standard log to the base 10 in most cases but there may be situations when the natural log (or indeed any base) seems more appropriate. There is a potential problem with this as any zero values will give an error (as $\log_{10} 0 = \infty$). To get over this possible problem, it is usual to add +1 to all our data, like so:

$$x_t = \log_{10}(x + 1)$$

We then carry on with our *Data Analysis* using the new values. We can work out logs using Excel or R quite simply and the commands are very similar. In Excel we use LOG(data, base) and in R we use *log*(data, base). In both cases we replace the base part with a numerical value that corresponds to the base we want (commonly 10). If we do need to add 1 to take care of zero values we simply modify the command, e.g. *log*(data + 1, base).

In Excel the default is to use base 10 and if we omit the base part of the command that is what we get. In R the default is to use the natural log and that is what we get if we omit the base part of the command. To get natural logs in Excel we use a different command LN(data). If we had negative numbers we can add an appropriate amount to every value so that we eliminate all negatives.

4.6.2 Square root transformation

The square root transformation can be useful in certain cases. It often seems to work in cases where you are counting things, the abundance of an invertebrate in kick samples for example. Applying the transformation is simple enough but usually we add 0.5 to each value before the square root to improve the move to normal distribution:

$$x_t = \sqrt{(x + 0.5)}$$

In Excel the SQRT function will do the trick. In R we use the *sqrt*() command.

4.6.3 Arcsine transformation (angular transformation)

When we are dealing with items that have some natural boundary, we often find that because the ends of the distribution are constrained we do not get normal distribution. Examples would include percentages and proportions, where we usually have boundaries of 0–100 and 0–1 respectively. In such cases the arcsine transformation can improve the distribution. To carry out the transformation, we take the arcsine of the square root of the proportion:

$$x_t = \text{arcsine}\ \sqrt{p}$$

where p is a proportion. If we have percentage data, we divide by 100 to create the proportion and then apply the square root and take the arcsine of that. The result is an angle and will range in value from 0–90°. Table 25 shows some percentages that have been transformed using this method.

Table 25. Examples of percentage values transformed using the arcsine method

Percentage	Arcsine $\sqrt{p}$ (°)
0	0.00
10	18.43
25	30.00
33	35.06
50	45.00
75	60.00
90	71.57
100	90.00

Once we have our new values we can perform our various analyses. Of course if we need to report the original means then we convert the values back to their proportions:

$$p = [\sin(x_t)]^2$$

Obviously if we are dealing with percentages then the final step is to multiply by 100.

Both Excel and R use radians by default and to convert to degrees we need to multiply by $\pi/180$. In Excel π is given using PI() and in R it is simply *pi*. For example sin 45° = $\sqrt{2} \div 2$ = 0.707. To do this in R we use the *sin()* command:

```
> sin(45 * pi/180)
```

In Excel we use: SIN(45 * PI()/180).

To go from degrees to radians we multiply by $180/\pi$. To carry out the angular transformation of 10% for example we would use the *asin()* command in R:

```
> asin(sqrt(0.1))*180/pi
```

In Excel we type:

```
ASIN(SQRT(0.1))*180/PI()
```

In both Excel and R we can convert a range of values and we would replace the 0.1 with the range of cells (in Excel) or the data name (in R) that we want. We must not forget that percentage values need to be divided by 100 first.

4.6.4 Reciprocal transformation

The reciprocal transformation might prove useful in certain cases where you have various dilutions for example. In these cases the original values are simply transformed thus:

$$x_t = 1/x$$

4.6.5 Reporting results from transformed variables

When we use transformations on our analyses we are improving our chances of getting a realistic result but making it harder to understand the resulting numbers. What we must do at some point is to un-transform the data to report mean values for example; however, the act of transforming the data is generally asymmetric and measurements of standard error would be misleading. What we need to do is to determine the confidence limits using the transformed values and report these in their un-transformed form.

4.7 When to stop collecting data? The running average

We have seen how to summarise data using averages and spread (e.g. standard deviation, quartiles). In Section 1.6.3 we thought about how much data we ought to collect and introduced the idea of the running mean. However, we did not go into any detail at that point because we had not even considered dealing with averages.

The idea behind the running mean is to give us an idea of how close to the true mean we might be. We start with a single value, our first measurement. Then as we add more data we calculate the mean as we go along, recalculating each time we obtain a new measurement. We can plot these means in a simple scatter graph as we proceed. In Figure 72 we see an example of a scatter plot of running means. The data are taken from our leaf size example shown in Table 22.

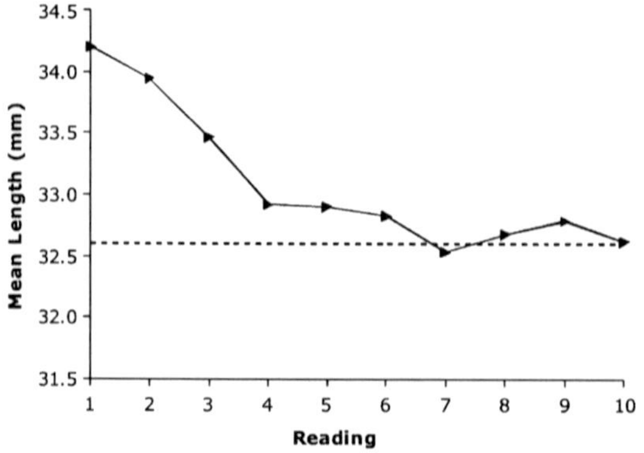

Figure 72. Example of a running mean. Each reading is a mean of ten values (leaf sizes in mm). The dashed line shows the overall mean of 100 leaves (32.6 mm)

We can see that early on the mean leaf size is somewhat larger than the mean for the 100 leaves eventually sampled (32.6 mm). As we collect more data the running mean levels out. We can do this sort of graph in our field notebooks as we go along. At other times we are likely to want to check that we have collected enough data by doing the running means after we've returned from the field (we assume that we could go out again and collect more data another time). If we are creating a report or simply exploring the data we will probably want to do this on a computer. We can use our spreadsheet or the R program to work out and display running means quite easily.

4.7.1 Running means in Excel

We can calculate running means easily using a spreadsheet. Excel provides us with the AVERAGE function for example and this can be pressed into service quite simply. First of all we assemble our replicate data. In Figure 73 we see some leaf size data from Table 22; in fact it is the final column of ten.

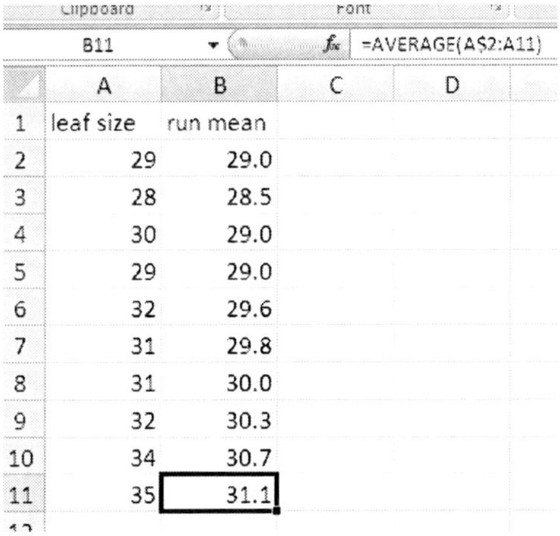

	A	B	C	D
1	leaf size	run mean		
2	29	29.0		
3	28	28.5		
4	30	29.0		
5	29	29.0		
6	32	29.6		
7	31	29.8		
8	31	30.0		
9	32	30.3		
10	34	30.7		
11	35	31.1		

Figure 73. Calculating a running mean using Excel. The $ sign forces the cell reference to remain absolute even when copied to another location

The formula to calculate the mean is AVERAGE and we supply the range of cells corresponding to our data in the brackets. For the first cell we use =AVERAGE(A2:A2). A2 is the cell where our first value lies. We could have used =A2 as we simply need the value; however, we are going to use a useful feature of cell referencing in Excel (and other spreadsheets).

First of all we need to edit the formula and insert a dollar sign ($) between the A and the 2: =AVERAGE(A$2:A2). Note that we do not do this for the second A2, only the first. Now we can copy and paste (or drag the corner handle of the cell) the formula down our list.

As the formula is copied down the column the cell reference will alter. The final formula will read =AVERAGE(A$2:A11). The $ sign forces Excel to keep the value after it; this will not change as we copy the formula down the column. We could also have put a $ before the A but since we are staying totally in column A this is not necessary.

Now we have a list of means running through the series of data. We get the impression that the mean is steadily rising but it is always a good idea to plot this out in a graph. The simplest way is to highlight the data (the means) and insert a chart. Figure 74 shows the finished result in our spreadsheet.

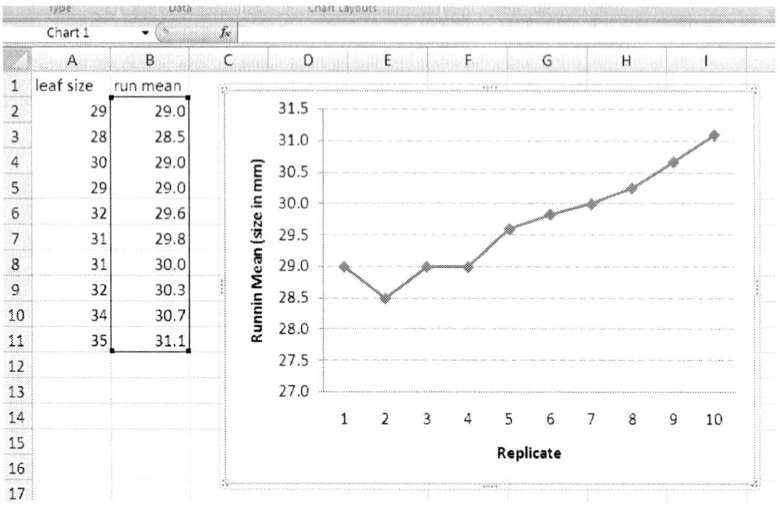

Figure 74. Plotting a running mean using Excel 2007

In our Excel graph (Figure 74) we selected the data and inserted a line plot. We only have one set of figures so Excel generates an index. This suits our purposes and all that remains is for us to add and edit suitable axis titles and to make the plot look sensible (see Section 12.4.3 for example).

4.7.2 Running means using R

Once we have our data in R we can use a variety of commands to explore and investigate. We have already met a few such as *mean*(), *sd*() and *hist*() commands for example. There is not a built-in function to determine the running mean; however, there are several functions that relate to cumulative data. The one we will use is *cumsum*(). This determines the cumulative sum of a vector of values.

```
> lf
 [1] 26 29 33 33 28 38 27 35 34 30
> cumsum(lf)
 [1]  26  55  88 121 149 187 214 249 283 313
```

In the example above we see a series of values from our leaf size data that we met in Table 22 when discussing standard error (Section 4.5.1). Here we have ten sizes (in mm) that represent a single sample. We have called the sample of 10 leaves *lf* and apply the *cumsum*() command by putting its name in the brackets. We see the result as a new list of values; each one is the sum of all the values up to that point. So, the third value is 88, which is the sum of the first three values (26, 29 and 33).

To get a running mean we need to divide each of our cumulative values by how far along we are. For example we would divide the first value (26) by one, as it is the first. We divide the second value (55) by 2 because it is the second (giving 27.5). R provides a command allowing us to deal with sequences, *seq*(). There are several ways it can be used but the one we want here allows us to step along a series of values.

```
> seq(along = lf)
 [1]  1  2  3  4  5  6  7  8  9 10
```

To generate a running mean we need to divide the cumulative sum by how far along the series of numbers we are. The following shows what we get with our leaf (*lf*) data:

```
> seq(along = lf)
 [1]  1  2  3  4  5  6  7  8  9 10
> cumsum(lf) / seq(along = lf)
 [1] 26.00000 27.50000 29.33333 30.25000 29.80000
31.16667 30.57143 31.12500
 [9] 31.44444 31.30000
```

Our running mean consists of ten values. Just by examining the numbers we can see that the last few seem to be evening out however, it might be useful to draw this as a graph. We can make a primitive plot using the following command (our data are the *lf* part):

```
> plot(cumsum(lf)/seq(along= lf), type= 'l')
```

The plot command makes a simple scatter graph. The y-data are the means and the x-data are a simple numerical index. We could plot just the points but it is more useful to see them joined up, so we add the type = "l" (letter "l") to tell R to create lines.

Figure 75 shows the basic plot that is produced. We can see that the line is starting to level out. It is often down to personal judgement as to when the graph is flattening out and when to stop.

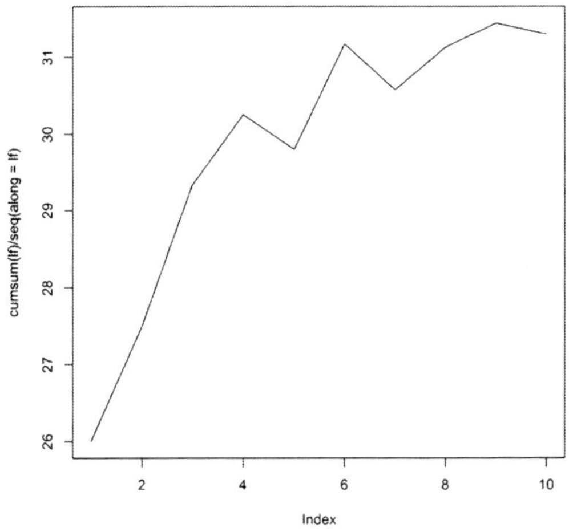

Figure 75. A simple lineplot of the running mean for ten leaves (size in millimetres)

We can add more commands tao our plot to make it look better. We'll deal with these commands later (Section 12.4) but here is a running mean graph for all 100 leaves that we had in Table 22.

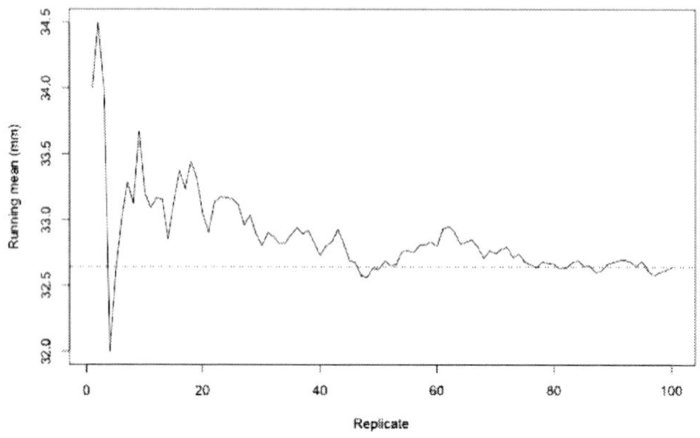

Figure 76. A lineplot of the running mean for 100 leaves (size in millimetres). Axis labels have been added as well as a dashed line indicating the overall mean

In this case, the graph in Figure 76 was produced from all the leaf data (which was stored under the name *leaves*) using two commands typed into R.

```
> plot(cumsum(leaves) / seq(along = leaves), type='l',
xlab = 'Replicate', ylab = 'Running mean (mm)')
> abline(h= mean(leaves), lty= 2, col= 'blue')
```

The plot command has three extra parts: type = "l" (draws lines rather than points); *xlab* = (labels the *x*-axis); and *ylab* = (labels the *y*-axis). The second command, *abline()*, is a special way of adding straight lines to existing graphs. It is often used for lines of best fit but can also add horizontal or vertical lines. Here we set the horizontal line to correspond to the final mean value. The two extra bits make the line dashed and blue.

4.8 Statistical symbols

We often need to use symbols in scientific writing. These symbols are used to convey some important information in a concise manner. In statistics there are a number of commonly used symbols and some of these we have already met.

In Table 26 we can see some of the most commonly used symbols and abbreviations. The mean is usually written with an overbar but you may also see the Greek letter μ (mu). Strictly speaking, overbar is used for your estimate and μ is used for the absolute mean; in most life sciences we are dealing with estimates. The standard deviation is given a simple letter *s* but you may also see the Greek letter σ (sigma). Generally *s* is used to show your estimate of standard deviation whilst σ is used to represent the actual standard deviation.

Table 26. Some commonly used symbols in maths and statistics

Symbol	Meaning
$\bar{x}$ μ	Arithmetic mean
σ s	Standard deviation
Σ	Sum of
n	Number in sample
s^2	Variance
t	Result of a *t*-test
U	Result of a *U*-test
z	Result of a *z*-test
χ^2	Result of a Chi-squared test
r_s	Spearman rank correlation coefficient
r	Pearson's product moment
$\lvert x \rvert$	Modulus, the absolute value of *x*
x^2	*x* to the power of 2
*x*E+3	*x* times 10 to the power of +3 (1000)
$p < 0.05$	Probability is less than 5%

The uppercase Greek letter sigma, $\sum$ is used to mean "sum of", i.e. everything added up. The number of items in your data list (sample) is given the designation n. A commonly used measure of dispersion in statistics is variance and this is written as s^2, which is useful as it really is the standard deviation squared.

Different stats tests usually have their own letter, especially the fairly simple ones: t for the t-test and U for the U-test. The two major types of correlation have similar letters so Pearson's coefficient is r and Spearman's rank coefficient is r_s. The Chi-squared test uses the uppercase Greek letter chi, χ.

The level of significance of a stats result is of great importance so the probability of this result happening by chance is given the letter p. Generally it is best to quote $p < 0.05$, $p < 0.01$ rather than the exact value (this is perceived as showing off so don't put $p = 0.0000023$); if not significant you either write $p > 0.05$ or n.s.

In some cases we are not interested in the sign of a difference but merely its size, so $x_1 - x_2$ might give a positive or negative result but we only want the magnitude. We indicate this using the modulus, a pair of upright brackets so we write $\lvert x_1 - x_2 \rvert$ to indicate this. This is sometimes called the absolute value and in Excel and R we obtain the modulus with similar commands, ABS and *abs()* respectively.

If you only have plain text and cannot use superscripts or subscripts then you may want to indicate more complex numbers using a basic notation. For example 10^2 equals 100 but how to write it without the superscript? Common notation uses a "hat" (properly called a caret): 10^2 = 100.

Really large or really small numbers can be given scientific notation so 10000 (which ought more properly to be written 10 000) becomes 1×10^4 and 347348 becomes 3.47×10^5. On the other hand 0.0000035 would be 3.5×10^{-6}. Without superscripts we can write the right-hand part (the exponent) as an "e". So we get 1e+4, 3.47e+5 and 3.5e–6. Excel tends to use uppercase E, and R uses a lowercase e. As long as you are consistent it does not matter.

Logarithms are not shown in the table but $\log_{10}$ or $\log_2$ (log by itself implies base 10) are usual ways to write logs. The natural log ($\log_e$) is often given its own symbol ln (or Ln).

5. Exploring data – which test is right?

Now that we have an idea of how to get our data into a format suitable for analysis and have looked at basic summary statistics, we need to move on to look at hypothesis testing. The hypothesis is the thing we were aiming at right from the start and we should have worked out our basic hypothesis at the planning stage (recall Section 1.4) but this is a convenient place to undertake a brief review.

5.1 Hypothesis testing

One of the first things you will need is an idea of what you are trying to show/determine. Examples might be:

"There are more buttercups (*Ranunculus repens*) in the south field compared to the north field."

or:

"There is a positive link between stream flow and abundance of stonefly (*Perla bipunctata*) larvae."

In broad terms, a hypothesis should be something that you can put to the test, something you can prove or disprove. In reality we can rarely be 100% certain so we talk about supporting or rejecting our hypothesis.

In scientific hypothesis testing, we will have biological data; this will be variable in nature and so our conclusions are generally based on the likelihood that we are correct rather than absolute certainty. In science we will accept our conclusions only if there is less than a 5% chance that the result could have happened due to chance. This is where the statistical tests come in; they allow us to put a value on this likelihood and so give us a tool to decide if we are above or below this 5% threshold.

In practice, it is a lot easier to disprove something rather than to prove it; this concept is carried into biological hypothesis testing. We switch the emphasis around and create a null hypothesis. In the case of the examples above our null hypotheses would be:

"There is **no difference** in buttercup (*Ranunculus repens*) abundance between the north and south fields."

and:

"There is **no link** between stream flow and stonefly (*Perla bipunctata*) larvae abundance."

So, the null hypothesis is **not** just the opposite of the original idea but a new idea in its own right. The null hypothesis iss what is actually tested. Once you have your null hypothesis (often abbreviated to H_0) you can move on to choosing the correct test to apply to your data. Ideally you do this **before** you collect any data at all as the type of test might influence the data collection. Thus the original hypothesis (often called the alternative hypothesis H_1), the H_0 and the selection of the stats test become part of the planning process. Once you know what you are aiming for, this will help you to decide which statistical test is right for your situation.

5.2 Choosing the correct test

Choosing the correct test should be part of your project planning process. Different tests have different requirements, e.g. some need a minimum amount of data so you need to know what you are aiming for. In simple terms, there are two broad categories of stats tests: those that split data and those that link data. If you are looking for differences in buttercup abundance between two fields/sites (as in the example above) then you are looking to split the data. In the second example you are clearly looking for a link (between flow and abundance).

Some books and computer programs provide a decision tree to enable you to select the correct test.

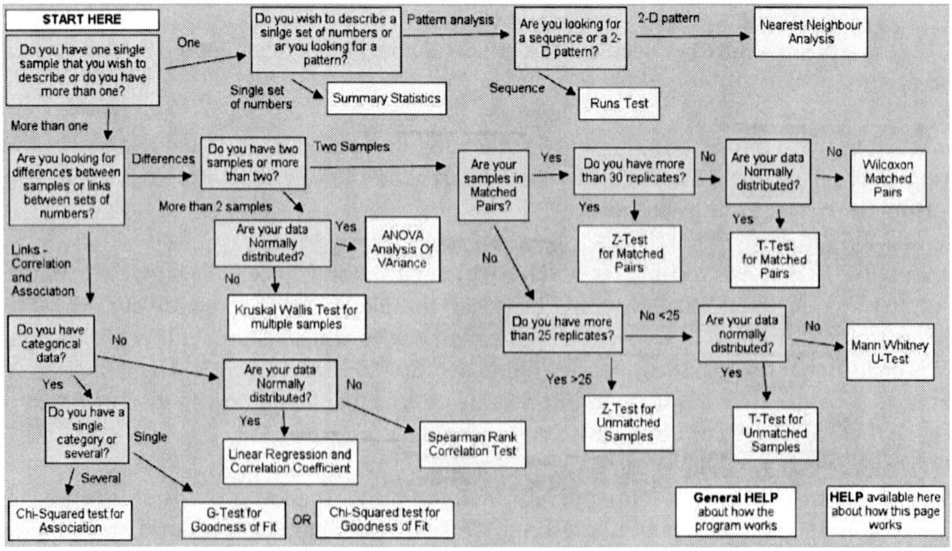

Figure 77. Decision tree for statistical analysis. This guides you towards the correct test for your data/situation

In the example shown in Figure 77, the decision tree provides a number of questions and you branch off accordingly. One question that crops up in several places is "are your data normally distributed?" This is a very important consideration because the distribution (Section 4.2) of the data affects quite profoundly the statistical approach. You can also see that in some case you need a lot more data to proceed (25–30 replicates). Once you have made your initial choice you should then determine what the requirements and limitations of your test are because this will influence your sampling strategy (you may of course alter your strategy in order to use a more suitable stats test). After looking at ways to explore our data graphically (Chapter 6), the next few sections will deal with specific types of statistical test. There are various kinds of test and each has a particular speciality.

In Table 27 we see a formalised method to help us decide which test is the right one. We start at number 1 and compare the pair of statements (each has *a* and *b* options). Then we select the most appropriate and read the option at the end of the line, Eventually we come to a final decision. This is somewhat simplified, as we will see as we go along, but it gives us a good framework and a place to start thinking about our analyses.

Table 27. Aid to choosing the correct statistical test

1a	Looking for... differences between items/samples.	Goto 2
1b	Looking for... links between items.	Goto 5
2a	Data are normally distributed (parametric).	Goto 3
2b	Data not normally distributed (i.e. skewed).	Goto 4
3a	There are only two samples to compare.	*t*-test (Section 7.1)
3b	There are more than two samples.	ANOVA (Section 10.2)
4a	There are only two samples to compare.	*U*-test (Section 7.2)
4b	There are more than two samples.	Kruskal–Wallis (Section 10.3)
5a	The data are categorical.	Goto 6
5b	The data are continuous variables.	Goto 7
6a	There are two sets of measured categories.	Chi-Squared Association (Section 9.1)
6b	There is one set of categories.	Goodness of Fit (Section 9.2)
7a	Data are normally distributed (parametric).	Goto 8
7b	Data not normally distributed (i.e. skewed).	Goto 9
8a	There is one dependent variable and one independent.	Pearson product moment (Section 8.2)
8b	One dependent variable and several independent.	Multiple Regression (Section 11)
9a	There is one dependent variable and one independent.	Spearman's rank Correlation (Section 8.1)
9b	One dependent variable and several independent.	This is tricky and beyond the scope of this book

We should ideally carry out this process as part of our original planning. Once we have determined the approach we need to take, we can begin to plan the data collection. Eventually we collect our data and can set about analysing it. A good start would be to draw graphs, histograms, and box–whisker plots of the individual samples as well as comparing samples, perhaps with bar charts or scatter plots. We looked at distribution graphs like histograms and stem–leaf plots in Section 4.1.4. We will discuss graphs briefly in Chapter 6 where we will start to think about which types of graph can be useful in which circumstances. We will examine graphs in much more detail later (Section 12.4 *et seq.*), including details on how to use our software to make the graphs we require. After the introductory chapter on graphs, we will focus on the mechanics of the various statistical tests. As we progress, you should bear in mind that producing a graphical summary of the situation is always a good idea.

6. Exploring data – using graphs

Graphs are useful for several reasons. They can help us to visualise the data and decide which statistical test is the best. We may spot patterns in the data and gain a better understanding of what we are dealing with. Graphs are also useful for summarising our final results, especially when we present our findings to other people.

We can think of graphs as being useful for two purposes: firstly to help us decide how to tackle the data, and secondly to present results. We will look at details of graphs and how to produce them in Excel and R in Section 12.4 *et seq.* where we examine ways to present our findings. We will also mention graphs throughout the text as we look at the various analytical methods to examine our data. Indeed we have already seen some examples in Chapter 4. In this short chapter we will summarise the graphs we might use to help us explore our data.

6.1 Exploratory graphs

One of the most common analysis of sample of data is to determine if they are normally distributed or not. This affects the kind of statistical analysis we are able to perform on the data. There are several ways we can illustrate the distribution of a data sample. We may use a simple tally plot or a stem–leaf plot; we can even do this right from our notebook in the field. The following example shows a stem–leaf plot.

```
1 | 679
2 | 112334
2 | 5666678899
3 | 01124
3 | 6
```

In this example, the data are sorted in numerical order in each row but we can still gain insights into the data distribution if the numbers are not sorted.

```
1 | 967
2 | 143123
2 | 9568667869
3 | 40121
3 | 6
```

A simpler version of a stem–leaf plot is the tally plot, and in this case we enter the data as a simple tally mark. In Table 28, we see a tally plot of the same data as our stem–leaf plot.

Table 28. A tally plot to show data distribution

Tally	Bin
X	16
X	18
X	20
XXX	22
XXX	24
XXXXX	26
XXX	28
XXX	30
XXX	32
X	34
X	36

These are simple plots but nevertheless can be extremely helpful. When we return from the field we may decide to use a more formal histogram to illustrate the distribution (Figure 78).

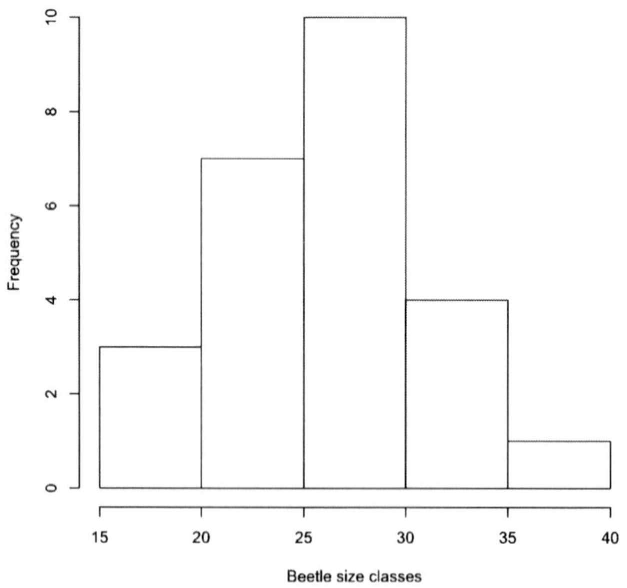

Figure 78. A histogram to illustrate the distribution of a data sample

The size of the bars in our histogram shows us the number of items (the frequency) of our dataset that lie within each size class, represented on the x-axis. We may decide to use a line instead of bars and the result is a density plot (Figure 79).

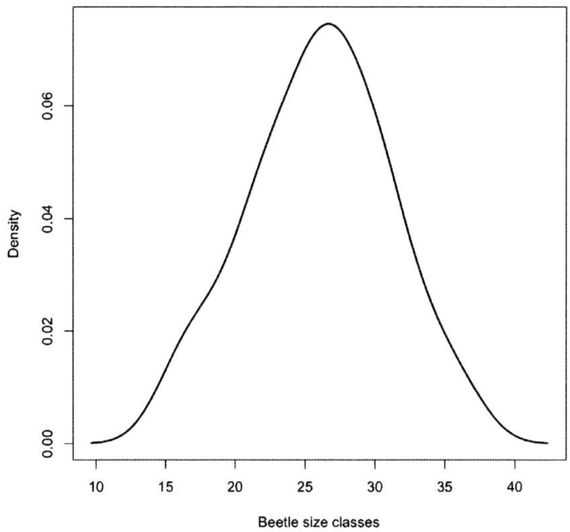

Figure 79. A density plot to illustrate the distribution of a data sample

Some types of graph are useful because they show a lot of information in a compact manner such as the box–whisker plot. A box–whisker plot shows us five pieces of information: median, maximum, minimum and both quartiles (Figure 80).

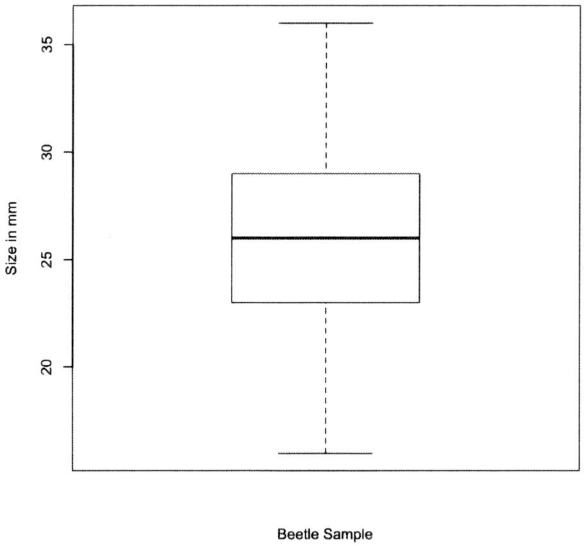

Figure 80. A box–whisker plot can be used to illustrate data distribution as well as providing other information, e.g. median, inter-quartiles and max/min

In Figure 80, we can see that the data appear normally distributed as the box–whiskers are symmetrical about the median stripe. We can use the box–whisker plot to look at several samples and illustrate not only differences between samples but their distribution as well (Figure 82).

Another way we can visualise our data is by using a line graph to show the running average (mean or median). We met this earlier in Section 4.7 where we used the idea to help determine if we had collected enough data. In Figure 81, we see an example of a running mean.

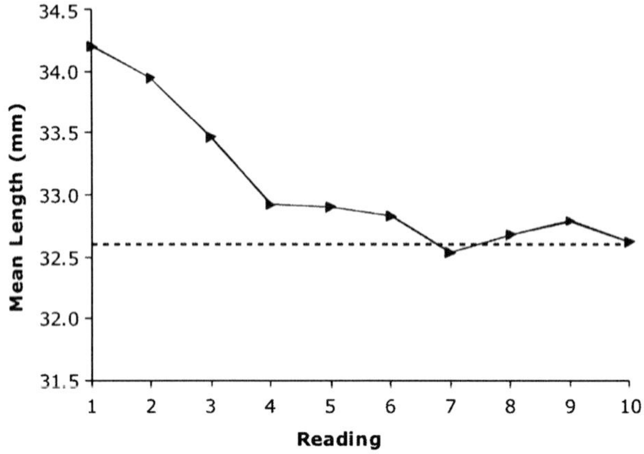

Figure 81. A line graph illustrating the running mean

This is another example of a graph we can sketch whilst out in the field. We do not have to be quite so exact when we are out in the field; the graph is simply a tool to help us make a decision.

6.2 Graphs to illustrate differences

When we have a project that is centred on looking at differences between samples we can illustrate the situation using bar charts or box–whisker plots. We met the box–whisker plot previously (Figure 80) when we used it to view a sample and check its distribution. In Figure 82 we look at three samples.

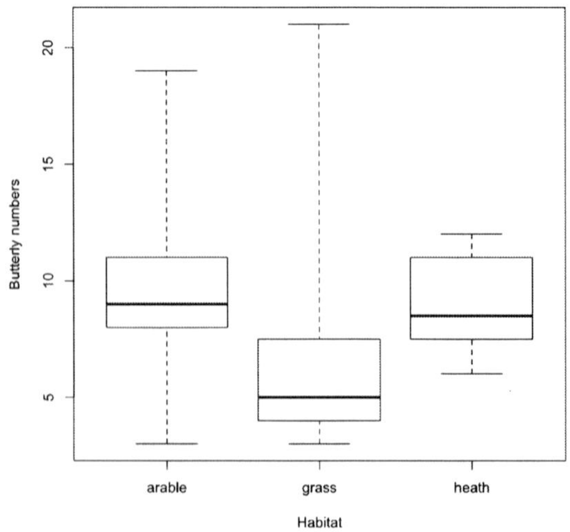

Figure 82. A box–whisker plot illustrating differences between three samples

We can see the differences between the three samples fairly easily and in addition we can gain some insight into the distribution. A common alternative to the box–whisker plot is the bar chart. This is useful to show differences between items in different categories and is therefore suitable to illustrate differences in samples. In Figure 83 we see the same data as in Figure 82 but here we use a bar chart with standard error bars to show the variability within each sample.

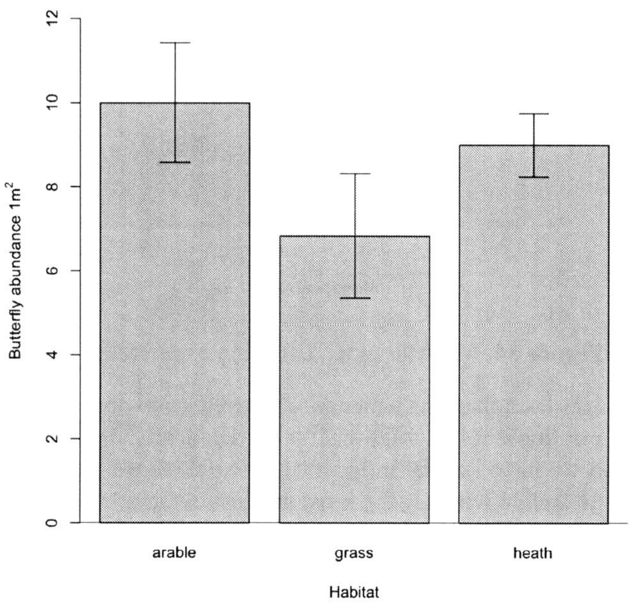

Figure 83. A bar chart illustrating differences between three samples

We can see from Figure 82 that there are differences between the three samples unlike in Figure 83 where we cannot tell anything about the distribution.

6.3 Graphs to illustrate links

When we think of ways to link data together there are two main approaches. In one approach, we have two sets of values, both are numeric and one represents a dependent variable and the other an independent variable. We are looking for a correlation. In the other kind of approach, we have categories of items and we are looking to associate one set of categories with the other.

6.3.1 Graphs to illustrate correlations

When we are looking for correlations, we can best illustrate the situation using a scatter plot; this allows us to see how one variable is related to the other. In Figure 84 we see a scatter plot showing how the abundance of a freshwater invertebrate is related to the speed of the water in which it lives.

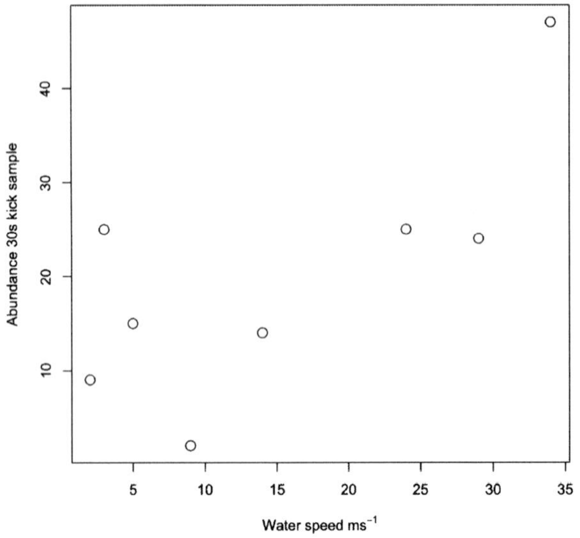

Figure 84. A scatter plot illustrating a correlation

In this case, it appears as though as the water speed increases so does the abundance of the invertebrate. We do not know if this relationship is statistically significant but it gives us an impression. When we have several independent variables we can plot several scatter plots; this may help us decide which is the most important factor to consider (Figure 85).

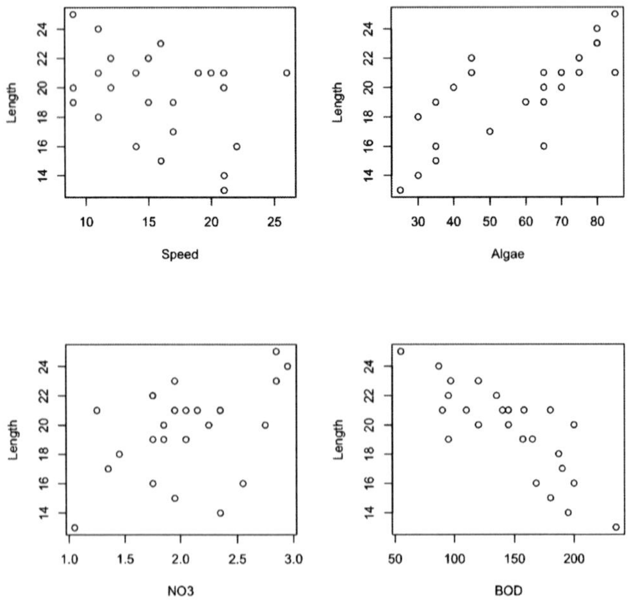

Figure 85. Multiple scatter plots showing one dependent variable plotted against several independent variables

In Figure 85 we can see that two of the independent variables show a more definite trend than the others; one shows a positive correlation and the other a negative one (although at this point we do not know if either is statistically significant).

6.3.2 Graphs to illustrate associations

When we have categorical variables, we have various choices. We can display the data for each row or column category as a pie chart (e.g. Figure 86); this will usually require several pie charts to be produced (one for each row or column category, depending on how we want to look at the data). The pie chart shows the data proportionally, each slice of pie shows the contribution as a proportion of the total.

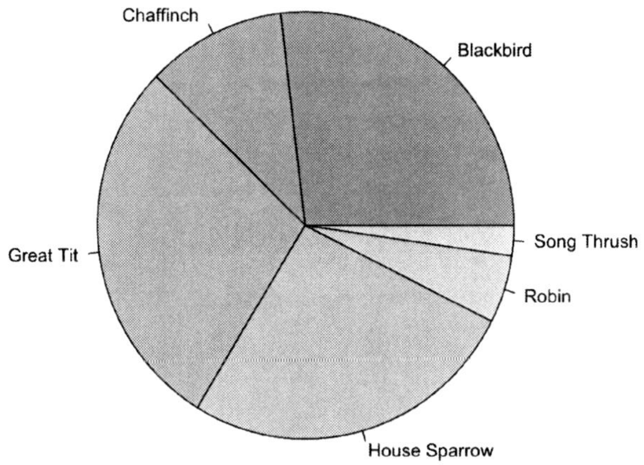

Figure 86. A pie chart illustrating categorical data. The proportions of common bird species in a garden habitat

When we have this kind of data we can always represent it in the form of a bar chart instead. The advantage of the bar chart is that we can show several categories at one time (Figure 87).

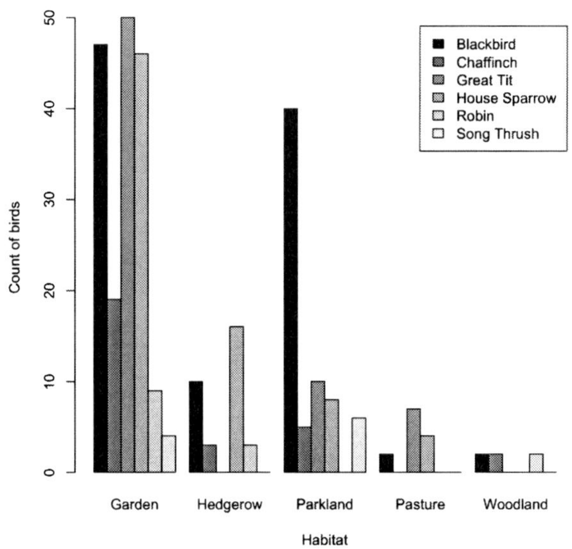

Figure 87. A bar chart illustrating categorical data. The number of common garden birds in various habitats

In Figure 87 we can see various bird species and various habitats; in this case we have also included a legend on the graph so the reader can identify the various bars more easily.

6.4 Graphs – a summary

There are quite a few different sorts of graph that we can utilise to help visualise our data and make important decisions about the analytical approach (Table 29). We should also use graphs to illustrate our data, which can make them more comprehensible to readers. When we present graphs we should ensure they are fully labelled and as clear as possible. Even when we use graphs for our own use it is good practice to label and title them fully.

Label axes and include the units.

Do not include too many different elements on a single graph – avoid clutter and if necessary produce two graphs rather than one.

Give a main title explaining what the graph shows. Usually this is done as a caption in a word processor. The caption should enable a reader to understand what the graph shows without having to read the main text. If your graph is in your field notebook then make sure you describe the graph so that someone else can understand it.

Table 29. Summary of graph types to use for different purposes

Purpose	Types of graph
Illustrating distribution	Stem–leaf plot, tally plot, histogram, density chart, box–whisker plot
Illustrating differences between samples	Bar chart, box–whisker plot
Illustrating correlations	Scatter plot
Illustrating associations	Pie charts, bar charts
Illustrating sample sizes	Line plot of running average (mean or median)

We will examine graphs in more detail in Chapter 12, which will also cover he presentation of results. Sections 12.4.1 and 12.5 will deal with producing graphs in R and Section 12.4.3 will cover producing graphs in Excel. We will also make some references to graphs in each of the sections dealing with the details of the various analytical methods. It is important to remember that our graphical analysis should go alongside the mathematical one.

7. Tests for differences

In tests for differences we generally have two or more samples that we wish to compare. We usually want to know if they are different, i.e. are their means (or medians) different? In this chapter we will consider three options and in all cases we will compare just two samples. In the first case, where we have two normally distributed samples we will examine the *t*-test, and where we have skewed data we will use the *U*-test. The last section will look at special cases of both tests where we have matched pairs of data. When we have more than two samples the situation becomes a bit more complex and we will cover these situations in Chapters 10 and 11.

Of course we should draw a graph to illustrate the situation. When we are looking at differences, we should be thinking in terms of bar charts and box–whisker plots (see Section 6.2); we will look at the details of constructing such plots later (Section 12.4 *et seq.*).

7.1 Differences: *t*-test

The *t*-test is a widely used test to determine if two samples are different. It is usually called Student's *t*-test after the nom de plume of the author. The test relies on the properties of the normal distribution (called a parametric test) and tests for differences in sample means. Figure 88 shows two samples (we will call them *a* and *b*) and the frequency distribution. We can see they have different means and that there is virtually no overlap between the two samples.

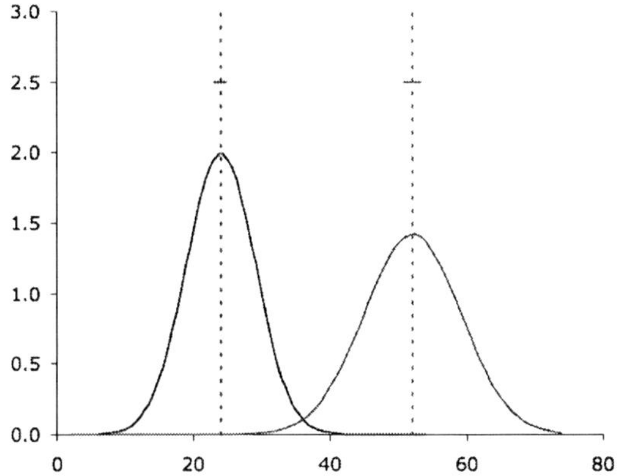

Figure 88. Frequency distribution of two samples. We can see very little overlap. Dotted lines represent the mean values and the bars are the standard error

In this case, when there is hardly any overlap we can be pretty certain that there is a real difference between the two samples and it is unlikely that they are really just part of one larger population/sample. Now look at the next example.

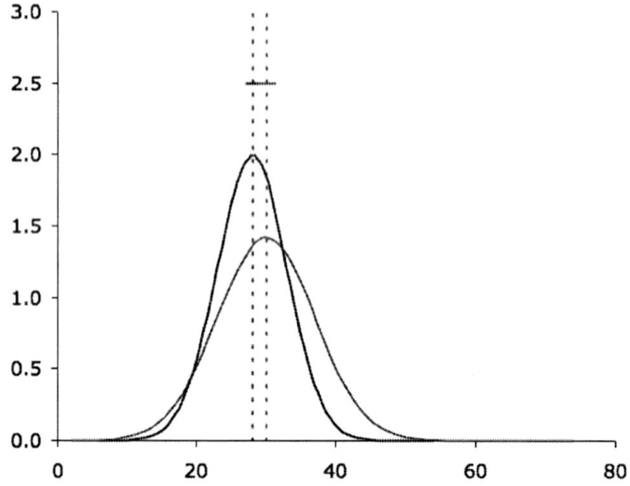

Figure 89. Frequency distribution of two samples. This time we see a large overlap, the means are close together and the standard error bars overlap

In this case, the means are very close together and there is a great deal of overlap. We might conclude that there is no real difference between the samples. In the two preceding examples, the situation appears fairly clear. In the next example however, the case is less certain.

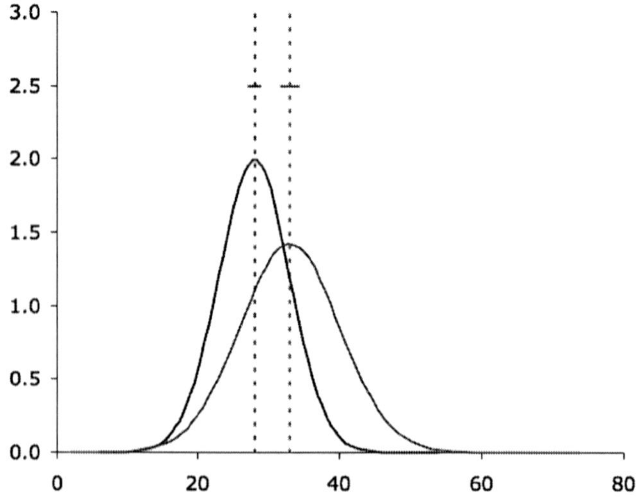

Figure 90. Frequency distribution of two samples. There is quite a bit of overlap but the standard error bars do not overlap

Figure 90 shows the frequency distributions of two samples. Their means are moderately far apart and there is a fair amount of overlap. Are these two samples different? This is where a statistical test will help because it will put a value to our uncertainty.

We are looking for differences between two samples and they are both normally distributed (the distribution graphs look symmetrical around the middle). The *t*-test is a suitable test to use to examine these data. This parametric situation (normal distribution) allows us to summarise the data using the mean and standard deviation (for each sample).

The *t*-test uses three pieces of information to determine if the samples are different. The formula for the test is shown in Figure 91:

$$t = \frac{|\bar{x}_a - \bar{x}_b|}{\sqrt{\frac{s_a^2}{n_a} + \frac{s_b^2}{n_b}}}$$

Figure 91. The formula for the *t*-test

On the top we have the means of the two samples. On the bottom we have the variance, which is the standard deviation squared, and the number of items in our sample (the replicates). In this version of the *t*-test, the variance of the two samples are assumed to be more or less equal. The vertical brackets on the top signify that we ignore the sign of the difference; we are interested only in the magnitude of the difference, i.e. the absolute difference. In Excel and R there is a function that ignores the sign, *abs()*.

If we look at the top and bottom separately we can see what is happening. If the difference between the means is large, then we have the situation akin to Figure 88 where the samples were widely separated. The bigger the difference, the larger our calculated value of *t* will be.

If the standard deviation of the samples is large, then the distributions will be wide and fat and there will likely be more overlap. This situation will be more like Figure 89. Since the standard deviations are on the bottom of the equation, the result is that larger *s* will lead to smaller *t*.

So do we want a large *t* or a small one? We want a large one to show that there is a real difference between samples (i.e. with widely separated means like in the Figure 88).

It is simple enough to calculate the value of *t* using the formula. Once we have a value, what does this tell us? We need to compare our calculated value with a table of critical values (see Table 30).

Table 30. Critical values for the Student's *t*-test. Reject the null hypothesis if your calculated value is greater than the tabulated value

	Significance level			
Degrees of freedom	**5%**	**2%**	**1%**	**0.1%**
1	12.706	31.821	63.657	636.62
2	4.303	6.965	9.925	31.598
3	3.182	4.541	5.841	12.941
4	2.776	3.747	4.604	8.610
5	2.571	3.365	4.032	6.859
6	2.447	3.143	3.707	5.959
7	2.365	2.998	3.499	5.405
8	2.306	2.896	3.355	5.041
9	2.262	2.821	3.250	4.781
10	2.228	2.764	3.169	4.587
11	2.201	2.718	3.106	4.437
12	2.179	2.681	3.055	4.318
13	2.160	2.650	3.012	4.221
14	2.145	2.624	2.977	4.140
15	2.131	2.602	2.947	4.073

In the table of critical values we see several columns. The first is labelled degrees of freedom and is related to how many data there are ($df = n_a - 1 + n_b - 1$, i.e. the total number of data items we have –2). We look down this column until we reach the one we want and then we read across. The following columns give the critical values of t for various levels of significance. We are prepared to accept 5% so if our calculated value is greater than the critical value we can reckon that there is a difference in the sample means. What we are saying is that there is only a 5% chance (or less) that we could have a t-value as high as we did by chance. We may get a value greater than the next column; this would then represent only a 2% chance.

Formally we say that we reject our null hypothesis and accept the alternative hypothesis. If our calculated value of t was lower than the critical value then we know that there was a greater than 5% probability that the result was due to chance so we have to accept our null hypothesis and conclude that there is no significant difference in sample means. We will look at ways of reporting our results in more detail in Section 12.3.

In this version of the t-test, we assume that the variance of the two samples is equal but this is obviously not always the case. When we have unequal variance we can modify the formula to take this into account. The final value of t is slightly different and the degrees of freedom are usually modified too. Most of the time we will be using a computer to carry out the test and it is generally safer to assume that the variance is unequal.

7.1.1 Using Excel for the *t*-test

Excel has routines built-in that can calculate the t-test for us. The basic form of the command is:

TTEST(*range1*, *range2*, 2, 3)

We type in the range of cells where our first sample is to be found; the next part is where we put the reference to the second sample. Finally we enter two numbers. The first is a 2 – this tells Excel to look at both ends of the normal distribution (which is the most common situation). The next value (3) tells Excel that the samples are independent of one another and that we should **not** assume that the variance is the same for both samples (which is what we did when using the formula shown previously in Figure 91). If we want to run the basic (equal variance) version, we replace the 3 with a 2.

The TTEST formula result gives us the probability that the two samples are from the same sample. In Figure 92 we see some data on the abundance of a plant species on ridges and in furrows of a field. We have used the TTEST formula and assumed that variance is unequal.

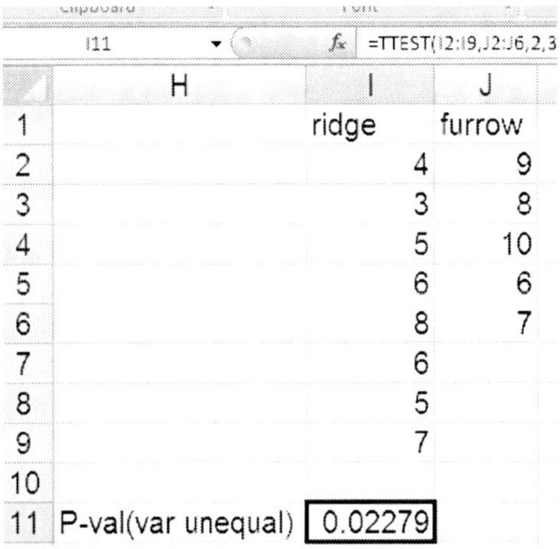

I11 =TTEST(I2:I9,J2:J6,2,3

	H	I	J
1		ridge	furrow
2		4	9
3		3	8
4		5	10
5		6	6
6		8	7
7		6	
8		5	
9		7	
10			
11	P-val(var unequal)	0.02279	

Figure 92. The TTEST command in Excel determines the probability that two samples are not different

The result is shown in cell I11 and the *p*-value of 0.02279 shows us that the probability of the two samples being the same is very low. In other words we can reject the *null hypothesis*; the samples have different means. The TTEST formula does not display the value of *t*, only the *p*-value. If we wish to work out what *t* is (and we probably should) then we need to do something else.

We could use the spreadsheet to calculate variance for each sample and then go on to solve the *t*-test formula but there is another way.

The TINV function works out the *t*-statistic given the probability (*p*-value) and the degrees of freedom:

TINV(*p-value, degrees_of_freedom*)

In Figure 93 we can see the result of determining the *t*-statistic for our ridge and furrow data.

I12 =COUNT(I2:J9)-2

	H	I	J
1		ridge	furrow
2		4	9
3		3	8
4		5	10
5		6	6
6		8	7
7		6	
8		5	
9		7	
10			
11	P-val(var unequal)	0.02279	
12	df	11	
13	T	2.645	
14	P-val	0.02279	

Figure 93. We can use the COUNT function to determine degrees of freedom in the *t*-test

The calculated result is $t = 2.645$. In Figure 93 the degrees of freedom are shown as 11 ($\text{sample}_1 - 1 + \text{sample}_2 - 1$). We can use the COUNT function to work this out as df = total # replicates – 2. At the very bottom we also see another *p*-value. This was generated using another command TDIST. This function works out the *p*-value given a *t*-statistic and degrees of freedom:

TDIST(*t-value, degrees_of_freedom*)

However, there is a slight problem. If we calculate *t* the long way, we end up with a different value (of 2.749). The reason is that the command assumes that the original *p*-value was calculated by the formula that assumes both samples had equal variance. In Figure 94 we can see the result of calculating the *p*-value assuming the variance is equal.

I13 =TINV(I11,I12)

	H	I	J
10			
11	P-val(var unequal)	0.02279	
12	df	11	
13	T	2.645	
14	P-val	0.02279	
15			
16	P-val(var equal)	0.01894	
17	T	2.749	

Figure 94. The TINV formula works out the *t*-statistic given a probability and degrees of freedom; however, it only works correctly if the *p*-value has been calculated assuming variance is equal

We can see (row 16) that the *p*-value is somewhat smaller (0.01894). This does not alter our conclusion in this case but it does show that when variance is equal we are able to be more definite about our results. When we apply the TINV formula to this new *p*-value, we see a *t*-statistic of 2.749. This is the value we get if we went the long way round and calculated *t* for ourselves.

It so happens that the variance of these two samples is quite close (2.57 and 2.50) so we are justified in using the equal variance version of the *t*-test. What we should probably do is to check the variance of each sample and if they are close we can simply run the basic *t*-test. If they are different then we should run the *t*-test twice, once using the proper unequal variance version to generate a *p*-value, then again with equal variance. We then use the equal variance *p*-value to work out the value of *t*. This is a bit tedious but still quicker than calculating *t* by the regular mathematical method! Alternatively we might elect to use the *Analysis ToolPak* (Section 1.9.1), which we describe next.

Using the *Analysis ToolPak* for the *t*-test

The *Analysis ToolPak* is a special set of routines that come as an add-in for most versions of Excel. In Section 1.9.1 we described how to install this add-in.

In order to run the *t*-test we need to include some data; we can see what form our data need to be by using the *Analysis ToolPak*, which will show us what is required. We can also use the *Help* button from the *Analysis ToolPak* screen. We begin by using the *Data* menu on the Excel ribbon (older versions of Excel use the same menu) and selecting the *Data Analysis* button. This brings up a menu box for the Analysis Tools and allows us to select the option we require (Figure 95).

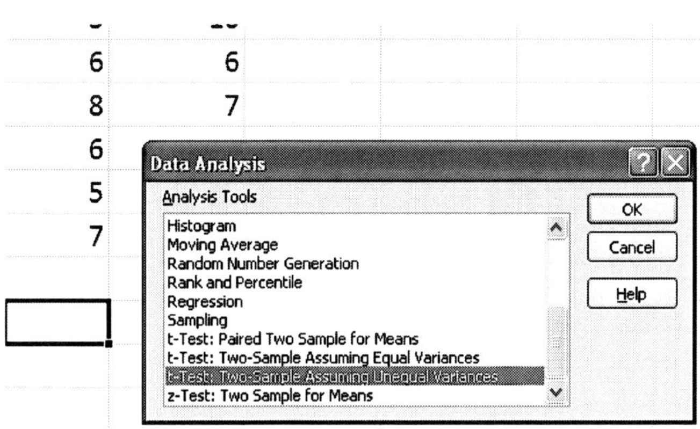

Figure 95. The *Data Analysis* options panel allows us to select a range of statistical routines

In our example we select the two-sample test for unequal variances and once we click the *OK* button we are prompted to select the data (Figure 96).

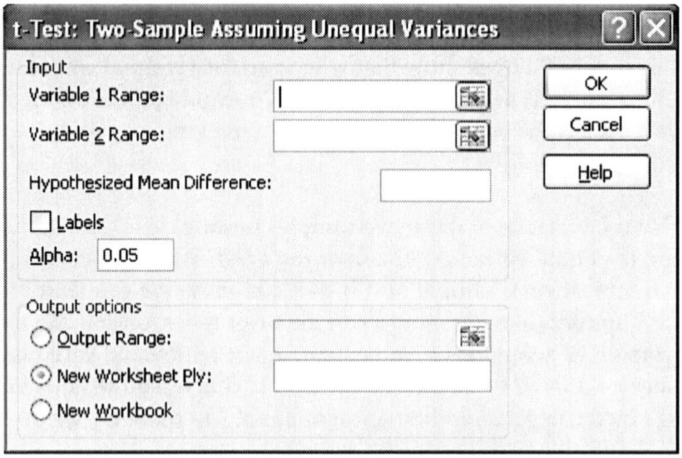

Figure 96. The selected *Data Analysis* prompts for the required inputs in the *Analysis ToolPak*

We can see that we need to input the ranges of cells for the two samples as well as the place where we wish the results to be placed. If we check the *Labels* box, we can use the column headings for our data. In Figure 97 we see the input boxes filled in; we have selected to place the output in the existing sheet in this case.

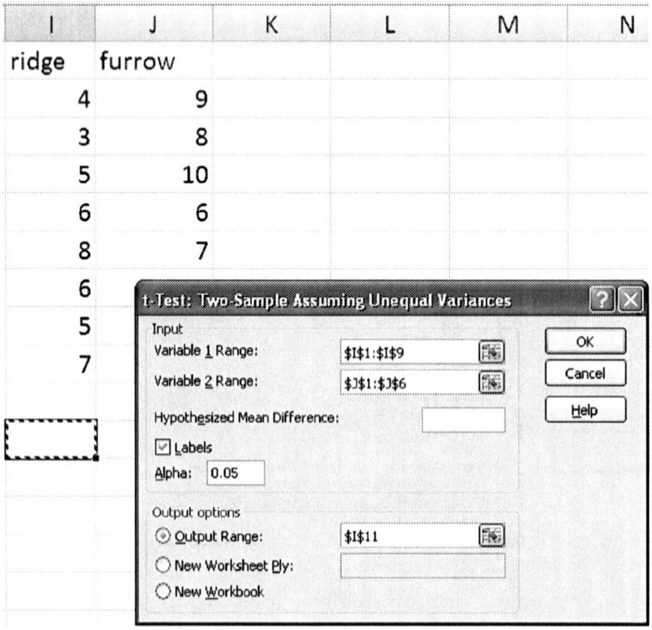

Figure 97. Selecting the data input for the *Analysis ToolPak*

Once we click the *OK* button, the *t*-test results are calculated and placed onto the spreadsheet at the spot we selected (Figure 98).

t-Test: Two-Sample Assuming Unequal Variances		
	ridge	*furrow*
Mean	5.5	8
Variance	2.571429	2.5
Observations	8	5
Hypothesized Mean Difference	0	
df	9	
t Stat	-2.75839	
P(T<=t) one-tail	0.011085	
t Critical one-tail	1.833113	
P(T<=t) two-tail	0.022171	
t Critical two-tail	2.262157	

Figure 98. The results from the *Analysis ToolPak* are placed in the spreadsheet in a neat tabular form

We see that we have two sets of results: one-tailed and two-tailed. We choose the two-tailed results. Essentially this refers to the normal distribution; the ends of the bell-shaped curve are called the "tails". We cannot assume *a priori* that one sample will have a larger mean than the other, so we must consider both ends of the normal distribution. In the very vast majority of cases we will use the two-tailed version of tests that we conduct.

7.1.2 Using R for the *t*-test

The basic distribution of R comes with a variety of standard analytical routines built-in. In fact we can perform quite complicated tests with very little effort.

T test, *t.test()* command

For normally distributed data, the *t*-test is the one when you wish to examine differences between two samples. We use the *t.test()* command, for example:

```
> t.test(ridge, furrow)
```

R produces something like the following:

```
Welch Two Sample t-test

data:  ridge and furrow
t = -2.7584, df = 8.733, p-value = 0.02279
alternative hypothesis: true difference in means is not equal to 0
95 percent confidence interval:
```

```
 -4.5598309 -0.4401691
sample estimates:
mean of x mean of y
      5.5       8.0
```

The *t*-test that R works out assumes that variance in unequal as default. This is probably a good thing but you may wish to run the standard version and to do so we add an extra part:

```
> t.test(ridge, furrow, var.equal = TRUE)
```

Here we are telling R to assume that the variance of the two samples is the same. The output is similar:

```
Two Sample t-test

data:  ridge and furrow
t = -2.7486, df = 11, p-value = 0.01894
alternative hypothesis: true difference in means is not equal to 0
95 percent confidence interval:
 -4.5018958 -0.4981042
sample estimates:
mean of x mean of y
      5.5       8.0
```

The results of the two versions of the *t*-test are broadly similar. Usually the effect of assuming unequal variance is to reduce the *p*-value slightly. This makes the test a little bit more conservative.

7.2 Differences: *U*-test

When the data are not normally distributed (i.e. they are skewed or non-parametric), we cannot use the mean and standard deviation. Instead we summarise the data using the median and range (and usually the quartiles). We cannot now use the properties of the normal (parametric) distribution and the *t*-test is **not** appropriate. Instead we use a test that is based on the ranks of the data, i.e. we replace the original values with their rank.

It would be a good idea to represent the situation with some type of graph. With non-parametric data, the box–whisker plot is a very good idea as this conveys a great deal of information about each sample as well as the differences between samples (see Chapter 6). We will cover graphs in more detail in Section 12.4 *et seq*. but for now we will focus on carrying out the mathematical parts of the analysis.

The following example in Table 31 shows some (non-parametric) data; we have counts of flour beetles in samples taken from two locations.

Table 31. Non-parametric data. Counts of flour beetles at two sites. The data are arranged in numerical size to make comparison easier. It is evident that there is no overlap

Site a	9	12	12	13	14	18	19	21	21	21										
Site b											23	23	24	25	28	32	33	34	34	45

In Table 31 the data have been arranged in numerical order to make the situation clearer. Here we can see that the values for *site a* are all smaller than the values for *site b*, there is no overlap at all. In the next example (Table 32) there is much more overlap:

Table 32. Non-parametric data. Flour beetle counts at two sites. Here the data overlap considerably

Site a	9	12		14		19	21		23	24		28		33	34	
Site b		12	13		18		21	21	23		25		32		34	45

Once again the data are arranged in numerical order so we can see the amount of overlap. In the first example with no overlap, we can be certain that there is a difference between the two samples. In the second example with almost complete overlap, we can be fairly certain there is no difference between the two samples. What we need is a way to determine how likely there is to be a real difference between two samples when there is some overlap.

Table 33. Non-parametric data. Counts of flour beetles at two sites. Here there is some overlap

Site a	9	12	12	13	14	18		21	21	23	24							
Site b							19	21		23		25	28	32	33	34	34	45

The example in Table 33 shows us two samples and there is some overlap. Are the two samples different? This is why we need a statistical test to tell us how certain we may be that there is (or is not) a difference. We start by putting the data into numerical order and changing the original value to a rank. This is simplest to illustrate using the first example.

Table 34. Flour beetle count data. The values are first converted to ranks, starting with the smallest. All the ranks of *site a* are smaller than for *site b*

Ranks R_a	**1**	**2**	**3**	**4**	**5**	**6**	**7**	**8**	**9**	**10**										
Site a	9	12	12	13	14	18	19	21	21	21										
Site b											23	23	24	25	28	32	33	34	34	45
Ranks R_b											**11**	**12**	**13**	**14**	**15**	**16**	**17**	**18**	**19**	**20**

Here we can see clearly that all the ranks of the first site are smaller than all the ranks of the second site. When there is a lot of overlap the ranks overlap too.

Table 35. Flour beetle count data. After converting to ranks, we see that the ranks also overlap considerably

Ranks R_a	**1**	**2.5**		**5**		**7**	**9**		**11.5**	**13**		**15**		**17**	**18.5**	
Site a	9	12		14		19	21		23	24		28		33	34	
Site b		12	13		18		21	21	23		25		32		34	45
Ranks R_b		**2.5**	**4**		**6**		**9**	**9**	**11.5**		**14**		**16**		**18.5**	**20**

When working out the ranks, we treat all the values as **one set** and start at the lowest (so the lowest value gets rank 1). The highest value gets the highest rank, so if there were 20 items of data then the highest rank would be 20. Where there are tied values the rank is averaged. We then step to the next available rank; in Table 35 for example we have two values of 12. These occupy ranks 2 and 3 so we get the average of 2.5. The next available rank is 4 and this is allocated to the 13 (from *site b*).

We carry on adding ranks until they are all done. For our third example (the one with the moderate overlap) we end up with:

Table 36. Flour beetle count data. After ranking the values, we see some overlap in the ranks

Ranks R_a	**1**	**2.5**	**2.5**	**4**	**5**	**6**		**9**	**9**	**11.5**	**13**						
Site *a*	9	12	12	13	14	18		21	21	23	24						
Site *b*							19	21		23		25	28	32	33	34	34	45
Ranks R_b							**7**	**9**		**11.5**		**14**	**15**	**16**	**17**	**18.5**	**18.5**	**20**

We can now add up the ranks for each sample (this gives R_a and R_b values). We now calculate our *U*-values, think of this as a measure of overlap. Because we have two samples, we get two *U*-values using the Mann–Whitney formula shown in Figure 99:

$$U_a = n_a n_b + \frac{n_a(n_a+1)}{2} - R_a$$

Figure 99. The formula to calculate the Mann–Whitney *U*-test

We calculate our first *U*-value then simply switch the subscripts and calculate the second *U*-value. We select the **smallest** one as our test statistic.

For our example above, we get $R_a = 63.5$, $R_b = 146.5$ and $U_a = 91.5$, $U_b = 8.5$ so we select 8.5 as our final *U*-value. We need to compare our test statistic against a table of critical values (Table 37).

Table 37. Critical values for *U*, the Mann–Whitney *U*-test at 5% significance level. Reject the null hypothesis of your value of *U* is equal or less than the tabulated value

n	1	2	3	4	5	6	7	8	9	10
1	-	-	-	-	-	-	-	-	-	-
2	-	-	-	-	-	-	-	0	0	0
3	-	-	-	-	0	1	1	2	2	3
4	-	-	-	0	1	2	3	4	4	5
5	-	-	0	1	2	3	5	6	7	8
6	-	-	1	2	3	5	6	8	10	11
7	-	-	1	3	5	6	8	10	12	14
8	-	0	2	4	6	8	10	13	15	17
9	-	0	2	4	7	10	12	15	17	20
10	-	0	3	5	8	11	14	17	20	23

We look down the first column until we reach the number of data for the first sample, then look across until we reach the column headed with the number of data items in our second sample (since the table is symmetrical we could read across the header row and then look down). The value in the cell is the *U*-value that we need to equal or be smaller than in order for the difference to be statistically significant. Note that in most statistical tests, you are looking for a large test value to be significant; however, in some (like the *U*-test) you are looking for a low value so it is important to check the requirements.

In this case, we have ten data in both samples so we look in the table and find the critical value to be 23. Since our calculated value is 8.5 we can reject our null hypothesis and accept the alternative hypothesis; there is a statistically significant difference between the two sites. Now look back at the medians and see which site has the most flour beetles (*Tribolium confusum*). Put another way, there is less than a 5% chance that the result we obtained could happen by random chance.

How the *U*-test works

If we look at the first example, where there was no overlap, we can see that the ranks add up to $R_a = 55$ $R_b = 155$. This is as good as it gets, with no overlap we will always get a *Rank* sum of 155 (the largest value) when we have $n_a = 10$, $n_b = 10$. If we take our last example, where we had some overlap, we got a *Rank* sum of $R_b = 146.5$. If we take this from our perfect result we get 8.5, which was the *U*-value we calculated using the equation! So we can summarise by saying that:

$$U = \textit{perfect rank sum} - \textit{highest actual rank sum}$$

If we look at our middle example we get rank sums of $R_a = 99.5$ $R_b = 110.5$. This would give us a *U* of 155 – 110.5 = 44.5, much higher than the critical value (of 23) which we would expect because there was almost complete overlap.

It is tedious to add up the ranks for the perfect situation every time but it so happens that there is a simple formula that will do it (Figure 100):

$$Perfect \sum R = n_a n_b + \frac{n_a(n_a+1)}{2}$$

Figure 100. The perfect rank sum is the basis for the *U*-test

This is of course the basis of the Mann–Whitney formula; all we do is to subtract the actual rank sum. We do this for both samples as we cannot always tell which will give us the smallest *U*-value. What we are doing in effect is to measure the overlap based on the ranks of the numbers in our two samples.

Checking the sums

Calculating the ranks and doing the maths is simple enough but when you have lots of data it is easy to make a mistake. There is a quick way to check that you have performed the calculations correctly (usually it is the ranking that is in error):

$$U_a + U_b = n_a \times n_b$$

In other words, the two *U*-values add up to the size of the two samples multiplied together. In our case, n is 10 for both samples so $n_a \times n_b = 100$. Our two *U*-values should add up to 100 also.

7.2.1 Using Excel for the *U*-test

There is no built-in formula to work out the *U*-test in Excel. It is possible to assign ranks to samples of data but Excel does not rank data is quite the manner we would like. When we get tied values, they need to end up with the same rank and we give them an average rank based on the ranks available. Unfortunately Excel gives the tied values equal ranks but it uses the lowest of the ranks available. It is possible to work out a formula that correctly ranks the data and we see an example of this in Figure 101.

fx =RANK(I2,$I$2:$J$9,1)+COUNTIF($I$2:$J$9,I2)*0.5-0.5

I	J	K	L	M
ridge	furrow	ridge_r	furrow_r	
4	9	2	12	
3	8	1	10.5	
5	10	3.5	13	
6	6	6	6	
8	7	10.5	8.5	
6		6	#N/A	
5		3.5	#N/A	
7		8.5	#N/A	

Figure 101. Calculating ranks in Excel requires some lateral thought

Even when we have successfully created the ranks we cannot automatically get the critical values for the *U*-test. The *Analysis ToolPak* is no help to us either as the routines to carry out the analysis are not built-in.

7.2.2 Using R for the *U*-test, the *wilcox.test()* command

When data are not normally distributed we need a *U*-test (Mann–Whitney). This was developed by Wilcoxon originally and R (perhaps confusingly) calls the *U*-test the *wilcox.test()*. We run it so:

```
> wilcox.test(data1, data2)
```

This will run the *U*-test and compare medians for our two specified samples. You may well get error messages if you have tied ranks. This is generally okay unless you are close to $p = 0.05$ when a correction to the *p*-value should be applied. The program to do this is not part of the basic package but the routines are available in a separate library (called *exactRankTests*), which can be easily loaded (see Section 10.1.2 on loading additional packages).

```
  Wilcoxon rank sum test with continuity correction

data:  data1 and data2
W = 19.5, p-value = 0.7925
alternative hypothesis: true location shift is not equal to 0

Warning message:
In wilcox.test.default(data1, data2) :
  cannot compute exact p-value with ties
```

We will not consider the correction factor any further; modification of the resulting *p*-values is usually quite small and mostly makes little practical difference to our conclusions.

7.3 Paired tests

The *t*-test and the *U*-test described in the previous sections look for differences between two samples of data. Each sample is independent of the other. There are occasions however, when the two samples may not be totally independent. For example, you may be interested in the difference in lichen abundance on different sides (aspect) of trees. In order to see if there are more lichens on the north compared to the south, you set out with your quadrat and measure abundance on trees. You place your quadrat on the north side of a tree and then move around to the south side. Then you repeat with lots of other trees.

Now you end up with two samples, one from the north and one from the south; however, the two sets of numbers are not completely independent. We can match up the first measurement on the north list with the first measurement of the south list because they come from the same tree. We have a set of matched pair data.

Other examples might include the abundance of a plant species before and after some management treatment. You might set up some quadrats and determine abundance. Then some management is performed. You come back later and as long as your quadrats are in exactly the same spot(s) you could count the data as a matched pair. Each individual quadrat has two values, one for the before and one for the after. You are measuring something over time (but in exactly the same spot).

A further example might be to examine the heart rate of undergraduate students before and after administration of large quantities of caffeine. There will be natural variation in heart rate amongst students and it is logical to pair up measurements and so control variability.

In reality, the occasions where you get matched pair data are fairly limited and if you are not absolutely certain then assume the data are not. To be sure, you must be able to match up one measurement in one sample with a specific measurement in the other sample.

There are versions of the *t*-test and of the *U*-test that can test for differences in matched pair data. The *t*-test for matched pairs looks for differences when the data are all normally distributed. Wilcoxon matched-pairs test looks at differences when the data are not normally distributed (i.e. skewed).

7.3.1 Paired *t*-test for parametric data

When we have normally distributed data we can use a version of the *t*-test to compare matched pair data. Essentially we are looking at differences between each pair of values. If all the values were positive then we could be sure that there was a significant difference. If all the differences were negative then we expect the same result. Often however, there will be some positive differences and some negative. If the positive differences balance out the negative differences then we can be certain that there is no significant difference.

Table 38 shows some paired data. Here we have used plastic squares to create targets in a greenhouse. Each target was bi-coloured, with a white half and a yellow half. The data show the number of whitefly (a greenhouse pest, actually a hemipteran bug and not a fly at all!) trapped on the respective halves of the targets (which were sticky). We can see fairly easily that we really have paired data; it makes perfect sense to compare the two halves of each target. If we had a set of white targets and a set of yellow ones then we would not have matched data, simply two samples.

We are interested to know if there is a difference in the number of whitefly attracted to yellow or white targets. The difference column shows the result for white–yellow. We can see that there are some positive values and some negative values as well as a zero difference.

If we look at the original formula for the *t*-test (Figure 91), we can see that we have summary statistics for each of the two samples: mean, variance and the number of items in each sample. Because we are looking at differences between pairs of values, then the top of the formula can be modified. We can take the difference between each pair of values and calculate the mean of the differences. Similarly we can replace the bottom of the formula

Table 38. Matched pair data. Counts of whitefly attracted to coloured targets in a greenhouse. Each target is bi-coloured with white and yellow halves

White	Yellow	Difference
4	4	0
3	7	–4
4	2	2
1	2	–1
6	7	–1
4	10	–6
6	5	1
4	8	–4

with an expression that relates to the variance of the differences. We end up with something that looks like Figure 102:

$$t = \frac{\bar{D}}{\sqrt{\frac{s_D^2}{n}}}$$

Figure 102. The formula to calculate the matched pairs *t*-test

Effectively this is half of the original *t*-test formula (Figure 91). If we look at the whitefly data from Table 38 we can work out the summary values for all the columns; these results are shown in Table 39.

Table 39. Mean and variance for samples and differences for whitefly matched pairs data

	White	Yellow	Difference
Mean	3.88	5.63	–1.75
Variance	2.98	8.27	8.5

Of course there are eight pairs of data here and therefore eight differences (although one difference is zero). If we substitute these values into the formula we get:

$t = –1.75 \div \sqrt{(8.5 \div 8)} = 1.698$

We look this value up in our table of critical values (Table 30) and use 7 degrees of freedom, i.e. the number of pairs of observations –1 (8 – 1 = 7). We find that in this case there is no statistical difference between the white and yellow targets.

Our original data were normally distributed; generally this means that the differences between the matched pairs will also be normally distributed but we really should check before we run the test.

7.3.2 Wilcoxon matched-pairs test for skewed data

When our data are not normally distributed, or the differences between the pairs are not normally distributed, the *t*-test is not applicable. We need a non-parametric alternative. Wilcoxon's matched-pairs test is based upon the ranks of the data (like most non-parametric tests). What we do is to look at the differences and give them a rank based on the magnitude of the difference. We can then split them up according to whether they relate to a positive difference or a negative difference.

If most of the ranks are due to positive differences then we will get a fairly large sum of ranks. The other sum will be quite small. Similarly if it is the other way round and most of the differences are negative we will get a large sum of ranks due to the negative differences. Let us consider our whitefly data again. This time we have determined the ranks of the differences. Table 40 shows the data and the ranks.

Table 40. Matched-pair data. Counts of whitefly attracted to coloured targets in a greenhouse. Each target is bi-coloured with white and yellow halves. Here the non-zero differences have been ranked

White	Yellow	Difference	All ranks	+ Ranks	– Ranks
4	4	0			
3	7	–5	5.5		5.5
4	2	2	4	4	
1	2	–1	2		2
6	7	–1	2		2
4	10	–6	7		7
6	5	1	2	2	
4	8	–4	5.5		5.5

The sum of all the ranks will always come to the same value for a given number of observations. Here we have eight pairs but one of them has a zero difference so we do not count that (this is slightly different to the *t*-test). For seven pairs of observations we will always get a sum of ranks of 28. The bigger the difference between the rank sums due to the positive differences and the rank sums due to the negative difference, the more likely it is that we have a significant difference.

This is how the Wilcoxon matched-pairs test works. It compares these rank sums. We can write this formally as:

$$W = \sum R^{+/-}$$

We always end up with two values, one for each rank sum (i.e. one for the positive and one for the negative differences). We take the smaller of the two values as our test statistic. If we add up the ranks for positive and negative differences from Table 40 we get 6 and 22 (which adds up to the maximum 28 for seven pairs of differences). We use 6 as our test statistic and compare to a table of critical values (e.g. Table 41).

Table 41. Critical values for Wilcoxon matched pairs. Compare your lowest rank sum to the tabulated value for the appropriate number of non-zero differences. Reject the null hypothesis if your value is equal to or less than the tabulated value

	Significance level		
N_D	5%	2%	1%
5	-	-	-
6	0	-	-
7	2	0	-
8	3	1	0
9	5	3	1
10	8	5	3
11	10	7	5
12	13	9	7
13	17	12	9
14	21	15	12
15	25	19	15
16	29	23	19
17	34	27	23
18	40	32	27
19	46	37	32
20	52	43	37
25	89	76	68
30	137	120	109

If our smallest rank sum is less than or equal to the critical value, we can reject the null hypothesis and accept the alternative that there is a significant difference between the two samples. In this case our lowest rank sum was 6 and the critical value is 2. We therefore accept the null hypothesis; there is no difference between the white and yellow targets.

7.3.3 Using Excel for paired tests

In Section 7.1.1 we looked at how Excel could perform a *t*-test. Excel can take paired data too and the formula only has to be modified slightly:

TTEST(*range1*, *range2*, 2, 1)

Here we have our two samples, of course they must be in matched pairs and so each sample will be the same length. We insert a 2 to tell Excel to look at both ends of the normal distribution curve and a 1 to indicate that the data are in matched pairs as opposed to independent samples.

The *Analysis ToolPak* (Section 1.9.1) has a paired *t*-test and we may use this to carry out the analysis.

There is no equivalent of the Wilcoxon matched-pairs test built-in to Excel. It is possible of course to work out a method to rank the data and carry out the maths but it is easier to use R in this case.

7.3.4 Using R for paired tests

The R program has routines for calculating paired tests built in. If you have paired data then both the *t*-test and *U*-test may be run in paired form simply by adding paired = TRUE to the command, e.g.

```
> wilcox.test(data1, data2, paired=TRUE)
```

If the two samples are unequal in length then R gives us a helpful error message:

```
Error in wilcox.test.default(data1, data2, paired = T) :
  'x' and 'y' must have the same length
```

If the samples are the same length but are not really paired, R will carry on regardless – it is a computer program, not a mind reader!

8. Tests for linking data – correlations

When we wish to examine a link between two samples, we are generally looking for a correlation. In this chapter, we will look at the simplest situations: finding links (correlations) between two samples. Generally these samples are a biological variable and an environmental variable but this is not always the case. Where our data are not normally distributed we will use the Spearman's rank test (Section 8.1). Where we have normally distributed data we will use Pearson's product moment (Section 8.2).

When we have one factor and think that there are several other factors that may influence it, we have a more complex situation and this will be dealt with in Chapter 11. If our data are categorical then we will look for an association; this will be the subject of Chapter 9.

8.1 Correlation: Spearman's rank test

So far our examples have been concerned with differences between things. Imagine you were interested in mayfly (Ephemeroptera) and how their abundance altered with stream flow. Alternatively you might be interested to know if different density of sheep (grazing pressure) affected the abundance of a particular plant. Another example might be the number of sites containing a species (e.g. Japanese knotweed) over time. These are examples of linking together sets of data. In the first case we link the stream flow (speed) to mayfly abundance. In the second case we link the density of sheep to plant abundance. In the third case we are looking to link number of sites with time (probably a year). We are looking for a correlation. If we collected some data we would represent it graphically using a scatter plot (see Section 6.3.1). We will cover graphs in a lot more detail in Section 12.4. One axis would be for the abundance (or number of sites) and the other for the stream flow (or sheep density or year).

Our finished graph might look something like the following example (Figure 103).

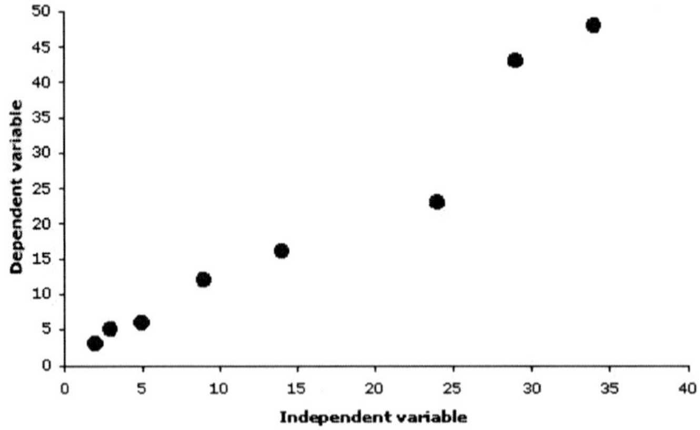

Figure 103. Scatter plot. This shows a perfect positive correlation between the independent and dependent variables (the points do not have to be a perfectly straight line)

Here the axes have been labelled as *Independent variable* and *Dependent variable*. They are always this way round, with the *y*-axis showing the dependent variable. Which axis is which? In our examples above, the stream flow, sheep density and year are the independent variables. This independent variable is what we think controls the abundance. The dependent variable is the abundance data or number of sites. These variables can also be called the *response* and the *predictor* variables respectively. It is the abundance that we are interested in and we want to know if this changes in some predictable manner when the independent variable changes.

There are three possible outcomes. In Figure 103 we see an example of a perfect positive correlation. As we move up the *x*-axis, the next largest value corresponds to the next largest value on the *y*-axis. As one axis increases so does the other, a positive correlation. We might also get a graph that looks like Figure 104.

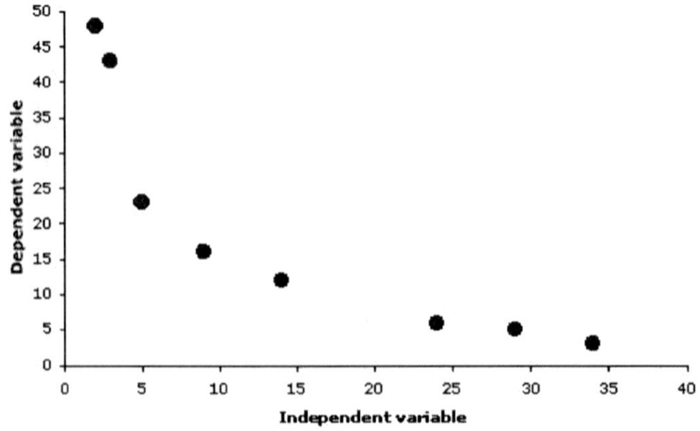

Figure 104. Scatter plot showing perfect negative correlation (note that this does not have to be a straight line)

In this case, we see that as we move up the x-axis we go from large y-values to smaller ones. We have a negative correlation. As we move up the x-axis we correspond exactly to steps down the y-axis so we have a perfect negative correlation. There is another possibility:

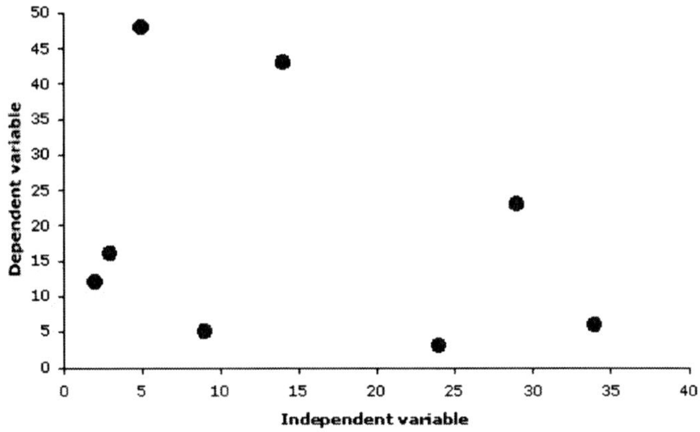

Figure 105. Scatter plot showing a perfect mess! There is no evident relationship between the two variables here

In Figure 105 we cannot determine any pattern at all. As we move up the x-axis we do not encounter any regular pattern.

The Spearman's rank Correlation test is used when you wish to determine the strength of the link between two sets of values. The link does not have to be in the form of a straight line (look at Figure 104 for example) but it should be in the form of a general trend up or down (a U-shape is not good). The test is designed so that you get a value of +1 when you have a perfect positive correlation, –1 with a perfect negative correlation and 0 when you have a perfect mess. As you may guess from the name, the test converts the actual values to ranks; it is a non-parametric test and is used on data that is not normally distributed (i.e. skewed).

Table 42 shows data collected about mayflies from a stream and for each sample we measured the stream speed and the number of mayfly. In this case the pattern is not as clear as our perfect illustrations.

Table 42. Stream speed (m s^{-1}) and mayfly (*Ecdyonurus dispar*) abundance

Speed	Abundance
2	6
3	3
5	5
9	23
14	16
24	12
29	48
34	43

If we create a scatter plot of the data in Table 42 we see the pattern more clearly (Figure 106).

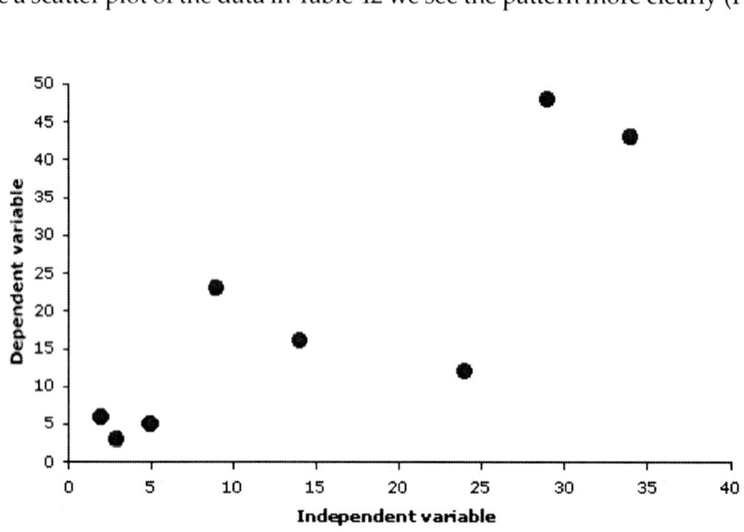

Figure 106. Scatter plot of mayfly (*Ecdyonurus dispar*) and stream speed data. The stream speed is the independent axis and the mayfly data the dependent axis

It looks like there may be a relationship between speed and abundance but there are a number of inconsistencies, it is not a perfect correlation. Is it statistically significant? We use the Spearman's rank test to determine.

We begin with a hypothesis to test. In this case we expect a greater abundance of mayfly as stream speed increases. We would write this formally (our alternative hypothesis H_1) as:

"There is a positive correlation between stream speed and mayfly (*Ecdyonurus dispar*) abundance."

We are saying that the relationship is a positive one because of the previous research we have done (reading papers and so on). Our null hypothesis (H_0) now becomes:

"There is no correlation between stream speed and mayfly abundance."

Note that H_0 is not the opposite of H_1; we did not say that the correlation is negative (look back at Sections 1.4 and 5.1).

8.1.1 Calculating the Spearman Coefficient

The next thing we must do is to assign ranks to our data. In the *U*-test (looking for differences, Section 7.2) we ranked the whole lot as one set. Here we are looking to compare the ranks of the *y*-data with the ranks of the *x*-data so we rank each set separately. The smallest value gets the smallest rank. Any tied values get the averages ranks (see rules for ranking in Section 7.2). Table 43 shows our data with the ranks added in.

Table 43. Mayfly and speed data ranked as part of the Spearman's rank correlation coefficient calculation

Speed	Rank speed	Abundance	Rank abundance
2	*1*	6	*3*
3	*2*	3	*1*
5	*3*	5	*2*
9	*4*	23	*6*
14	*5*	16	*5*
24	*6*	12	*4*
29	*7*	48	*8*
34	*8*	43	*7*

In this case it is pretty simple, there are no tied ranks. We now want to compare ranks; we are looking for differences between the two sets. We can start by subtracting one from the other.

Table 44. Differences in rank of mayfly (*Ecdyonurus dispar*) and speed data

Speed	Rank speed	Abundance	Rank abundance	Difference
2	*1*	6	*3*	**2**
3	*2*	3	*1*	**−1**
5	*3*	5	*2*	**−1**
9	*4*	23	*6*	**2**
14	*5*	16	*5*	**0**
24	*6*	12	*4*	**−2**
29	*7*	48	*8*	**1**
34	*8*	43	*7*	**−1**
			Sum:	**0**

Of course now we have some positive and negative values and they add up to zero. Not terribly helpful at this point; however, a common trick in statistics is to get rid of negative values by squaring. We now take the square of the differences.

Table 45. Squaring the difference in ranks for the mayfly (*Ecdyonurus dispar*) and speed data

Speed	Rank speed	Abundance	Rank abundance	Difference	D^2
2	*1*	6	*3*	**2**	*4*
3	*2*	3	*1*	**−1**	*1*
5	*3*	5	*2*	**−1**	*1*
9	*4*	23	*6*	**2**	*4*
14	*5*	16	*5*	**0**	*0*
24	*6*	12	*4*	**−2**	*4*
29	*7*	48	*8*	**1**	*1*
34	*8*	43	*7*	**−1**	*1*
			Sum:	**0**	*16*

Lastly we use the Spearman's rank formula (Figure 107) to calculate our final test statistic:

$$r_s = 1 - \frac{6\sum D^2}{n(n^2-1)}$$

Figure 107. The formula for the Spearman's rank correlation coefficient

In this case n is the number of pairs of data (since each y-data item is paired with an x-data). D^2 is the sum of the squares of the differences of the ranks (the final column in our calculation from Table 45). We have eight replicates so $n = 8$ and $D^2 = 16$.

Once we have evaluated our r_s value (don't forget that the 1-part applies to the whole of the fraction on the right of the formula) we get $r_s = 0.810$. Okay so far. Already we know one thing, that the correlation is positive (we could of course tell that from the graph) but we do not yet know if it is statistically significant of not. We need to compare to a table of critical values (Table 46).

Table 46. Critical values for r_s the Spearman's rank correlation coefficient. Reject the null hypothesis if your calculated value is equal to or greater than the tabulated value

	Significance level		
No. of pairs, *n*	5%	2%	1%
5	1.000	1.000	-
6	0.886	0.943	1.000
7	0.786	0.893	0.929
8	0.738	0.833	0.881
9	0.683	0.783	0.833
10	0.648	0.746	0.794
12	0.591	0.712	0.777
14	0.544	0.645	0.715
16	0.506	0.601	0.665
18	0.475	0.564	0.625
20	0.450	0.534	0.591
22	0.428	0.508	0.562
24	0.409	0.485	0.537
26	0.392	0.465	0.515
28	0.377	0.448	0.496
30	0.364	0.432	0.478

We look down the first column and find the number of data (in this case 8) then we read the critical value from the next column (in this case it is 0.738). If our value is greater or equal to this then we can reject our null hypothesis and accept the alternative. In this instance, we can say that there is a statistically significant positive correlation between stream flow and abundance.

We could look over to the next column (headed 2%) but we see that we do not exceed this value. Our correlation is significant at the 5% level so we assume that there is a less than 5% probability that the result was due to random chance. If we get a negative correlation coefficient, then we simply ignore the sign when looking up the table of critical values. We can see that it is much easier to cross the finishing post when you have more data. This is something to consider when designing your project.

How the Spearman's rank test works

When we have a perfect positive correlation, the ranks of the *x*- and *y*-values will all agree perfectly. Look at the example in Table 47.

Table 47. Example of perfect positive correlation. The ranks for the independent factor agree perfectly with the ranks of the dependent factor

Speed	Rank speed	Abundance	Rank abundance	Difference	D^2
2	*1*	3	*1*	**0**	*0*
3	*2*	5	*2*	**0**	*0*
5	*3*	6	*3*	**0**	*0*
9	*4*	12	*4*	**0**	*0*
14	*5*	16	*5*	**0**	*0*
24	*6*	23	*6*	**0**	*0*
29	*7*	43	*7*	**0**	*0*
34	*8*	48	*8*	**0**	*0*
			Sum:	**0**	*0*

In Table 47 we see that the ranks agree perfectly and as a result the sum of the rank differences squared ($\sum D^2$) is bound to be zero. Now let's look at an example of perfect negative correlation (Table 48).

Table 48. Example of perfect negative correlation. The ranks for the independent factor disagree perfectly with the ranks of the dependent factor

Speed	Rank speed	Abundance	Rank abundance	Difference	D^2
2	*1*	48	*8*	**7**	*49*
3	*2*	43	*7*	**5**	*25*
5	*3*	23	*6*	**3**	*9*
9	*4*	16	*5*	**1**	*1*
14	*5*	12	*4*	**–1**	*1*
24	*6*	6	*3*	**–3**	*9*
29	*7*	5	*2*	**–5**	*25*
34	*8*	3	*1*	**–7**	*49*
			Sum:	**0**	*168*

In Table 48 (perfect negative correlation) we see that the ranks all disagree as much as possible so the difference of the ranks squared will be as large as possible. Exactly how large will depend on how many data there are. It so happens that the maximum the $\sum D^2$ can be determined using the following equation (Figure 108):

$$Max \sum D^2 = \frac{n(n^2-1)}{3}$$

Figure 108. Maximum differences of squared ranks form the basis for the Spearman's rank correlation test

Now we want to get to the point where we have values ranging from +1 to –1. At present we have 0 to Max $\sum D^2$. We can now create a new equation using Max $\sum D^2$ that will give us the required –1 to +1 range (Figure 109).

$$r_s = 1 - \frac{2\sum D^2}{Max \sum D^2}$$

Figure 109. The Spearman Rank correlation can be expressed in terms of the maximum differences in squared ranks

If we substitute in the equation for Max $\sum D^2$ we end up with the final Spearman's rank formula we saw earlier.

Correlation may not mean cause

It is important to keep in mind that a statistically significant correlation does not necessarily mean that one factor causes the other. There may be some other factor that a actually responsible that you have not determined. For example the stream velocity itself may not be responsible for the abundance of mayfly but the speed of flow may influence the number of food particles washed along or the oxygenation.

We can use R to carry out the Spearman's rank test quite easily but before we do let us look at another type of correlation.

8.2 Pearson's product moment

The Spearman's rank correlation test mentioned above (Section 8.1) examined for links between dependent and independent factors. The data do not have to be normally distributed (parametric) and the correlation itself does not have to be in the form of a straight line (although it should be generally heading in one direction). Look again at Figures 103 and 104; in the first case the scatter plot shows a positive correlation and in the second case a negative one. We used the Spearman's rank test to determine of the correlation was significant.

There may well be occasions where you suspect that the link between two factors can be described as a straight line. To begin with the two factors must themselves be normally distributed. Assuming this is the case, we can describe the relationship using the following formula:

$$y = mx + c$$

In this case the y variable is our dependent factor. The x is the independent factor. The c represents the point on the y-axis where the line of best-fit crosses. Finally the m represents the slope of the line. Look at Figure 110. We see a scatter plot with a series of points representing our x and y measurements, the dependent and independent variables.

We have added the line of best fit (we will see how to do that later, in Sections 11.2 and 12.4) and we can see that it crosses the y-axis at around 15. The slope of the line can be worked out using the values on the x and y-axes and we get something a bit over one and a half (1.74). In this case the equation is displayed on the graph for reference.

Once we know what the equation is, then if we are given a value for x we could calculate the corresponding value of y.

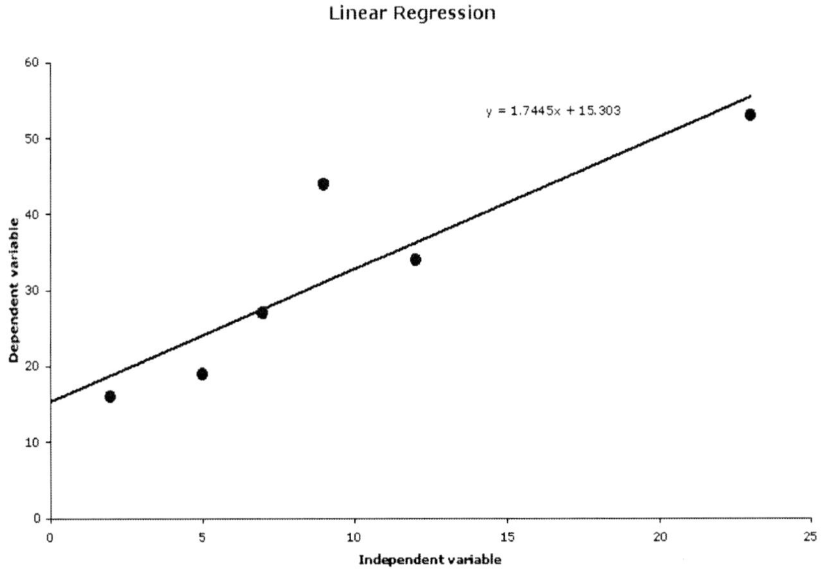

Figure 110. A scatter plot to show the elements of a straight-line relationship

You might expect a straight-line relationship in many cases, for example plant growth and nitrogen fertiliser treatment, butterfly abundance and percentage cover of food plant.

What Pearson's product moment does is to calculate not only the strength of the relationship (like Spearman's rank test) but also the equation that describes the linear fit between the two variables.

To work out the slope, *m*, we need to evaluate the following formula (Figure 111):

$$m=\frac{\sum(x-\bar{x})(y-\bar{y})}{\sum(x-\bar{x})^2}$$

Figure 111. The formula to calculate the slope of a straight line

It looks a little scary but actually it is quite simple (just tedious to do). We start by evaluating the means for *x* and *y* (the dependent and independent factors). Then we take the mean away from each value. On the top line we multiply differences between *x* and between *y*, then add these all together. On the bottom line we examine the *x* only but square them before adding them together.

The intercept is quite easy to calculate using the formula shown below (Figure 112):

$$c=\bar{y}-m\bar{x}$$

Figure 112. The formula to calculate the intercept of a straight line

Once we know the slope, we are there as we already worked out the means of *x* and *y*. Now we can reconstruct the formula that represents the line of best fit and could draw it on a graph if we wanted. The line must pass through a point on the graph that is represented by the mean of *x* and the mean of *y*. We also know the intercept. If we marked these two points and joined them up, we would make the best-fit line. We could then check the slope to ensure it was correct.

This is only part of the story because we still do not know if our putative linear relationship is statistically significant. In order to achieve this we set up our hypothesis (H_1), which will be that there is a correlation between the two factors. Recall from the Spearman's rank test that we also defined the direction of the correlation; we should do the same here. The null hypothesis (H_0) is that there is no correlation. The formula that Pearson came up with is in Figure 113:

$$r=\frac{\sum(x-\bar{x})(y-\bar{y})}{\sqrt{\sum(x-\bar{x})^2\sum(y-\bar{y})^2}}$$

Figure 113. Formula to calculate the Pearson correlation coefficient

It is more tedious than really difficult and most often you would use a computer to do it for you. In fact there is a function built into Excel to do this. Because the linear relationship is so commonly used, Excel also has functions that will evaluate the slope and intercept of a set of paired values (see Section 8.3).

So, we end up with a statistic called r. The final stage is to compare this to a critical value. Table 49 gives critical values for the Pearson correlation coefficient. The degrees of freedom are calculated by subtracting 2 from the number of items of data you have. So, if you have three pairs of data, the df = 1 (that is 3 y-values and 3 x-values – you cannot have an x-value without a y-value, they are always in pairs).

Table 49. Critical values for the correlation coefficient r, Pearson's product moment. Reject the null hypothesis if your value is equal to or greater than the tabulated value

	Significance	
Degrees of freedom	**5%**	**1%**
1	0.997	1
2	0.95	0.99
3	0.878	0.959
4	0.811	0.917
5	0.754	0.874
6	0.707	0.834
7	0.666	0.798
8	0.632	0.765
9	0.602	0.735
10	0.576	0.708
12	0.532	0.661
14	0.497	0.623
16	0.468	0.59
18	0.444	0.561
20	0.423	0.537
22	0.404	0.515
24	0.388	0.496
26	0.374	0.478
28	0.361	0.463
30	0.349	0.449
35	0.325	0.418
40	0.304	0.393
45	0.288	0.372
50	0.273	0.354
60	0.25	0.325
70	0.232	0.302
80	0.217	0.283
90	0.205	0.267
100	0.195	0.254
125	0.174	0.228
150	0.159	0.208
200	0.138	0.181
300	0.113	0.148
400	0.098	0.128
500	0.088	0.115
1000	0.062	0.081

If your calculated value is greater than or equal to the tabulated value, the relationship is statistically significant. If your *r*-value is negative just ignore the sign when looking it up in the table. As with the Spearman correlation, a negative value indicates that the correlation is negative, that is as *y* increases, *x* decreases.

With the Spearman correlation, we did not make any assumptions regarding the cause and effect. The same is true here – just because there is a linear link between the two variables does not mean necessarily that there is cause and effect.

8.3 Correlation tests using Excel

We can use Excel to carry out tests for correlation; there are several built-in functions as well as the *Analysis ToolPak* (Section 1.9.1).

8.3.1 Parametric tests of correlation

It is possible to conduct parametric correlation using Excel. There is a function built-in that will work out the Pearson correlation coefficient:

CORREL(*range1, range2*)

Here we replace the *range1* and *range2* parts with the cell references that contain our data (it does not matter which is the dependent and which the independent variables). The result given is the coefficient.

We may also determine the slope and intercept:

SLOPE(*range_y, range_x*)
INTERCEPT(*range_y, range_x*)

In these two cases we need to specify which range of data is the dependent (response) and which is the independent (predictor). It is important to get these the right way around!

We can calculate the slope, intercept and correlation coefficient but we really need to know if the correlation is statistically significant. It is possible to use a function to do this in Excel but it is a little more involved than a simple formula. We start by using the LINEST formula, which works out a variety of statistics based on the normal distribution and fitting lines of best fit (which is what the slope and intercept are). The formula looks like this:

LINEST(*y-data, x-data, 1, 1*)

The structure of the formula is fairly straightforward. The *y-data* and *x-data* parts are the ranges of cells where the dependent and independent variables are found. The next bit is the number 1. This tells Excel to fit a standard line; if we use a 0 here then the line would be forced through the origin. There are few occasions when this is necessary. The final number 1 tells Excel to show a range of summary statistics. If we make this a 0 we get the slope and intercept only.

The formula is an array formula; we met this previously (Section 4.2) when looking at histograms. The result of the formula will be spread over several cells. Unfortunately Excel is not smart enough to put the answer in the blank cells for us so we must highlight them in advance. The result(s) will be spread across two columns and five rows in the spreadsheet so we need to highlight the appropriate number of blank cells before we type the formula.

Figure 114 shows data and the LINEST function in action. In the spreadsheet we have two columns of data. The abund column is the abundance of a freshwater invertebrate and the flow column the water speed. We have already checked that these data are normally distributed and have calculated the correlation coefficient, slope and intercept using the methods described above.

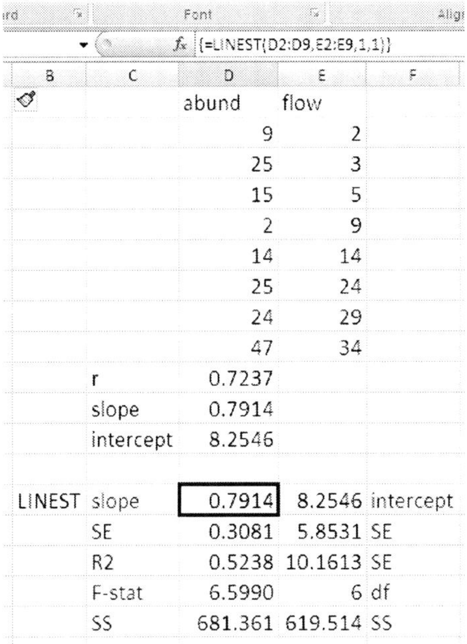
{=LINEST(D2:D9,E2:E9,1,1)}

B	C	D	E	F
		abund	flow	
		9	2	
		25	3	
		15	5	
		2	9	
		14	14	
		25	24	
		24	29	
		47	34	
	r	0.7237		
	slope	0.7914		
	intercept	8.2546		
LINEST	slope	0.7914	8.2546	intercept
	SE	0.3081	5.8531	SE
	R2	0.5238	10.1613	SE
	F-stat	6.5990	6	df
	SS	681.361	619.514	SS

Figure 114. Using the LINEST function to determine the significance of a correlation in Excel

To get the LINEST function to work, we need to highlight ten blank cells where we want the answers to appear. We can then type in the formula in the regular manner. When we come to enter the formula though, we must remember to press CTRL+SHIFT+ENTER as this is an array formula. The results appear as plain values. In Figure 114 the text labels have been added so we can see which bits are of interest because Excel does not do this for us.

We get a variety of results. The top row shows the slope and intercept, and underneath we get values relating to the standard error of each of these. The third row shows us the R^2 value, we will cover this in more detail when we look at multiple regression (Chapter 11). For now let us just say that it is related to how much variability in the data (the dependent or response data) is accounted for by our independent (predictor) variable. It is not quite the same as the correlation coefficient. In this case we see a value of 0.5238, which informs us that only about half the variability in abundance is accounted for by the flow.

The value to the right of the R^2 value is the standard error of the residuals, essentially how far the points lie from the line of best fit. The very bottom line shows the sums of squares of the deviations of the data from their means and from the estimated values of y (from the line of best fit). We do not need to worry about these but they are used to calculate the value in the fourth row, which has been labelled *F*-stat.

We are going to use the *F*-stat value (6.599 in this case) to determine the *p*-value and so the statistical significance of the correlation. In order to do that we will need to use a new function that calculates probabilities based on the *F*-statistic.

The function we need is called FDIST and works like this:

FDIST(*F-value, df-top, df-bott*)

The *F-value* is the value of the *F*-statistic we just calculated. The *df-top* part relates to how many columns of data we have and the *df-bott* relates to the number of rows of data. Figure 115 shows the formula in action.

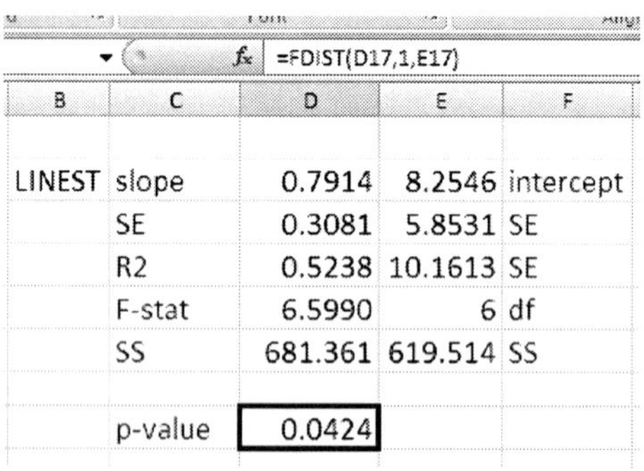

Figure 115. After the LINEST function is used, we determine the significance from the *F*-statistic

Since we have the *F*-statistic calculated for us, we can point to this in the formula and enter the cell where the value lies (D17). The next part is a number 1 and this is the number of columns – 1. This is not shown as a result so we need to type a 1 for ourselves. The next part relates to how many rows there are. It is slightly more complex and is worked out as:

df = (number of rows – 1) – (number of columns – 1)

In this case we get (8 – 1) – (2 – 1) = 6. The answer has already been worked out for us and the result is to the right of the *F*-stat value (E17 in our example). So, we can enter the cell reference of this into our FDIST formula.

The result we get is a probability. In this case we get 0.0424 and this tells us that the correlation is a statistically significant one. In other words there is only a 4% chance that we have this high correlation by random chance. We are able to reject the null hypothesis and accept the alternative, which is that there is a significant correlation (positive) between flow and abundance.

You will also notice that we still do not get the correlation coefficient and need to run the CORREL function. We will look at producing a graph of our result with a scatter plot in Section 12.4.3.

8.3.2 Non-parametric test for correlation in Excel

Simply put, there is no built-in function for determining the correlation coefficient of non-parametric data. It would be possible to rank the data and calculate Spearman's rank coefficient but your answer would still need to be compared to a table of critical values.

8.3.3 Using the *Analysis ToolPak* for correlation in Excel

We can carry out correlation using the *Analysis ToolPak*. We can only determine the correlation using parametric data. We can determine the basic Pearson correlation coefficient by selecting the Correlation routine from the *Data Analysis* window (Figure 116).

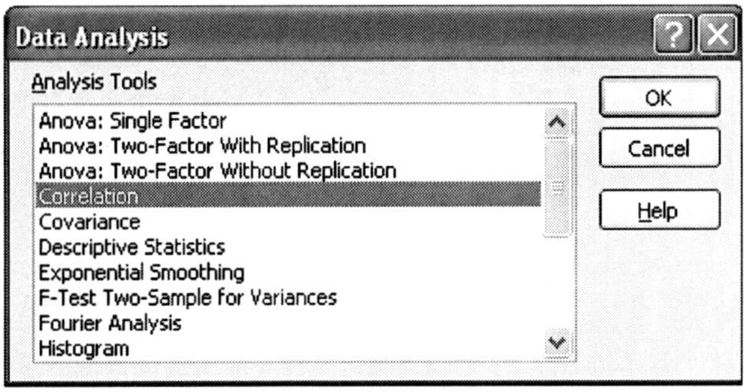

Figure 116. Selecting Correlation from the *Data Analysis* window

Once we have chosen this routine, we can select our data from the input window that appears (Figure 117).

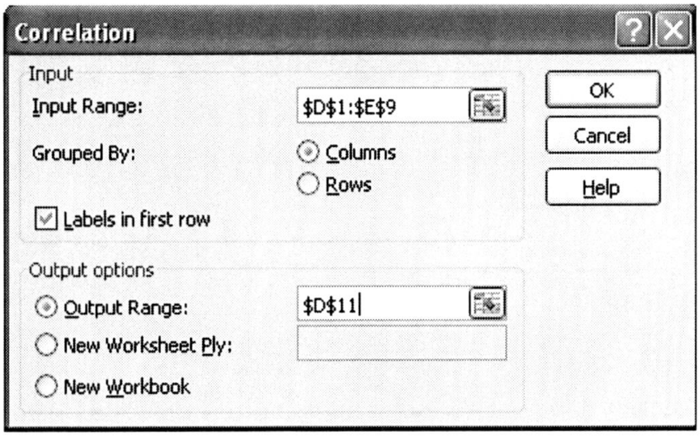

Figure 117. Choosing the data for correlation from the *Data Analysis* input window

The Pearson coefficient is displayed in the spreadsheet where we chose to have it placed. Here we selected the existing sheet but we might have chosen to use a new workbook to hold our result (Figure 118).

	abund	*flow*
abund	1	
flow	0.723721	1

Figure 118. The results from Correlation are placed in a simple table

If we wish to carry out a correlation test then we must use the Regression option from the *Data Analysis* window (Figure 119).

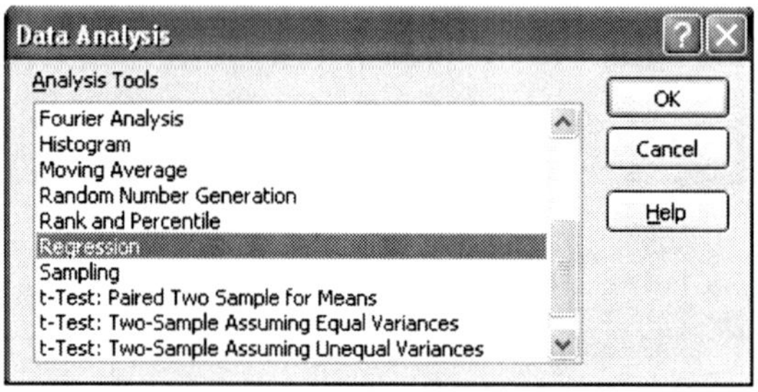

Figure 119. Selecting Regression to carry out a significance test for correlation

This time we get to select our y-values and x-values separately. It is important we get these the correct way round! In our case the abundance values are the y-values and the flow forms the x-data (Figure 120).

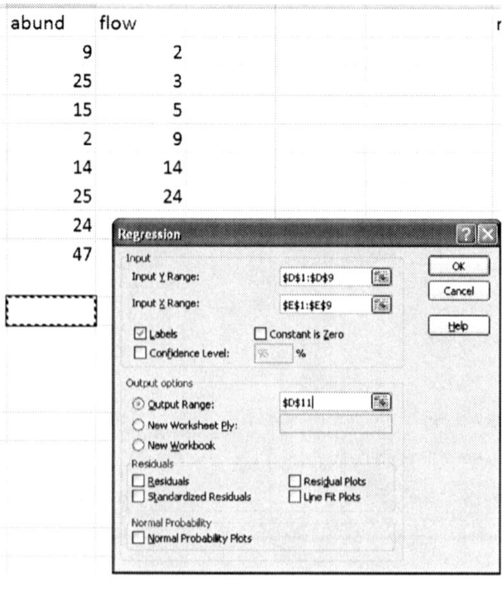

Figure 120. Regression data are selected with *y* and *x* data separately

We have a variety of additional options but we will look at these in Chapter 11. Our results are placed at the location we select in the input window (Figure 121).

SUMMARY OUTPUT

Regression Statistics	
Multiple R	0.72372058
R Square	0.52377147
Adjusted R Square	0.44440005
Standard Error	10.1613138
Observations	8

ANOVA

	df	*SS*	*MS*	*F*	*Significance F*
Regression	1	681.3612132	681.3612	6.5989932	0.0424005
Residual	6	619.5137868	103.2523		
Total	7	1300.875			

	Coefficients	*Standard Error*	*t Stat*	*P-value*	*Lower 95%*	*Upper 95%*	*Lower 95.0%*	*Upper 95.0%*
Intercept	8.25459559	5.8531414	1.410285	0.2081327	-6.0675254	22.576717	-6.06752544	22.576717
flow	0.79136029	0.308060074	2.568851	0.0424005	0.0375645	1.5451561	0.03756445	1.5451561

Figure 121. The regression results from the *Analysis ToolPak* are extensive

The results show us quite a few things; we can see the basic intercept and slope listed under Coefficients in the lower table. In the upper table we can see the R^2 value and the number of observations (replicates) that we used. The middle table shows us the statistical significance of our correlation; of particular note are the *F* and Significance items. The lower table also shows us the significance of the regression and we can see that the *p*-value for the flow is the same as the Significance shown in the middle table (0.042). We also see values for the 95% confidence intervals. When we come to examine multiple regression in Chapter 11 this bottom table would contain more information.

8.4 Correlation tests using R

We can carry out correlation in R quite easily using the *cor.test()* command. The command will work with parametric or non-parametric data.

8.4.1 Non-parametric tests of correlation

The basic *cor.test()* command will work out Spearman's rank or Pearson coefficients. It will also calculate Kendall's coefficient; this is a non-parametric test and broadly similar to Spearman's rank test. Here we will stick to Spearman's rank for non-parametric data.

Correlation tests *cor.test()* command

Simple correlations may be run using the *cor()* command. This simply calculates the correlation coefficient (you may select Spearman, Pearson or Kendall's coefficients).

```
> cor(data4, method= "pearson")
```

In this case we have two variables in a single file and R displays a matrix of results:

```
          Abundance     Speed
Abundance 1.0000000 0.8441408
Speed     0.8441408 1.0000000
```

This may be useful if you have a set of data with several variables because R will calculate a matrix of them all. If you have two separate variables you can alter the command to include them both, e.g.

```
> cor(data4, data5, method = "pearson")
```

The result is now displayed as a single item:

```
[1] 0.8441408
```

It is helpful to select two variables and to calculate the coefficient and the statistical probability as well. The command needs only a tiny bit of tweaking:

```
> cor.test(data4, data5, method= "spearman")
```

This will work out the Spearman's rank correlation coefficient and determine the significance:

```
 Spearman's rank correlation rho

data:  data4 and data5
S = 16, p-value = 0.02178
alternative hypothesis: true rho is not equal to 0
sample estimates:
      rho
0.8095238
```

The output is sparse but covers all we really need to know. The results give *S*, which is the sum of the differences in ranks squared; *p*-value, which tells us if the correlation is significant; and rho, the value of the Spearman's rank correlation coefficient itself.

8.4.2 Parametric tests of correlation

The basic *cor.test()* command will calculate the parametric Pearson's product moment, indeed the default command does this and if we need a non-parametric version we must explicitly state that is what we want using the *method* = part of the command; however, as we saw in dealing with correlation in Excel (Section 8.3.1) we may well want to know the slope and intercept of the line of best fit to tell us about the linear relationship between the two variables.

Let us start by using the same data that we used when looking at correlation in Excel (Section 8.3.1). We will run a simple Pearson correlation using the *cor.test()* command. The results are shown below:

```
> abund
[1]  9 25 15  2 14 25 24 47
> flow
[1]  2  3  5  9 14 24 29 34
> cor.test(abund, flow, method = 'pearson')

  Pearson's product-moment correlation

data:  abund and flow
t = 2.5689, df = 6, p-value = 0.0424
alternative hypothesis: true correlation is not equal to 0
95 percent confidence interval:
 0.03887166 0.94596455
sample estimates:
      cor
0.7237206
```

We have also shown the data, which were saved as two items, abund and flow. Here we added *method = 'pearson'* to the command but this is the default in R so we need not have done the additional typing. We get a variety of results but the two we are interested in are the *p*-value and the cor. The *p*-value (0.0424) is the same as that we calculated using Excel. The correlation coefficient (0.7237) is at the bottom and is of course also the same as that calculated in Excel via the CORREL function.

We can extract just the correlation coefficient from our result by adding a bit to the end of the command:

```
> cor.test(abund, flow, method = 'pearson')$estimate
      cor
0.7237206
```

We could also extract the *p*-value alone like so:

```
> cor.test(abund, flow, method = 'pearson')$p.val
[1] 0.04240053
```

Although this is not especially helpful here it illustrates the fact that many results in R are in fact comprised of several parts and that it is possible to retrieve just those parts you are interested in.

In order to get the slope and intercept for our data we need to use a slightly different command. R has a range of functions bundled into the *lm()* command (*lm* stands for linear modelling). Below we see the *lm()* command used to provide a summary of the linear relationship between abundance and flow:

```
> summary(lm(abund ~ flow))

Call:
lm(formula = abund ~ flow)
```

```
Residuals:
    Min       1Q   Median       3Q      Max
-13.377   -5.801   -1.542    5.051   14.371

Coefficients:
            Estimate Std.  Error  t value  Pr(>|t|)
(Intercept) 8.2546         5.8531 1.410    0.2081
flow        0.7914         0.3081 2.569    0.0424 *
---
Signif. codes:  0 '***' 0.001 '**' 0.01 '*' 0.05 '.' 0.1 ' ' 1

Residual standard error: 10.16 on 6 degrees of freedom
Multiple R-squared: 0.5238,     Adjusted R-squared: 0.4444
F-statistic: 6.599 on 1 and 6 DF,   p-value: 0.0424
```

The *lm()* command requires a slightly different look to that we have met previously. We need the names of our response and predictor variables and put a squiggle between them (the ~ character is called a tilde). It is important to get the variables the right way around, e.g.

```
> lm(y-variable ~ x-variable)
```

The results of the linear modelling show much the same as the LINEST command in Excel (Section 8.3.1) except that there is some labelling to help us. The slope and intercept are found under the Estimate heading. Here we see the intercept as 8.25 and the slope underneath as 0.79. The statistical significance of the relationship is shown at the end of the row called flow. In this case we see that $p = 0.0424$, which is the same as before and of course leads to the same conclusion: that there is a statistically significant correlation between abundance and flow.

We also see the p-value repeated along the bottom where the F statistic and the degrees of freedom are. The multiple R^2 value gives us an idea of how good the flow is at predicting the abundance. We will cover this area in more detail when we come to look at multiple correlations in the section on multiple regression (Chapter 11).

If we are just interested in getting the slope and the intercept, we can shortcut some of this and type in a modified command to extract only those values:

```
> lm(abund ~ flow)$coefficients
(Intercept)         flow
  8.2545956    0.7913603
```

Here we have used the *lm()* command (we do not need to full summary this time) but have added a bit to the end. This displays only the coefficients (intercept and slope) of the linear relationship as described by the $y = mx + c$ equation.

We will look at methods of producing graphs using R in Sections 12.4.1 and 12.5.4.

8.5 Curved linear correlation

In the examples of correlation we have looked at so far, the relationship between the two variables was one where broadly speaking the link was in the form of a straight line (or at least heading in one direction). In cases where the data are not normally distributed we use the Spearman's rank correlation test (Section 8.1). When the data are normally distributed we can use Pearson's product moment and assume the relationship is in the form $y = mx + c$ (Section 8.2). This assumes that the relationship is in a straight-line form.

There are plenty of times when we will get a relationship that is not in the form of a straight line. For example in Figure 122 we see a relationship where the response variable goes up and then down again in a definite inverted U-shape.

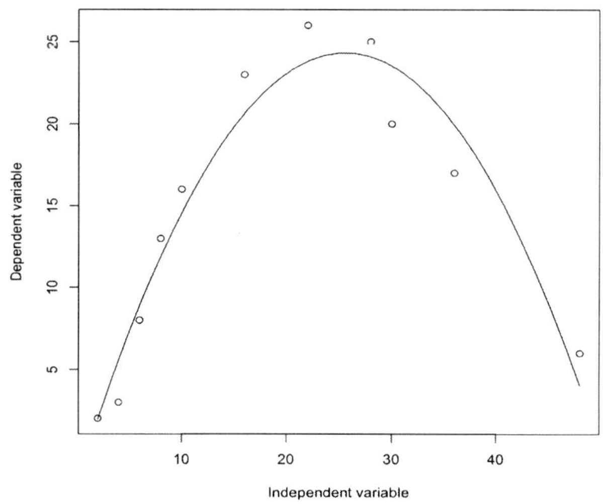

Figure 122. The inverted U-shape is an example of a correlation that is not a straight line. A polynomial equation is required

The standard correlations that we have looked at would not describe this relationship adequately as it is not in the form of a straight line. We get this relationship when looking at things that have an optimum in terms of an environmental factor. For example, a species might be most happy when soil moisture is at a certain level. When the soil becomes too dry or too wet, the abundance of the species declines. The result is this inverted U-shape.

Figure 123 shows a relationship where the general trend is upwards; however, we can see a definite bend and the line of best fit starts off steep and becomes gradually flatter.

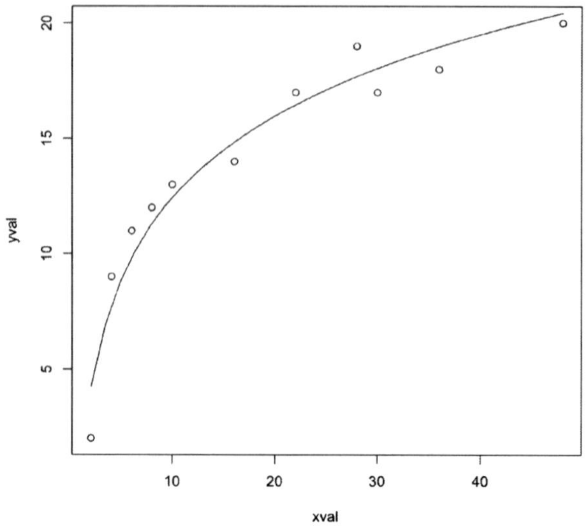

Figure 123. This curved shape is another example of a non-linear correlation. Here a logarithmic correlation best describes the relationship

In this instance, the response variable is dependent not on the value of the predictor variable directly but to the logarithm of the variable. Logarithms crop up quite a bit in natural sciences, for example pH, Richter scale and sound measured in decibels.

The main point is that rather than $y = mx + c$ we have another mathematical formula that could represent the relationship we observe. It is therefore important to present our data graphically as part of the analysis (see Section 12.4 *et seq.*). By drawing a graph we can guess the most correct form of relationship. For the two we have represented here, the relationships would be:

$$y = ax + bx^2 + c$$
$$y = a\,\mathrm{Ln}(x) + c$$

In the first example x is represented twice. This is a polynomial equation and produces the inverted U-shape. The terms a, b and c are fixed values. The a and b are coefficients that are analogous to the slope m in the straight-line equation. The c is the intercept.

In the second example we have c again, the intercept. We also have a, which is analogous to the slope. This time x is represented once but we use the natural logarithm rather than the plain value. This form of equation produces the curve that starts steeply and then flattens out.

There are many other potential equations but we will illustrate how to approach these two forms and the rest will be easy to deal with if you come across them. The way we carry out these correlations is more akin to multiple regression than to simple regression between two factors so we will provide a quick overview now and a more comprehensive look in Chapter 11, where we examine multiple regression.

8.5.1 Polynomial correlation

The equation that describes a polynomial is:

$$y = ax + bx^2 + c$$

We now have a more complex situation compared to before as we have an extra part to deal with. We can use our spreadsheet to carry out the analysis using the LINEST function but we need to add an extra column containing the values for x^2. In R we can carry out the analysis without adding anything but we need a modified version of the *lm()* command we used earlier.

```
> my.lm = lm(yvals ~ xvals + I(xvals^2), data = my.data)
```

Here we create a new item to hold our result. We use the *lm()* command that we met previously but we add an extra bit to take into account the x^2 part. The *yvals* part is of course our response variable and the *xvals* part is the predictor variable. In the command as shown the two variables are held in a separate container, a CSV file that we read previously (using the *read.csv()* command). We therefore add the *data* = part so that R can find the two variables we are interested in (we will look at this later in Section 10.1.1).

We want to take into account x^2 but we need to put it in a special form. The *I(xvals^2)* part does this. The reason is that ^2 in a formula of this type usually means something other than raise to the power of two. The point is that whatever is in the brackets following the *I* is a mathematical operation. We always use *I*, which is short for Inhibit (or insulate). Think of it as meaning "inhibit the regular formula and carry out a mathematical operation".

We will look at polynomial regression again when we examine multiple regression in Chapter 11.

8.5.2 Logarithmic correlation

The equation that describes a logarithmic fit is:

$$y = a\,\mathrm{Ln}(x) + c$$

Although we have a different looking formula to the one for a straight line, we can see it has essentially the same elements. We have a slope/coefficient (a), an intercept (c) and a bit relating to x. We can carry out the correlation using LINEST in the spreadsheet just as we did previously (Section 8.3) but with one small modification: we need to create a new column of data and use LN(x) to get the natural log of the values. We then run the LINEST function on the new (logarithmic) values instead of the original values.

In R we can simply modify the *lm()* command and create the logarithmic values as needed.

```
> my.lm = lm(yvals ~ log(xvals), data = my.data)
```

The command will generate log values (the default in R is to use natural logs) directly and carry out the regression appropriately. In the above example we create a new item to hold our result. The *yvals* part is our response variable and the *xvals* part is the predictor variable. In the command as shown, the two variables are held in a separate container, a CSV file that we read previously (using the *read.csv()* command). We therefore add the *data* = part so that R can find the two variables we are interested in.

We can also use the regular correlation test *cor.test()* with a tiny bit of tweaking. We have two options:

```
> cor.test(log(xval), yval)
> cor.test(~ yval + log(xval), data = my.data)
```

In both cases the Pearson correlation is worked out using the standard method. The difference is that in the first instance the *xval* and *yval* data are available to R and exist as separate data objects. In the second case we have a data frame (called *my.data*), which contains two columns (or possibly more) where the data are held. We tell R to look inside to find the items we want (more on this in Section 10.1.1).

We will look at logarithmic regression again when we examine multiple regression in Chapter 11.

9. Tests for linking data – associations

When we have categorical data we need to use tests for association. This chapter deals with two versions. The basic *Chi-squared* test examines two sets of categories. When we have a single set of categorical data we can compare these data to a theoretical set. This crops up in genetic studies for example and are called "goodness of fit" tests (see Section 9.2).

9.1 Association: Chi-squared test

Imagine the following situation. You are interested to know if certain plant species are found growing in proximity with one another. In other words you wish to know if there is an association between certain species. You could take up a quadrat and go to your field site and look for the two plants. One way to do this would be to note down the presence or absence of the species in each quadrat. When you are done you could generate a table of data similar to Table 50, which shows data for two species of heather.

Table 50. Presence–absence data for two heather species (*Calluna vulgaris, Erica cinerea*). Here we looked at 137 quadrats and noted the presence of two plant species. From the results we can determine how many quadrats contained both species and so on

	C. vulgaris present	*C. vulgaris* absent	Total
E. cinerea present	3	23	26
E. cinerea absent	35	76	111
Total	38	99	**137**

From the quadrat data we can work out how many of the quadrats contained the first species; this is the column headed *C. vulgaris* present. The number of quadrats that contained both species together goes in the top left cell; the next one down is how many quadrats contained the first species (*C. vulgaris*) by itself. The margins of the table show the totals and we can see that there were 137 quadrats in total.

This sort of table is called a contingency table. We can apply this kind of arrangement to other situations. For example, imagine that we are interested in certain invertebrate groups: beetles (Coleoptera), bugs (Hemiptera) and ants (Formicoidea) and want to know if they are associated with a particular habitat. Our data look like Table 51.

Table 51. Habitat selection amongst some invertebrate groups. These data were collected by examining a number of plants; a tally of the number of taxa were recorded against each part of the plant

	Ant	Bug	Beetle	Total
Upper leaf	15	13	68	96
Lower leaf	12	11	15	38
Stem	65	78	5	148
Bud	3	21	3	27
Total	95	123	91	**309**

Here we can see that each column is the number of each invertebrate group that we encountered. Each row represents a habitat and each cell in the table shows how many of that particular invertebrate taxa were found at a particular habitat. The margins show the totals and we can see that our sampling survey found 309 invertebrates in total.

In both of our examples we have categorical data. In both examples our categories are represented in the titles of the rows and columns. Because of the way we have collected our data (by a simple count or tally) we do not have replicated data. We cannot describe any of the cells in the table in terms of an average or spread using means, medians, standard deviation or rang, as we would do with replicated data. Each cell of the table represents a unique combination of two categories. We can represent these data graphically (see Section 6.3.2) and create a series of pie charts showing the proportion of items in each column or row. We may also use a bar chart to visualise our data (see for example Figure 87 in Section 6.3.2).

We need to use the Chi-squared test to analyse these data. We first need to make a hypothesis. In our invertebrate example we might say:

"There is an association between invertebrate group and habitat."

This is our alternative hypothesis H_1. We now need a null hypothesis H_0 to test.

"There is no association between invertebrate group and habitat."

In this case we have not stated the direction of the association, it could be positive or negative. Because of the complexity (some associations are likely to be positive others negative), we merely state that there will be associations. In essence, our null hypothesis is stating that the categories are randomly associated with respect to one another. Our alternative hypothesis states that this is not the case and that there are associations of some kind.

Now we need to work out the number we would expect in each cell of our table if there were no associations (i.e. if H_0 were true). We can do this by taking each cell in our contingency table in turn. Consider our invertebrate/habitat example.

How likely is it that we will capture an ant? We can see from the total that there were 95 ants and that was out of a complete invertebrate haul of 309. The probability of getting an ant will therefore be 95/309 = 0.307.

What is the likelihood of finding anything at all on an upper leaf? We read the total invertebrates captured on upper leaves as 96 so the probability will be 96/309 = 0.311.

To determine the probability of getting an ant on the upper leaf we multiply the two together: 95/309 × 96/309; however, we actually want the number rather than the probability so we multiply by the total number of all captures (we end up with 30 rounded up to the nearest whole ant). We can see the general arrangement in (Figure 124).

$$Expected\ no. = \frac{row\,total}{grand\ total} \cdot \frac{col\ total}{grand\ total} \cdot grand\ total$$

Figure 124. Working out the expected count from a contingency table

We can do a bit of cancelling here and reduce our generic formula (Figure 125).

$$Expected\ no. = \frac{row\,total\,.\,col\,total}{grand\ total}$$

Figure 125. The expected count formula can be simplified

If we apply this to our data we end up with something that looks like Table 52.

Table 52. Observed and expected values for invertebrate habitat selection data

OBS	Ant	Bug	Beetle	Total
Upper leaf	15	13	68	96
Lower leaf	12	11	15	38
Stem	65	78	5	148
Bud	3	21	3	27
Total	95	123	91	**309**

EXP	Ant	Bug	Beetle
Upper leaf	*30*	*38*	*28*
Lower leaf	*12*	*15*	*11*
Stem	*46*	*59*	*44*
Bud	*8*	*11*	*8*

The original data are also shown so we do not have to look back at the previous table. We have rounded up to the nearest whole number (since you would not expect to get a fraction of an invertebrate) but in practice you would carry over your decimal places until the final result. Notice how the row and column totals for the expected values come to the same as for the actual values (try it and see for yourself).

Each of our expected values should be greater than five. If one or more are smaller than this, the Chi-squared test becomes less reliable. If we do have smaller values then we might have to rethink. It may be possible to get extra data or to combine (or ignore) certain categories. So, how do these expected values compare to the original observed values? We can see that there are differences but are they statistically significant or not?

We could start by taking the difference between the observed and expected values. The problem here is that there will be some positive differences and some negative, these will cancel out and we will end up with a sum of zero, which is not very helpful.

We could square the differences; this is an approach that is commonly used in statistics to get rid of negative values. Table 53 shows the differences and their squares.

Table 53. Observed – expected values are squared for the invertebrate and habitat selection data

O – E	**Ant**	**Bug**	**Beetle**	
Upper leaf	–14.51	–25.21	39.728	
Lower leaf	0.3172	–4.126	3.8091	
Stem	19.498	19.087	–38.59	
Bud	–5.301	10.252	–4.951	
			Total:	**0**
(*O* – *E*)²	**Ant**	**Bug**	**Beetle**	
Upper leaf	210.67	635.73	1578.3	
Lower leaf	0.1006	17.026	14.509	
Stem	380.19	364.33	1488.9	
Bud	28.1	105.11	24.517	
				4847.46

The problem now is that we have no idea how important a particular difference is. A difference of 2 for example will become 4 when squared. If this difference is based on an expected value of 4 then it is pretty large. If however the expected value is 400 then a difference of 4 is actually quite small. We need to take into account the size of the expected value; to do that we can simply divide by the expected value.

Our final Chi-squared formula is shown in Figure 126.

$$X^2 = \sum \frac{(O-E)^2}{E}$$

Figure 126. Formula to determine Chi-squared in a test for association

If we apply this to our invertebrate data we get something like Table 54.

Table 54. Final Chi-squared values for invertebrate and habitat selection data

Chi-sq	Ant	Bug	Beetle
Upper leaf	7.1379	16.636	55.827
Lower leaf	0.0086	1.1256	1.2965
Stem	8.3555	6.1842	34.159
Bud	3.3852	9.7801	3.0833

If we add the individual values we get Chi-squared = 147 (to three significant figures). This is great but how do we know if this is statistically significant or not? We need to look at a table of critical values.

Table 55. Critical values for the Chi-squared test (also used with Kruskal–Wallis test). Reject the null hypothesis of your value is greater than the tabulated value

	Significance level			
Degrees of freedom	5%	2%	1%	0.01%
1	3.841	5.412	6.635	10.830
2	5.991	7.824	9.210	13.820
3	7.815	9.837	11.341	16.270
4	9.488	11.668	13.277	18.470
5	11.070	13.388	15.086	20.510
6	12.592	15.033	16.812	22.460
7	14.067	16.622	18.475	24.320
8	15.507	18.168	20.090	26.130
9	16.909	19.679	21.666	27.880
10	18.307	21.161	23.209	29.590

First of all we look down the left column until we find our degrees of freedom. This is related to the amount of data. In the case of Chi-squared it is related to the original contingency table (Table 51). The margins contain the row and column totals (and we also have the grand total). If we had a blank table and only had these totals, how many cells would we need to be filled in before we could work out the rest (it is a bit like Sudoku)? The degrees of freedom is this value.

A simple way to determine degrees of freedom is to calculate it as:

$$\textit{degrees of freedom} = (\#\ \text{columns} - 1) \times (\#\ \text{rows} - 1)$$

Here we have three columns (invertebrate categories) and four rows (habitat categories) so df = (3 – 1) × (4 – 1) = 6. We look down to the 6 and read across. The second column is headed 5% and gives us the value we need to exceed. Here we have a value of 12.592 and we comfortably exceed that. We can now reject our null hypothesis and conclude that there is an association between invertebrate type and habitat (we accept the alternative hypothesis). Of course we have a much larger value so we can look across the table row further and find our calculated value is larger than the final column (headed 0.01%). We can now conclude that there is less than 0.01% probability that our result was due to chance.

It is all very well to have a statistically significant association but we are not necessarily saying that all the associations in our contingency table are significant. We must look back at the observed and expected values and see which associations are positive and which are negative. Then we need to look at the individual Chi-squared values. Recall that to get our final statistic we added all the values together. We can now look to see which of the individual values makes the greatest contribution to the total. We can see by looking back to Table 54 that the first row has fairly large values, the second row has fairly small values. We can thus gain an idea of the relative strengths of the associations by comparing the sizes of these individual Chi-squared values. In fact if we look at the table of critical values (Table 55) we can see that something of the order of 3.8 would be likely to be significant for a pairwise comparison.

Pearson residuals

We can look at the differences between observed and expected values in a slightly different way. This is called the Pearson residual (after Karl Pearson). In short, instead of using the standard way to calculate Chi-squared we use a different formula shown in Figure 127.

$$\frac{(Observed - Expected)}{\sqrt{Expected}}$$

Figure 127. Formula to calculate Pearson's residuals in a test for association

If we apply this simple formula to our data on invertebrates and habitat selection we end up with something like Table 56.

Table 56. Pearson residuals for invertebrate and habitat selection data

Residuals	Ant	Bug	Beetle
Upper leaf	*-2.67*	*-4.08*	7.47
Lower leaf	0.09	*-1.06*	1.14
Stem	2.89	2.49	*-5.84*
Bud	*-1.84*	3.13	*-1.76*

We can see at a glance which of the associations are positive and which are negative. In the table we can see negative associations as negative numbers. We can also see the relative sizes of these associations and gain some insight into which ones are most important. Any individual residual that is – 2 is likely to be a significant one.

Yates' correction for 2 × 2 contingency tables

When we have small contingency tables (2 × 2) like our heather example in Table 50, it is usual to apply a small correction factor to improve the reliability of the result. This is generally known as Yates' correction.

During the calculation, each observed – expected value is reduced by 0.5. This reduced value is then squared and divided by the expected value in the normal manner. We can write the correction down formally as:

$$|O - E| - 0.5$$

The effect is to subtract 0.5 if the value is positive and to add 0.5 if the value is negative.

9.2 Goodness of fit test

In the two examples above, the plant presence/absence data (Table 50) and the invertebrate/habitat data (Table 51), we collected data (the observed values) and worked out the expected values using the row and column totals. The Chi-squared test can also be used in situations where you already have these expected values. The most common example would be in genetic studies. In a simple single locus study you would expect offspring in the ratio of 2:1:1 and you can use these as your expected values in a Chi-squared calculation.

Imagine the situation shown in Table 57 where you are looking at the genetics of peas.

Table 57. Goodness of fit test on pea genetic data

	Green	Yellow	Total
Smooth	116	31	
Wrinkled	40	13	
Total			200

Here we have two categories of colour and two of the coat texture. Our knowledge of Mendelian genetics would lead us to expect ratios of 9:3:3:1 and we can now calculate expected numbers as shown in Table 58.

Table 58. Expected numbers in pea genetic study using goodness of fit

Ratio	Category	Calculation	Expected #
9	Green smooth	9/16 × 200	112.5
3	Green wrinkled	3/16 × 200	37.5
3	Yellow smooth	3/16 × 200	37.5
1	Yellow wrinkled	1/16 × 200	12.5

If we add up the ratios, we get 9 + 3 + 3 + 1 = 16 so each expected value is a certain number of 1/16ths of the total observed (200). We can now use the standard Chi-squared formula: $(O - E)^2/E$ to work out the significance. In Table 59 we see the results of the calculations. The first column shows the names of the categories, the second shows the observed numbers in each category. The third column shows the expected values determined from above. The final column shows the individual Chi-squared values.

Table 59. Goodness of fit test results for pea genetics study

	Obs	Exp	$(O - E)^2/E$
Green smooth	116	112.5	0.11
Green wrinkled	40	37.5	0.17
Yellow smooth	31	37.5	1.13
Yellow wrinkled	13	12.5	0.02
Total:	200	200	1.42

The total Chi-squared value is shown at the bottom. We see a value of 1.42. We need to compare this to a table of critical values (Table 55) and see if the result is statistically significant. We have four categories here and so the degrees of freedom are 3 (i.e. # categories – 1). The critical value from the table is 7.815 so we can say that the observed values are not statistically different from the expected values.

9.3 Using R for Chi-squared tests

R will calculate Chi-squared tests quite simply but your data need to be in an appropriate form. Since the test involves a contingency table your data also need to be in a table. The simplest way is to make a CSV file and read this into R using the *read.csv()* command:

```
> data6 = read.csv(file.choose(), row.names=1)
```

Here we make a new object (which we will do our Chi-squared test on) but we tell R to use the first column as row names. By default R takes the first row as column names so now we have an object, *data6*, that has both row and column names. We have our contingency table. To see the data we can type its name:

```
> data6
       Upper Lower Stem
Aphid    230   175  321
Bug       34    31   35
Beetle    72    23  101
Spider    11     3    5
Ant       12     9   15
```

Notice that we do not need the row and column totals. Here we have some more data on invertebrate taxa and habitat selection. Our contingency table shows three columns and five rows. The Chi-squared test is run easily using the *chisq.test()* command:

```
> chisq.test(data6)
```

The result is displayed quite simply:

```
  Pearson's Chi-squared test

data:  data6
X-squared = 25.1296, df = 8, p-value = 0.001478
```

```
Warning message:
In chisq.test(data6) : Chi-squared approximation may be incorrect
```

R actually creates more information but in this case it only displays the basic information. We can get additional information more easily if we make a new object to hold our Chi-squared test result.

```
> data6.chi = chisq.test(data6)
```

Now we have a new item called *data6.chi* and to see the result try:

```
> summary(data6.chi)
> names(data6.chi)
```

The second command shows us that there are a number of items we haven't seen. The most useful ones are observed, expected and residuals. To view them we type our object name, add a dollar sign and then the name of the bit we want:

```
> data6.chi$expected
```

This shows us a table of expected values:

```
             Upper      Lower       Stem
Aphid  242.000000 162.456825 321.543175
Bug     33.333333  22.376973  44.289694
Beetle  65.333333  43.858867  86.807799
Spider   6.333333   4.251625   8.415042
Ant     12.000000   8.055710  15.944290
```

Now we see the reason for the error message, one of our expected values is <5. Below we can see the results of looking at our Chi-squared object:

```
> names(data6.chi)
[1] "statistic" "parameter" "p.value"   "method"
"data.name" "observed"
[7] "expected"  "residuals"
> data6.chi$obs
       Upper Lower Stem
Aphid    230   175  321
Bug       34    31   35
Beetle    72    23  101
Spider    11     3    5
Ant       12     9   15
> data6.chi$exp
             Upper      Lower       Stem
Aphid  242.000000 162.456825 321.543175
Bug     33.333333  22.376973  44.289694
Beetle  65.333333  43.858867  86.807799
Spider   6.333333   4.251625   8.415042
Ant     12.000000   8.055710  15.944290
```

```
> data6.chi$res
           Upper      Lower        Stem
Aphid  -0.7713892  0.9840984 -0.03029148
Bug     0.1154701  1.8228842 -1.39588632
Beetle  0.8247861 -3.1496480  1.52324712
Spider  1.8543453 -0.6070112 -1.17724779
Ant     0.0000000  0.3327004 -0.23648449
```

We do not have to type the full name at the end (after the $ sign) as long as we type enough for R to work out what we want; typing $p for example would not work as there are two parts that begin with "p". If we want to display only the *p*-value we must add $p. as an absolute minimum.

By default R will apply a special correction factor (Yates' correction) to all the expected values if the contingency table is 2 × 2. You can turn it off by adding *correct = FALSE* (*or correct* = *F*) to the command.

Goodness of fit test

The *chisq.test()* command can also perform a goodness of fit test. To do this you will need a series of observations and a corresponding list of expected values. The test is run in a similar way to above but this time you add the list of expected values (a simple list of numbers is called a vector) like so:

```
> data7.chi = chisq.test(data7, p = data8, rescale.p = T)
```

In this case the *data7* object would be our observed values and *data8* our expected. We really need *data8* to be probabilities and to sum to 1; the *rescale.p* part ensures that this is the case. We get a similar looking output:

```
 Chi-squared test for given probabilities

data:  data7
X-squared = 40, df = 2, p-value = 0.2578
```

If we have just a few data we might consider typing the values directly into R rather than making a CSV file. The following example shows the pea genetics data we looked at previously:

```
> pea = c(116, 40, 31, 13)
> ratio = c(9, 3, 3, 1)
> chisq.test(pea, p = ratio, rescale = TRUE)

 Chi-squared test for given probabilities

data:  pea
X-squared = 1.4222, df = 3, p-value = 0.7003
```

We can see from the p-value at the end that there is no statistical difference between the observed and expected ratios.

9.4 Using Excel for Chi-squared tests

We can carry out Chi-squared tests using Excel fairly easily although we might have to do a bit of work. The program has the critical values encoded within it and there are some functions that make use of them. These are:

CHITEST
CHIDIST
CHIINV

The CHITEST function takes a set of observed values and a set of expected values (which you supply) and calculates the probability that the associations are due to random chance.

The CHIDIST command takes a value for Chi-squared and the degrees of freedom. It then gives you the p-value. The CHIINV command takes a p-value and degrees of freedom and works out what Chi-squared value corresponds.

Figure 128 shows the invertebrate data that we met previously when using R for looking at Chi-squared (Section 9.3). The CHITEST command does not calculate the expected values for us and expects them to be supplied. So our first task is to work out the row and column totals as well as the grand total. We use the SUM formula for this.

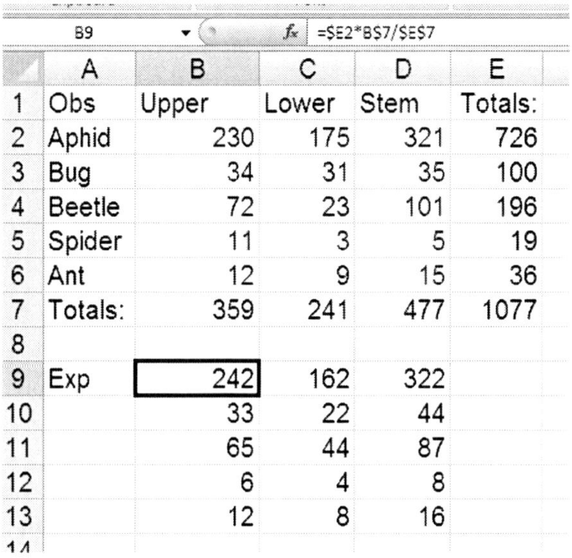

B9 fx =$E2*B$7/$E$7

	A	B	C	D	E
1	Obs	Upper	Lower	Stem	Totals:
2	Aphid	230	175	321	726
3	Bug	34	31	35	100
4	Beetle	72	23	101	196
5	Spider	11	3	5	19
6	Ant	12	9	15	36
7	Totals:	359	241	477	1077
8					
9	Exp	242	162	322	
10		33	22	44	
11		65	44	87	
12		6	4	8	
13		12	8	16	

Figure 128. Using Excel for Chi-squared tests. The CHITEST command requires the expected values to be ready calculated

We now need to create a table of expected values. This is simply the row total multiplied by the column total and divided by the grand total. In Figure 128 we can see the first of the values highlighted to show the formula. We can see that we have modified the formula and added $ signs at various places. This is so that when we copy the formula across the rows and down the columns that we retain the correct cell references. We met the $ sign before when dealing with running means (Section 4.7.1) and in making lookup tables (Section 3.2.6).

Once we have created our table of expected values we can use the CHITEST formula to give us the *p*-value, i.e. the chance that the associations are down to random chance. The basic form of the command is:

CHITEST(*observed*, *expected*)

We can now enter the formula and get a result. Figure 129 shows the final result.

B15 fx =CHITEST(B2:D6,B9:D13)

	A	B	C	D	E
1	Obs	Upper	Lower	Stem	Totals:
2	Aphid	230	175	321	726
3	Bug	34	31	35	100
4	Beetle	72	23	101	196
5	Spider	11	3	5	19
6	Ant	12	9	15	36
7	Totals:	359	241	477	1077
8					
9	Exp	242	162	322	
10		33	22	44	
11		65	44	87	
12		6	4	8	
13		12	8	16	
14					
15	P-val	0.001478			
16	Chi	25.1296			
17					

Figure 129. The CHITEST command in Excel calculates the probability of no association given a range of observed and expected values

We can now see that we have a *p*-value (of 0.001478), which tells us that the probability of the associations being random is very small. In other words, there are significant associations between the invertebrate taxa and the habitats where they were found. In Figure 129 we see that a Chi-squared value has also been given. This is not generated by the CHITEST formula and we need to work it out for ourselves using the CHIINV command as see in Figure 130.

B16 =CHIINV(B15,8)

	A	B	C	D
14				
15	P-val	0.001478		
16	Chi	25.1296		

Figure 130. The CHIINV formula determines the Chi-squared value associated with a particular probability and degrees of freedom

The CHIINV formula works out the Chi-squared value for a given probability and degrees of freedom. Remember that degrees of freedom is worked out from the size of the contingency table so:

df = # rows – 1 × # columns – 1

Here we have five rows and three columns so df = (5 – 1) × (3 – 1) = 4 × 2 = 8. This goes into the formula along with the *p*-value.

We can now relate the observed and expected values but we still do not know which of the associations contributes most to the overall Chi-squared value. We should work out the individual Chi-squared values for ourselves. If we do this then there is little point in using the CHITEST formula. We can use the CHIDIST formula instead. This works out the probability that the associations are random based on degrees of freedom and a Chi-squared value.

CHIDIST(*Chi_value, degrees_of_freedom*)

Since we have worked out a Chi value the CHITEST formula becomes redundant.

Goodness of fit and Excel

The CHITEST formula will work just as well for goodness of fit tests. Once we have a set of observed values, we need to create our expected values and then proceed from there. Recall the pea genetics data we looked at before; in Figure 131 we have entered these into the spreadsheet.

B26 =CHITEST(B20:B23,D20:D23)

	A	B	C	D	E
18					
19		Obs	Ratio	Exp	
20	GS	116	9	112.5	
21	GW	40	3	37.5	
22	YS	31	3	37.5	
23	YW	13	1	12.5	
24	Totals:	200	16		
25					
26	P-val	0.700334			
27	Chi	1.422			

Figure 131. The CHITEST formula in Excel can determine goodness of fit Chi-squared tests

In column C we have entered the ratios we expect. We convert these into expected values: the top one (D20) for example is 9/16 × 200. The probability that these ratios are different is given by the CHITEST formula.

Here we see a value of 0.700, which indicates that there is no significant departure from the expected ratio. If there were departures then we might perhaps wish to look further. We could fairly easily calculate the Chi-squared values for ourselves and see which of the individual Chi-squared values contributes most to the total.

Notice that, as before we do not automatically get an overall Chi-squared value. We need to use the CHIINV formula again. Here we work out the degrees of freedom by looking at how many items we have in our pea categories. There are four options and so df = # options – 1, i.e. 3.

The *Analysis ToolPak* (Section 1.9.1) will not perform Chi-squared tests for us. In reality this is hardly a big blow as it is pretty straightforward to undertake the analyses ourselves.

10. Differences between more than two samples

The differences and correlation tests that we have covered so far have looked at two samples. The *t*-test (Section 7.1) and *U*-test (Section 7.2) look for differences between two sets of data for example. The Spearman's rank test (Section 8.1) and Pearson's product moment (Section 8.2) look for a correlation between two variables.

It is quite common to want to examine more than two sites or to look at more than two factors. In most cases you will be looking at a single dependent variable, this might be the abundance of a plant or invertebrate species. You might look at differences in abundance between several sites or you may look to see how several environmental factors affect the abundance of a species. In these cases we need to use some form of analysis of variance (for looking at differences, see Sections 10.2 and 10.3) or multiple regression for correlations (see Chapter 11).

We may still use the same kinds of graphs to help us make sense of the data (see Chapter 6) using box–whisker plots for example; however, our graphs will be rather more complex, depending on the complexity of the situation we are describing.

When you have several independent factors (probably species abundances) and also several factors we talk about multivariate analyses. The approach for these multivariate analyses is rather more complex and won't be covered in this book (community analyses will be dealt with in a separate volume).

10.1 Using R for more complex statistical analyses

We can use R quite easily to perform quite complex analyses. In order to make this easier, R uses a slightly different notation to keep track of what is going on. In order to progress we will need to look briefly at how this works.

10.1.1 Complex statistical commands

When you are simply comparing two samples or performing a correlation, it is simple enough to type the names of your two samples; however, many statistical analyses look at more than two samples and R provides a slightly different approach to writing the commands.

Model syntax

R uses what is called model syntax (or formula notation). In brief this means that you specify both sides of your equation in the command. Recall the *t*-test where we compared two samples.

```
> t.test(data1, data2)
```

This type of command relies on you having two samples as separate items; however, it is much more sensible to write your data in columns. For example one column might be the abundance of a plant species in quadrats. The second column is the name of the site. Each row is a separate item (like a biological record). The first column contains a series of values and the second contains a set of names (in this case we have two sites). We could probably label the first column "*abund*" and the second "*site*".

We start by reading our CSV file of data:

```
> our.data= read.csv(file.choose())
```

In this example we call our data item *our.data* but it would probably make more sense to give it a more meaningful (but short) name. To remind ourselves of the variables we can try the *names()* command.

```
> names(our.data)
```

This will show us the two variables like so:

```
[1] "abund" "site"
```

To use these in an analysis, we type our command with a slight modification, e.g.

```
> t.test(abund ~ site, data= our.data)
```

This says in essence: take the variable *abund* and compare by *site*, look in the *our.data* object to find the two variables. The result looks pretty much the same as the previous *t*-test we did:

```
  Welch Two Sample t-test

data:  abund by site
t = 2.2423, df = 13.611, p-value = 0.04216
alternative hypothesis: true difference in means is not equal to 0
95 percent confidence interval:
 0.06819405 3.26513929
sample estimates:
mean in group Lower mean in group Upper
           5.444444            3.777778
```

We can run our *U*-test in similar fashion:

```
> wilcox.test(abund ~ site, data= our.data)
```

We may also use this to plot a graph as in Figure 132:

```
> boxplot(abund ~ site, data= our.data)
```

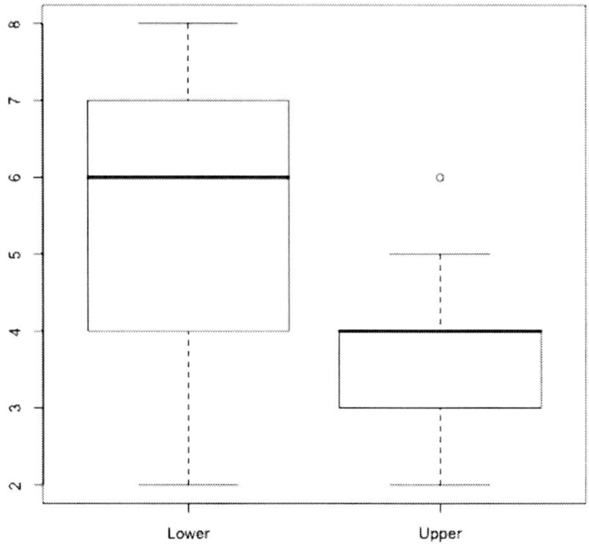

Figure 132. Box plot using model syntax "boxplot(abund ~ site, data= our.data)". Note how columns are automatically labelled. The box labelled "Upper" shows an outlier

This time our plot (Figure 132) shows labels automatically. Note also that there is an outlier on the right hand side. The boxplot command draws the whiskers to a limit equal to 1.5 times the inter-quartile distance. If we wish to make the whiskers extend to the extreme range (i.e. max and min) then we set *range = 0* as part of the command.

Variables within data objects

In the above examples we told R where to find the variables using the *data=* command. This is because the two variables were within the *our.data* object and do not appear in our list of items (remember the *ls()* command?).

One way to let R see these objects is to append the name to the object with the dollar sign. We did this previously with the Chi-squared test (Section 9.3). So if we want to see out abundance data we might use:

```
> our.data$abund
```

Attach() command

Using the $ is okay for one or two small items but it can get pretty tedious adding the dollar sign every time. We can use the *attach()* command to allow R to see inside another dataset and so read the variables within.

```
> attach(our.data)
```

The attach command now lets R see the variables in the *our.data* object. The *ls()* command will not see these variables but they are there. Now we can use abund and site as objects in their own right.

Detach() command

Once we are finished with these extra objects it makes sense to get rid of them. The *detach()* command does this.

```
> detach(our.data)
```

Now we have effectively removed the objects from memory (the *our.data* object is still present but R cannot see inside it unless we add the $ sign like before).

10.1.2 Additional analysis packages

The base distribution of R can do a lot but often you will find the need to perform some analysis that it cannot do. Because of the nature of R, there are a lot of people working on it and they have produced a wealth of packages. These are libraries of specialist routines that perform a huge variety of statistical analyses.

It is fairly simple to find and install these extra packages. One method is to search online, adding CRAN to your search term will generally work. You may be able to download a zip file. Then you can install the package from within R using one of the menu commands on the toolbar (under the *Packages* menu). Alternatively, if you are connected to the Internet you can use the menu to install packages directly.

Loading extra analysis routines using the *library()* command

To load a library of routines into R you use the *library()* command, e.g.

```
> library(vegan)
```

This loads the "vegan" library of analysis tools (vegan is short for Vegetation Analysis). Once they have been loaded, these additional packages may be used like any other. They have their own help files built-in to R and the syntax for usage follows the standard R format.

10.2 Analysis of variance

Analysis of variance (usually abbreviated to ANOVA); is a method of looking for differences between more than two samples at once. It is a parametric test and therefore requires that the data be normally distributed. If your data are not parametric then you cannot use analysis of variance and must find an alternative. You might use the Kruskal–Wallis test (Section 10.3), which is a non-parametric version or you may be able to transform your data by using a logarithm or another mathematical trick (see Section 4.6), and then run the analysis. Consider what we are dealing with. Figure 133 shows three samples, the data are represented by the three normally distributed frequency curves.

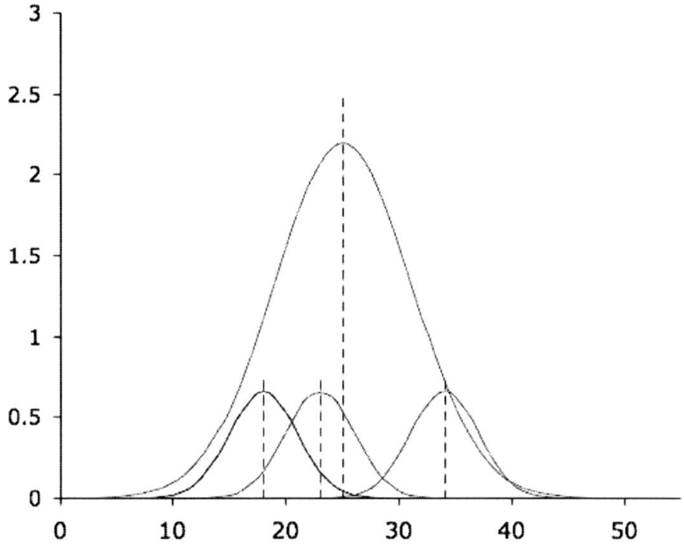

Figure 133. Three samples are shown as frequency distributions. Are they really all part of the single overall distribution or are they different? Analysis of variance can help us determine

The larger curve shows us the frequency distribution of the whole lot as if they were one big sample. What analysis of variance does is to look at the variability (the variance, s^2) and work out the likelihood that the samples are really all taken from this one big set. We can see that we have three means (one for each sample) and a further mean representing the overall dataset. In analysis of variance we measure two lots of variability. First of all we look at the variability of each of the samples, i.e. we look at the variability of each of the three small curves. This is called the variance within samples (also called the error term). We can see that if this variability is great then the three curves are more likely to overlap and it is less likely that the three samples are different from one another. We also measure the variability between samples. What we do is to see how far each sample is from the overall mean. If this between-samples variability is large then it implies that the samples are far apart from one another and it is likely that the samples are distinct (i.e. different).

Although we call the method analysis of variance, we mostly use something called sums of squares. The formula for calculating the variance, which is shown in Figure 134, is essentially the same as for standard deviation (Figure 65) except we do not take the final square root.

$$s^2 = \frac{\sum (x - \bar{x})^2}{n - 1}$$

Figure 134. The formula used to calculate variance. Essentially this is standard deviation squared

We work out the mean of our sample and then look at the difference between each individual value and the mean. Some are positive values and some negative so we square the result. Finally we divide by $n - 1$. The standard deviation, s, is simply the square root of the variance (Section 4.3.4).

In analysis of variance we calculate the sums of squares and use this value. You will see later that we take into account the size of the sample(s) and thus retain the variance. You can think of the sums of squares as variance before taking sample size into consideration. We can write a formula for the sums of squares as seen in Figure 135.

$$SS = \sum (x - \bar{x})^2$$

Figure 135. The formula used to determine the sums of squares

To begin with we determine the sums of squares (usually abbreviated to SS) for each sample. The total sums of squares then relates to how fat the samples are, how spread out the data are about their means. Looking back to Figure 133 this would represent how wide the three small curves are (these represent the three samples). This sums of squares of the samples is called the *within groups sums of squares* (also called the error term).

The next step is to work out the mean value of everything all added together, in other words looking back to Figure 133 we look at the large curve. In our example we have three samples so we would add all the values together and divide by the total number of observations. We now look to see how far away from the overall mean each sample is. The sums of squares for this are worked out using the formula shown in Figure 136.

$$SS_{between(indiv)} = n(\bar{x} - \bar{T})^2$$

Figure 136. The formula used to determine the sums of squares between samples

You can see that we take the mean of a sample and subtract from it the overall mean. This value is then squared and the result multiplied by the number of items in the sample. This is repeated for all samples and the sum of these is known as the *between groups sums of squares*.

We now have values for sums of squares but need to determine variance. In order to do this we work out the degrees of freedom for the various components. For our between groups sums of squares we use the number of samples – 1. If, as in our example, we have three groups, the degrees of freedom = 2. We simply divide the sums of squares by the degrees of freedom to get our variance. This variance is called a special name in analysis of variance: the mean square (MS). The degrees of freedom for the within groups sums of squares is simply the total number of observations minus the number of samples.

Once we have our two mean square values, we calculate our final *F*-value (Figure 137; it is called *F* after the statistician Fisher). The final *F*-value is determined by dividing the between mean square by the within mean square.

$$F = \frac{MS_{between}}{MS_{within}}$$

Figure 137. Calculating the *F*-value using mean squared values (sums of squares divided by degrees of freedom)

We now compare our calculated value of *F* to published tables of critical values; Table 60 shows some example values. If our final result is greater or equal to the critical value then we are in business and can reject the null hypothesis.

Table 60. Table of critical values for *F* at $p = 0.05$. The columns are degrees of freedom for between groups (no. groups = *k*, df = *k* – 1) and rows are degrees of freedom for within groups (df = *n* – *k*). Reject the null hypothesis if your calculated value exceeds the tabulated value

k/*n*	2	3	4	5	6
2	19.00	19.20	19.20	19.30	19.30
3	9.55	9.28	9.12	9.01	8.94
4	6.94	6.59	6.39	6.26	6.16
5	5.79	5.41	5.19	5.05	4.95
6	5.14	4.76	4.53	4.39	4.28
7	4.74	4.35	4.12	3.97	3.87
8	4.46	4.07	3.84	3.69	3.58
9	4.26	3.86	3.63	3.48	3.37
10	4.10	3.71	3.48	3.33	3.22
15	3.68	3.29	3.06	2.90	2.79
20	3.49	3.10	2.87	2.71	2.60
30	3.32	2.92	2.69	2.53	2.42

When we report the result, we use a shorthand method, e.g. $F_{2,20} = 4.75$, $p < 0.05$. This informs the reader that we used analysis of variance (because of the *F*) and that the degrees of freedom were 2 and 20. The degrees of freedom are always written down with the groups first, so in this case we can infer that there are three groups (since degrees of freedom = $k - 1$).

In fact the results of analysis of variance are often written down in the form of a small table. Table 61 shows results of an analysis.

Table 61. Example of analysis of variance table used to present results of analysis

Source of variation	SS	df	MS	*F*	*p*	Crit
Between groups	106.278	2	53.139	5.633	0.015	3.682
Within groups	141.5	15	9.433			
Total	247.778	17				

The analysis of variance table shows clearly what each element in the calculation is. For example we see the sums of squares for both between and within variation. The degrees of freedom are also shown. We see both mean square values (worked out by dividing the sums of squares by the degrees of freedom). We get F by dividing the upper mean square by the lower mean square. The analysis of variance table demonstrates some interesting properties. Look first at the degrees of freedom. We can see that there were three samples since the degrees of freedom = 2 (we know that degrees of freedom = $k - 1$). The degrees of freedom for within groups is worked out by taking the number of all the observations, n, and subtracting the number of samples (i.e. $n - k$). This means that there must have been 18 observations in total ($18 - 3 = 15$). The degrees of freedom of the overall data would be $n - 1$ (which is $18 - 1 = 17$), and we see the total degrees of freedom written as this value at the bottom. If we now look at the sums of squares (SS) we see values for between and within as well as a total. The total sums of squares is worked out by taking each observation and subtracting the overall mean, then squaring the result and adding them all up. You will notice that the total sums of squares is the sum of the two values above. This means that once we know two of the values we can determine the last by simple subtraction. It also provides a means to test our maths since we can calculate the total sums of squares by two methods.

The final result merely tells us if there is a significant difference between the samples (i.e. that they are not all from one big group) or not (i.e. that in reality they are all sub-sets of some larger sample). In the example above (Figure 133) we might get the result that there was a significant difference; however, when we look, it appears that the two samples on the left are quite close together and that the one on the right is different. How can we tell which is different from which? We will examine this shortly (Section 10.2.1).

Comparing analysis of variance and the *t*-test

The *t*-test (Section 7.1) is very similar to analysis of variance and can be regarded as a special case of it, where of course you only have two samples to compare. The top part of the equation for the *t*-test (Figure 91) is analogous to the top of the formula for F (Figure 137). Essentially this is dealing with how far apart samples are. The bottom of both equations is concerned with how "fat" the samples are; the more spread the samples the greater likelihood that they overlap and less likely that differences are significant. If you calculated t

and then went on to work out F for a pair of samples you would find that the similarity between t and F is such that $t^2 = F$.

10.2.1 Post-hoc testing in analysis of variance

When our analysis of variance result shows a significant result we know that the samples are not all from one big group and that there are differences; however, we might want to know if site *a* is different from site *b* and if site *c* is different again or more similar to site *b*. We could run *t*-tests on each pair of samples but this would not be an appropriate approach. The problem here is that the more tests you run, the more likely it is that you will get a significant result by chance (e.g. the more dice you roll the more likely you are to get a six. Alternatively the more lottery tickets you buy, the greater chance you have of winning, although it is still a very small chance).

Look at the example below shown in Figure 138. Here we see a box–whisker plot of three samples. The data are taken from the example used to make the analysis of variance table shown above (Table 61).

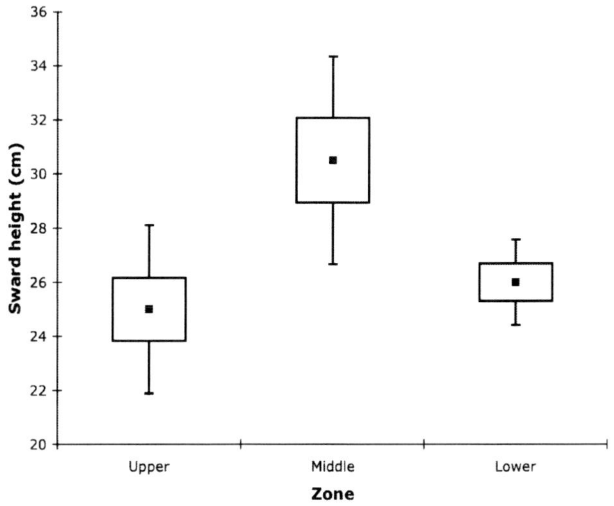

Figure 138. Results of ANOVA for mean sward height (cm) at three zones in a meadow in Shropshire. $F_{2,15} = 5.633$, $p < 0.05$. Boxes represent standard error; whiskers are standard deviation

We will look at graphs in more detail in Section 12.4; note how the caption is pretty complete and says much of what we need to know without having to refer to the text. The result of the analysis of variance tells us that there are significant differences between the sward heights at the different zones of this meadow; however, when we look at the graph it would appear that the *Upper* and *Lower* samples are quite close together; with the *Middle* zone looking like it might have a taller sward. Our analysis of variance does not tell us anything more than "there is a difference between samples".

The answer is to run as post-hoc test, meaning "afterwards" (literally "after this"). What we do is run a special version of the *t*-test but designed to take into account the fact that we are running several tests at once. Many statistical programs will run post-hoc tests (including R of course) and although there are different types they run more or less on the same principle. The Tukey Honest Significant Difference test (Tukey HSD) is a commonly used method (Figure 139).

$$t = \frac{|\bar{x}_1 - \bar{x}_2|}{\sqrt{\frac{MS_{within}(\frac{1}{n_1}+\frac{1}{n_2})}{2}}}$$

Figure 139. The Tukey post-hoc test is a modified version of the *t*-test

We can see that superficially this looks much like a regular *t*-test. On the top we have the difference between the means of the two samples we wish to compare. On the bottom we would normally have a measure of the variance and sample size of the two samples. In the Tukey test we use the mean square of the within sample variability (called the error term) as well as the sizes of the two samples. This version takes into account the variability of all the data as well as differences in sample size.

So for each pair of samples we evaluate this formula, the means and sample sizes vary but the error term remains the same for each comparison. This is one advantage in producing the ANOVA table (like Table 61 above) as we can readily see the value we need (9.433 in this example). Once we have our *t*-values we can compare them to a table of critical values for *t*.

Table 62. Table of critical values of *t* used in the Tukey HSD post-hoc test

5%	2%	1%	0.1%
4.303	6.965	9.925	31.598

There is only one row of values since we are always interested in pairs of comparisons (the degrees of freedom = 1). Once you have done your comparisons you need a method of conveying the result to potential readers. You might use a small table showing the various results (*t*-values, critical values, *p*-values) for example.

Table 63. Post-hoc *t*-values calculated by the Tukey HSD test for sward heights at three zones in a Shropshire meadow

	Post-hoc *t*-values		
	Upper	**Middle**	**Lower**
Upper	-	4.552	0.786
Middle		-	3.422

Table 63 shows the calculated *t*-values for three samples worked out using the Tukey HSD test. We can compare these to critical values and we can readily see that only one pair of samples has significantly different means (middle and upper). In Table 64 we have determined the *p*-values exactly.

Table 64. Post-hoc *p*-values calculated by the Tukey HSD test for sward heights at three zones in a Shropshire meadow

	Post-hoc *p*-values		
	Upper	**Middle**	**Lower**
Upper	-	0.045	0.514
Middle		-	0.076

The *p*-values show us where the differences really lie. You might have assumed by looking at Figure 138 that the middle sample was statistically different from the other two; you would have been incorrect! In the table we have shown the *p*-values as calculated but you might easily replace the non-significant results with n.s. and the significant one with $p < 0.05$. There is more on reporting of results below and in Section 12.3.

Tables are fine but can take up a lot of space so it is common to add the post-hoc information directly onto your graph. The simplest way is to add a symbol to each sample (letters are convenient). Where letters are the same the samples are not significantly different. Where letters differ then the samples are significantly different (as determined by your post-hoc test). Look at Figure 140 where our sward height box plot has been modified.

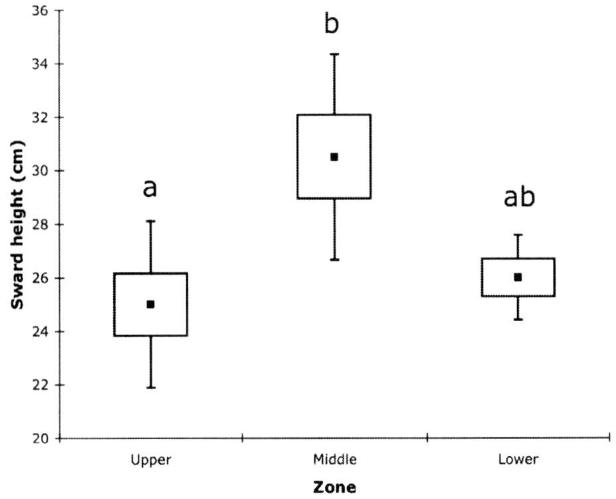

Figure 140. Results of ANOVA for mean sward height (cm) at three zones in a meadow in Shropshire. $F_{2,15} = 5.633$, $p < 0.05$. Boxes represent standard error, whiskers are standard deviation. Letters above samples represent differences as determined by the Tukey HSD post-hoc test. Different letters are significant at 5%

Now we have letters above the samples to show post-hoc differences. We can see that we start with sample upper and that gets an "a" (a good starting point). The next sample (middle) is significantly different from the first and so we give it a different label, "b" will do nicely. Now the third sample is not different from the first so we give it an "a", which shows it is not different; however, it is also the same as the middle sample so it needs to have the same letter as that. We add a "b" and now the third sample has "ab" as a label indicating that it is not significantly different from either of the other two samples. When you have a lot of comparisons, the labelling can be tricky to get right. Start at the left and work your way across and you will hopefully get it correct.

We can use R to carry out analysis of variance (Section 10.2.4) and also our spreadsheet (Section 10.2.5). Before we do though we will carry on and describe analysis of variance in more complex situations.

10.2.2 Analysis of variance and several factors

If we are looking for a simple difference in abundance of something over several sites we can run analysis of variance (assuming normally distributed data). What we have is a response variable (the abundance) and a single predictor variable (the site). We would call our analysis a one-way ANOVA to remind us that we have the single predictor variable; however, it may be that there are different management strategies at these sites. You might be interested to see if the management also has an effect. Now we have two predictor variables: site and management. We can use analysis of variance once more but now that we have two predictor variables we call it two-way ANOVA.

We might have even more factors to consider and then we could have three-way or just multi-way ANOVA; however, when you add more factors, things get complicated very quickly!

For example, take our site and management scenario. We may well find differences in abundance of our species between sites; we may also find that management affects the abundance; however, there is another thing to consider and that is the combined effect. In ANOVA-speak this is called an interaction and is often written *site * management* (i.e. the two factors look like they are multiplied by one another). If we had a significant interaction then this would indicate that our management strategy was having a different effect at different sites.

Look at Table 65, this shows some data on plant abundances under three different grazing regimes and at two different sites.

Now we are looking for differences between treatments and between sites but what analysis are we going for? If we examined each site in turn we could apply one-way ANOVA and determine if the grazing treatment had any effect on abundance; however, since we have all the data together we should use an analytical method that uses all the data. This is where two-way ANOVA comes in.

What we do is to split up the data so that we can see how each element contributes to the variability. Table 65 is already partitioned like this. We can see that we have two sites. We also have three treatments. In this case we see that we have three observations (replicates) for each

Table 65. Abundance of a plant species (per m^2) at two sites under three differing grazing treatments

	Site	
Grazing treatment	**Top field**	**Lower moor**
Lo	9	7
Lo	11	6
Lo	6	5
Mid	14	14
Mid	17	17
Mid	19	15
Hi	28	44
Hi	31	38
Hi	32	37

of the treatments. This gives a certain symmetry, which is nice. More importantly the properties of the variance are such that our calculations become more inexact the greater imbalance we have. We strive therefore to collect data in such a way as to achieve this symmetry.

Usually we would not have our data written out in this format; we should have one column for the response factor and one for each of the predictor factors. We will come back to this a bit later (Table 68).

Before looking at how the variance components are calculated, it is useful to look at a graph of the means. Figure 141 shows us the means for the six combinations.

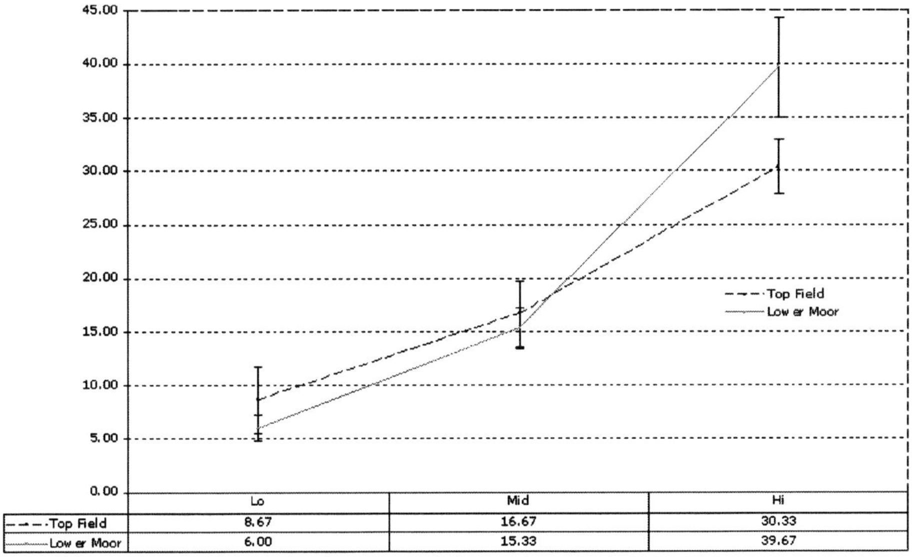

Figure 141. Abundance of a plant species (mean per m^2) at two sites under differing grazing treatments to illustrate interaction between treatments. Error bars are standard error

We should not join the treatments together with lines but in this case it has been done to highlight the data for each treatment. First of all, look at the three main treatments; we can see that as the grazing becomes more intense so the abundance increases. We do not yet know if this is statistically significant but since the error bars do not overlap we can assume that it is likely. Now look at each treatment in turn and compare the means for the sites. At the low treatment we can see that there is a great deal of overlap and indeed the data table (at the bottom of the graph) shows that the means are quite close. Similarly when we examine the middle treatment we see a lot of overlap. When we look at the high treatment however, we see that the means are some way apart and the error bars do not overlap.

Just by looking at the graph we can see that the grazing treatment appears to be affecting the abundance of the plant. In addition we can see that the plant is responding slightly differently at the two sites, with the high grazing treatment at lower moor giving us the highest abundance. What we say here is that there appears to be an interaction between the two factors (grazing treatment and site). At the moment all we have are the means and the graph; the next step is to analyse the data using ANOVA and see if our observations are backed up.

The first stage involves calculating sums of squares for the various elements. We start with the simplest, which is the sums of squares for each block. There are six of them in our example, the first being the low grazing treatment at the top field site, then the low treatment at the other site and so on. This will give us the error term (within groups sums of squares). We determine this "error term" by looking at each of the six blocks in turn. We take each observation in a block and deduct the block mean, then square the result. The sum of these squared values is our "within group sums of squares". The total error term is then the sum of all the "within group sums of squares" for all six blocks.

The next step is to take each column and look at the sums of squares. What we are doing is comparing the sites and disregarding the grazing treatments. We work out this "between groups sums of squares" as before by taking the overall mean away from the mean for each column, squaring and then multiplying by the number of observations (nine in this case as we are comparing columns). The formula is repeated below. We add these together (there are only two values, one for each column) to give an overall "between columns sums of squares" (Figure 142).

$$SS_{between(indiv)} = n(\bar{x} - \bar{T})^2$$

Figure 142. The between groups sums of squares in an analysis of variance

Now we do the same thing using the rows. That is to say we take each treatment and ignore the sites. We have three grazing treatments and so will get three sums of squares. Adding these together gives us a total for "between rows sums of squares".

So far we have the individual sums of squares (six of them), which represent the variability within groups. We have now also partitioned the variability by row and column. The next bit is to examine the interaction sums of squares. We could do this by determining the overall sums of squares and then subtracting the other sums of squares, which we have already calculated; however, as a proper check we should compute it directly. The formula to achieve this is shown in Figure 143.

$$SS_{ab} = \sum^{a} \sum^{b} n(\bar{x} - \bar{x}_a - \bar{x}_b + \bar{\bar{x}})^2$$

Figure 143. The sums of squares for the interaction in a two-way ANOVA

At first glance it looks pretty horrendous but actually it is not that hard, the double sigmas at the start make it look daunting. We have six blocks in our current example, each one representing the interaction between a grazing treatment and a site. So, we should end up with six interaction sums of squares values. The formula works like this: select a block and determine its mean, now subtract from that the mean for the row and then subtract the mean for the column. Now add the overall mean. Then we square the result and multiply by the number of observations in the block (three in this case for all blocks). The double sigma means that we add up the sums of squares for the rows and for the columns (two columns and three rows makes six in total in our current example).

It is useful to create ANOVA table to keep track of the results. We did this in Table 61 for the one-way ANOVA. This time we will have more rows. We start by writing the sums of squares. It is usual to put the two main treatments at the top followed by the interaction sums of squares and then the within sums of squares (the error SS) and finally the total sums of squares. In Table 66 we see the sums of squares entered in this fashion.

Table 66. Sums of squares part of two-way ANOVA table for abundance of a plant species at two sites under differing grazing treatment

Source of Variation	SS
Rows (grazing treatment)	2403.10
Columns (site)	14.22
Interaction	129.78
Within (Error)	69.33
Total	2616.40

The next part involves working out the degrees of freedom for the various components. We shall start with the rows (grazing treatments) because we have them written down first. Degrees of freedom is generally the number of things you have –1. In this case we have three treatments, so degrees of freedom = 3 – 1 = 2. We have two columns (sites) so the degrees of freedom = 2 – 1 = 1.

The degrees of freedom of the interaction is determined by how many rows and columns we have, i.e. the number of treatments and sites. We take rows – 1 and multiply by columns – 1. We have three rows (treatments) and two columns (sites) so we get degrees of freedom = (3 – 1) × (2 – 1) = 2 × 1 = 2.

The degrees of freedom for the error term is worked out as the total number of observations – (number of rows × number of columns). We have 18 observations, three rows (treatments) and two columns (sites) so we get:

$$\textit{degrees of freedom} = 18 - (2 \times 3) = 18 - 6 = 12.$$

The total degrees of freedom is simply the total number of observations – 1. We have therefore 18 – 1 = 17 degrees of freedom overall. If we add together the previously calculated degrees of freedom we ought to get the same total (i.e. 17), which we do. The completed table is shown in Table 67.

The final stage is to determine the mean squares and the *F*-values and significance of the various components. The results are summarised below in Table 67.

Table 67. Completed two-way ANOVA table for abundance of a plant species at two sites under differing grazing treatments

Source of Variation	SS	df	MS	*F*	*p*-value	*F* crit
Rows (grazing)	2403.1	2	1201.6	207.96	4.9E-10	3.89
Columns (site)	14.22	1	14.22	2.46	0.1426	4.75
Interaction	129.78	2	64.89	11.23	0.0018	3.89
Within (error)	69.33	12	5.78			
Total	2616.4	17				

The mean square values are simply calculated by dividing the sums of squares by the degrees of freedom. To calculate the *F*-values we take the mean square for the element we want and divide by the error MS (that is the within groups mean square). So, for example the interaction *F*-value is determined by dividing 64.89 by 5.78 to give 11.23. Since this is larger than the critical value we can determine that the interaction between grazing treatment and site is a significant one.

Notice that we have a mixture of significant and not significant results. The interaction is a significant one as is the difference between treatments (the rows), but the difference between sites (columns) is not. We can see by referring back to our graph in Figure 141 that the two sites are very similar with much overlap at two out of the three grazing treatments.

Reporting results of a two-way ANOVA

When reporting the results of a two-way ANOVA, it is common to present the whole table. Should you wish to highlight one element however, and present a result in words, you give the *F*-value and the appropriate degrees of freedom. For example if we wanted to mention in the text that the interaction was significant we would present thus: $F_{2,12} = 11.23$, $p < 0.01$. If we wanted to present the site result we would present thus: $F_{1,12} = 2.46$, n.s.

Recording data for a two-way ANOVA

Our original data (Table 65) were written out in a block design, which mimicked to some extent what the situation on the ground was. This is not a terrible way to go about recording the data but it is also not the best manner to set out the values. In Table 68 we have reorganised the data into three columns. The first column is the response variable, i.e. the abundance, which is what we actually measure. The next two columns contain the predictor variables, one for site and one for grazing. Many computer programs expect the data to be laid out in this manner. It resembles biological recording format more closely as each row contains all the required information.

Table 68. Abundance of a plant species (per m^2) at two sites under differing grazing treatments. Data are laid out in single column format, i.e. each column contains a single variable

Abund	Site	Grazing
9	Top field	Low
11	Top field	Low
6	Top field	Low
14	Top field	Mid
17	Top field	Mid
19	Top field	Mid
28	Top field	High
31	Top field	High
32	Top field	High
7	Lower moor	Low
6	Lower moor	Low
5	Lower moor	Low
14	Lower moor	Mid
17	Lower moor	Mid
15	Lower moor	Mid
44	Lower moor	High
38	Lower moor	High
37	Lower moor	High

If we have more complex situations, then we may not be able to set out our data in the on-the-ground layout but we will always be able to add extra columns representing additional predictor variables.

Post-hoc testing in two-way ANOVA

Much like one-way ANOVA, our final result only tells us if there are significant differences between samples. Our graph of the means (Figure 141) shows us that some of the differences are pretty close and so likely not significant. We can carry out Tukey HSD tests in the same manner as before (Section 10.2.1) but of course now we have more pair-wise comparisons to make.

10.2.3 Multi-way ANOVA

It is certainly possible to add more predictor variables to the mix. In this way you can get three-way ANOVA and more. Calculating the results becomes more and more difficult as does the interpretation. If you are going to attempt a multi-way ANOVA, then you are likely to want to use a dedicated computer program to carry out the calculations. Our obvious choice is to use R.

10.2.4 Using R for ANOVA

We can use R to carry out ANOVA quite easily – there is a generalised command for running ANOVA (the *aov()* command) and this will run the calculations we need. For post-hoc testing we have another command *TukeyHSD()*.

ANOVA, *aov()* command

ANOVA allows you to examine the differences between more than two samples at one go. It is not correct to run a series of *t*-tests. R uses the model syntax to drive the routines in the *aov()* command.

First of all we need a set of data. In this set each row would be a separate item of data. Each column represents a factor. One would be the response variable, this is the thing that changes and an example would be abundance. The rest of the columns are factors and are known as predictor variables. In this example we might have site and management.

We read in our data from the CSV file.

```
> data10 = read.csv(file.choose())
```

To remind us of the factors involved we can use the *names()* command:

```
> names(data10)
[1] "site"        "ssp"        "management"        "abundance"
```

We might be interested in how the abundance alters between the sites so we use:

```
> data10.aov = aov(abundance ~ site, data = data10)
```

If we had used the *attach()* command, then we would not need the *data* = part. We assign a variable name to our result and to look at it we use:

```
> summary(data10.aov)
```

Our result looks something like the following:

```
             Df Sum Sq Mean Sq F value    Pr(>F)
site          1    725     725  15.906 6.875e-05 ***
Residuals  2190  99797      46
---
Signif. codes:  0 '***' 0.001 '**' 0.01 '*' 0.05 '.' 0.1 ' ' 1
```

We get fairly sparse information but it contains all we really need; we have the degrees of freedom and the *F* value. The bottom line is the *p*-value, which in this case is highly significant. R also adds a code after the p-value as a quick visual aid. The codes are shown after the main result. Notice how the small value is represented using scientific notation.

Of course we actually have two factors here and both may be important. The first test we ran was a one-way ANOVA but we should use both variables in a two-way ANOVA. We can use a variety of models to perform ANOVA.

```
> aov(abundance ~ site + management)
> aov(abundance ~ site * management)
```

In the first example we are examining the two factors (site and management) and they are treated more or less independently. We see the result:

```
             Df Sum Sq Mean Sq F value    Pr(>F)
site          1    725     725  16.151 6.048e-05 ***
management    1   1557    1557  34.684 4.477e-09 ***
Residuals  2189  98240      45
---
Signif. codes:  0 '***' 0.001 '**' 0.01 '*' 0.05 '.' 0.1 ' ' 1
```

Notice that the *F* value for site is affected (a little) by adding the second factor.

There may also be a difference in the response to management between the sites. This is called an interaction and the * symbol tells R to look at both factors as well as the two in concert. If we summarise the relationship using the second model above we get:

```
                  Df Sum Sq Mean Sq F value    Pr(>F)
site               1    725     725 16.1572 6.027e-05 ***
management         1   1557    1557 34.6984 4.445e-09 ***
site:management    1     85      85  1.8996    0.1683
Residuals       2188  98155      45
---
Signif. codes:  0 '***' 0.001 '**' 0.01 '*' 0.05 '.' 0.1 ' ' 1
```

We now have a third row to the table. The row labelled site:management shows us the interaction between the two factors. In this case the interaction is not significant. We conclude that the abundance is different between the site and is affected by the management. The lack of interaction implies that the effects of management are similar across the sites.

Post-hoc testing in R using *TukeyHSD()*

We can carry out the Tukey post-hoc test quite simply in R using the *TukeyHSD()* command. First of all we need an analysis of variance result of some sort. We could run a two-way ANOVA like so:

```
> pw.aov = aov(height ~ water * species, data = pw)
> summary(pw.aov)
              Df  Sum Sq Mean Sq  F value    Pr(>F)
```

```
water           2 2403.11 1201.56 207.9615 4.863e-10 ***
species         1   14.22   14.22   2.4615  0.142644
water:species   2  129.78   64.89  11.2308  0.001783 **
Residuals      12   69.33    5.78
---
Signif. codes:  0 '***' 0.001 '**' 0.01 '*' 0.05 '.' 0.1 ' ' 1
```

In this case we have two different plant species (we can work this out from the degrees of freedom, Df) and three watering regimes. Our response variable is the plant growth. We can see that we have a significant effect of water treatment on the plant growth. There appears to be no significant difference between the species but the interaction between species and water treatment is significant. To look more closely we run a post-hoc test on the result (which we called *pw.aov*):

```
> TukeyHSD(pw.aov)
  Tukey multiple comparisons of means
    95% family-wise confidence level

Fit: aov(formula = height ~ water * species, data = pw)

$water
             diff        lwr       upr     p adj
lo-hi  -27.666667 -31.369067 -23.96427 0.0000000
mid-hi -19.000000 -22.702401 -15.29760 0.0000000
mid-lo   8.666667   4.964266  12.36907 0.0001175

$species
                      diff       lwr       upr    p adj
vulgaris-sativa -1.777778 -4.246624 0.6910687 0.142644

$`water:species`
                               diff        lwr          upr     p adj
lo:sativa-hi:sativa      -33.666667 -40.258930 -27.07440314 0.0000000
mid:sativa-hi:sativa     -24.333333 -30.925597 -17.74106981 0.0000004
hi:vulgaris-hi:sativa     -9.333333 -15.925597  -2.74106981 0.0048138
lo:vulgaris-hi:sativa    -31.000000 -37.592264 -24.40773647 0.0000000
mid:vulgaris-hi:sativa   -23.000000 -29.592264 -16.40773647 0.0000007
mid:sativa-lo:sativa       9.333333   2.741070  15.92559686 0.0048138
hi:vulgaris-lo:sativa     24.333333  17.741070  30.92559686 0.0000004
lo:vulgaris-lo:sativa      2.666667  -3.925597   9.25893019 0.7490956
mid:vulgaris-lo:sativa    10.666667   4.074403  17.25893019 0.0016201
hi:vulgaris-mid:sativa    15.000000   8.407736  21.59226353 0.0000684
lo:vulgaris-mid:sativa    -6.666667 -13.258930  -0.07440314 0.0469217
mid:vulgaris-mid:sativa    1.333333  -5.258930   7.92559686 0.9810084
lo:vulgaris-hi:vulgaris  -21.666667 -28.258930 -15.07440314 0.0000014
mid:vulgaris-hi:vulgaris -13.666667 -20.258930  -7.07440314 0.0001702
mid:vulgaris-lo:vulgaris   8.000000   1.407736  14.59226353 0.0149115
```

The first part of the post-hoc test looks to compare each of the watering treatments. In this case we are ignoring the species. We see that all three treatments are different from one another. The next part of the post-hoc summary considers the species only and ignores the watering treatments. Here we have two species so there is only one comparison and we see that there is no significant difference between the two species; however, we have several water treatments and lumping them all together does not seem very sensible.

The final part of the post-hoc analysis compares all the possible combinations of species and water treatment. The more water treatments and the more species we have, the more complicated this section will become! We are really most interested in differences between the two species at each of the watering treatments:

```
                              diff        lwr       upr        p adj
hi:vulgaris-hi:sativa   -9.333333 -15.925597 -2.741070 0.004813836
lo:vulgaris-lo:sativa    2.666667  -3.925597  9.258930 0.749095605
mid:vulgaris-mid:sativa  1.333333  -5.258930  7.925597 0.981008407
```

We can see now that there is no significant difference between the two species until we get to the high watering treatment. Our interpretation is that the two species both react in similar fashion, increasing growth as the amount of water is increased; however, once the highest level is reached, the *sativa* species is significantly taller than the *vulgaris* species.

10.2.5 Using Excel for ANOVA

Excel does not have functions built-in that will carry out ANOVA as it does for the *t*-test. We can use the *Analysis ToolPak* to carry out ANOVA, as we shall see shortly, but before that we will look at a couple of useful functions that can help us. The mathematics to carry out simple one-way ANOVA are not especially complex and it is easy enough to work out the variance components. Figure 144 shows a simple one-way ANOVA carried out comparing two samples.

Once you have determined the value of the *F*-statistic it is possible to use built-in functions to determine the *p*-value and the critical value in a similar manner to the *t*-test we saw earlier (Section 7.1.1). The two commands are FINV and FDIST. The first of these works like so:

FINV(*F-value, df_between, df_within*)

If we look at Figure 144 we can see how this operates in practice.

	A	B	C	D	E	F	G	H	I	J	K	L	M	N
1		Ridge	$x-\bar{x}$	$(x-\bar{x})^2$	Furrow	$x-\bar{x}$	$(x-\bar{x})^2$				$x-\bar{\bar{T}}$	$(x-\bar{\bar{T}})^2$	$x-\bar{\bar{T}}$	$(x-\bar{\bar{T}})^2$
2		4	-1.5	2.25	9	1	1				-2.462	6.05917	2.538	6.44379
3		3	-2.5	6.25	8	0	0				-3.462	11.9822	1.538	2.36686
4		5	-0.5	0.25	10	2	4				-1.462	2.13609	3.538	12.5207
5		6	0.5	0.25	6	-2	4				-0.462	0.21302	-0.462	0.21302
6		8	2.5	6.25	7	-1	1				1.538	2.36686	0.538	0.28994
7		6	0.5	0.25							-0.462	0.21302		
8		5	-0.5	0.25							-1.462	2.13609		
9		7	1.5	2.25							0.538	0.28994		
10									All data					
11	Count	8			5				13					
12	Sum	44	0	18	40	0	10		84		-7.692	25.396	7.692	21.834
13	Mean	5.5			8				6.462					
14	Variance	2.571			2.500				3.936					
15														
16	SS within	28		18			10							
17	SS bet(ind)	7.396			11.834									
18	SS between	19.231												
19	SS Total	47.231										25.396		21.834
20														
21	Source	SS	df	MS	F	P	Crit							
22	Between	19.231	1	19.231	7.555	0.019	4.844		$SS_{between(indiv)} = n(\bar{x}-\bar{\bar{T}})^2$					
23	Within	28	11	2.545										
24	Total	47.231	12											

Figure 144. A simple ANOVA carried out in a spreadsheet

The F-value has been worked out to be 7.555 and the result is in cell E22. To the right we see the *p*-value (0.019), which is calculate by the FINV formula. The degrees of freedom between samples is 1 and this is the number of samples – 1. Here we only have two samples so degrees of freedom = 1. The degrees of freedom within (the error term) is worked out as number of items in first sample – 1 + number of items in second sample – 1; in other words, 8 – 1 + 5 – 1 = 11. The total degrees of freedom is always the total number of values we have – 1.

The FINV formula gives us the probability that this result was due to random chance. Here we see that the small value indicates that the two samples are significantly different.

We can also use the FDIST command to work out the critical value of *F* for any given situation like so:

FDIST(*p-value, df_between, df_within*)

In most cases we want to compare to a *p*-value of 0.05 and in Figure 144 we see that the critical value has been calculated to be 4.844. Since our calculated value (of 7.555) is greater than this we can reject the null hypothesis and determine that the samples are statistically different (with 95% confidence). We could replace the *p*-value with smaller numbers (e.g. 0.01) and we would see a critical value of 9.646. Our calculated value does not exceed this so we cannot be 99% certain that there is a difference. We have already determined that the exact *p*-value is 0.019 and we can see that indeed this is greater than 0.01 but smaller than 0.05.

If we wish to carry out two-way ANOVA then things become more difficult and we may wish to use the *Analysis ToolPak* (Section 1.9.1).

Using Excel add-ins for ANOVA

Although Excel does not have a dedicated function to work out ANOVA, there are some routines built-in to the *Analysis ToolPak* (see Section 1.9.1). We can access the routines via the *Tools* menu on the ribbon and clicking the *Data Analysis* button. You can carry out one-way and two-way ANOVA.

In our first example, we will carry out a one-way ANOVA. Our data are arranged in separate sample columns and we call up the *Analysis ToolPak* by clicking the *Data Analysis* button on the *Data* ribbon menu. We can now select the type of analysis we require (Figure 145).

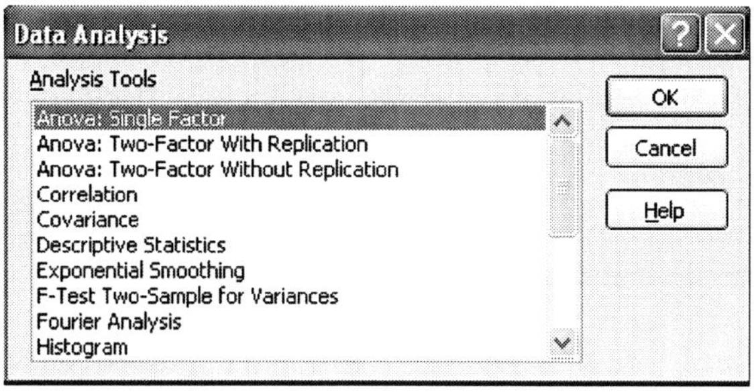

Figure 145. Selecting one-way ANOVA on the *Data Analysis* menu

Once we have selected our data and chosen the location for the results, we can carry out the analysis by clicking OK (Figure 146).

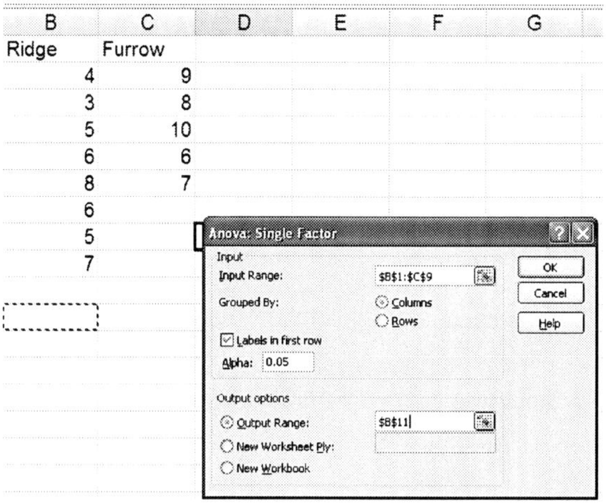

Figure 146. Choosing the data for a one-way ANOVA using the *Analysis ToolPak*

The results are sent to the location we selected (Figure 147). We see that we are presented with a summary table followed by the main analysis results.

Anova: Single Factor

SUMMARY

Groups	Count	Sum	Average	Variance
Ridge	8	44	5.5	2.571429
Furrow	5	40	8	2.5

ANOVA

Source of Variation	SS	df	MS	F	P-value	F crit
Between Groups	19.23077	1	19.23077	7.554945	0.018937	4.844336
Within Groups	28	11	2.545455			
Total	47.23077	12				

Figure 147. The results of a one-way ANOVA from the *Analysis ToolPak*

We get to see the classic ANOVA table showing the sums of squares and so on and culminating in the calculated *F*-value and the probability that our result was down to random chance.

If we wish to carry out two-way ANOVA we need to set out our data in a special way. In Figure 148 we see our data ready for analysis. This time we select "Two-Factor With Replication" because each of our treatments has several observations.

	Top Field	Lower Moor
Lo	9	7
Lo	11	6
Lo	6	5
Mid	14	14
Mid	17	17
Mid	19	15
Hi	28	44
Hi	31	38
Hi	32	37

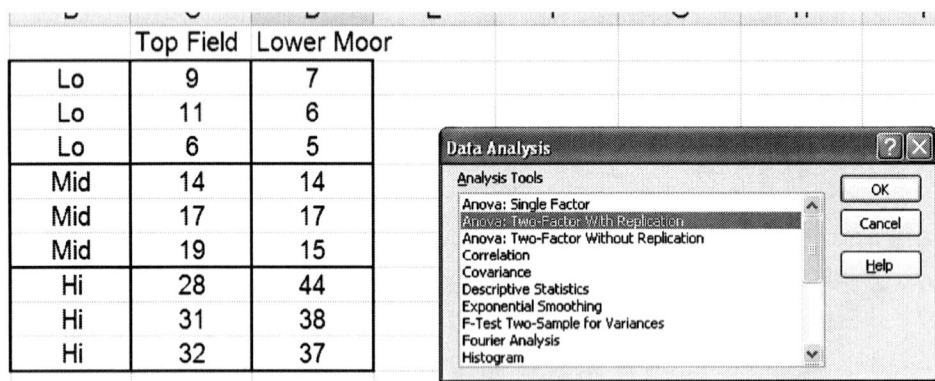

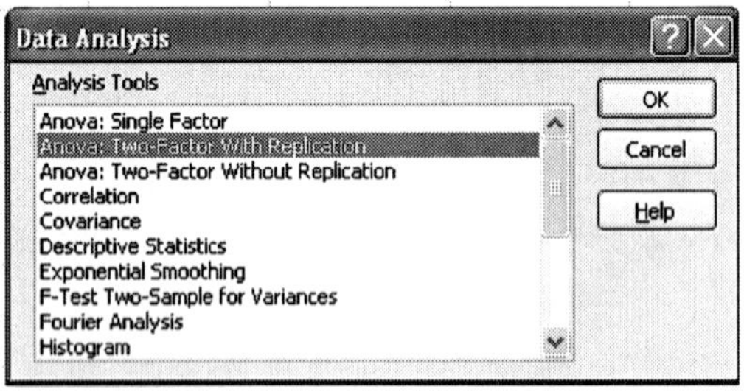

Figure 148. Selecting two-way ANOVA from the *Data Analysis* menu

Our data are set out with columns representing one treatment; in this case the site is the factor we are interested in. The rows represent the other factor (grazing treatment). Each row has three observations and these form the replicates. It is important that we have the same number of rows for each treatment. Once we click OK, we can select the data and the location for the results to be sent (Figure 149).

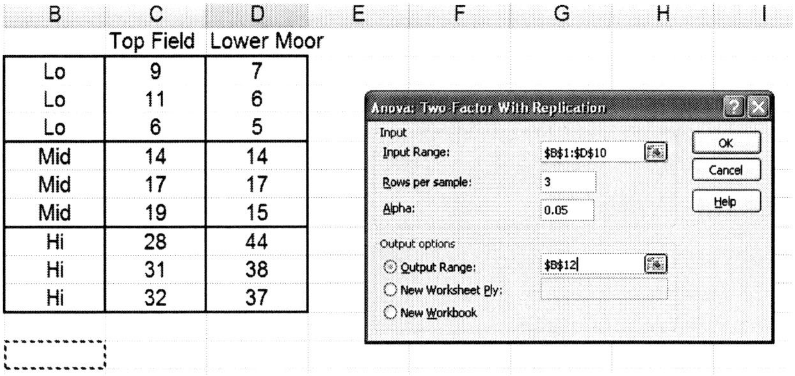

Figure 149. Two-way ANOVA data need to be set out in a special manner

The results are quite a bit more extensive that for the one-way ANOVA as we get a summary table for each of the levels of the row treatment (Figure 150). In this case we had three levels of grazing (low, mid and high) so we get three summary tables.

Anova: Two-Factor With Replication						
SUMMARY	Top Field	Lower Moc	Total			
Lo						
Count	3	3	6			
Sum	26	18	44			
Average	8.666667	6	7.333333			
Variance	6.333333	1	5.066667			
Mid						
Count	3	3	6			
Sum	50	46	96			
Average	16.66667	15.33333	16			
Variance	6.333333	2.333333	4			
Hi						
Count	3	3	6			
Sum	91	119	210			
Average	30.33333	39.66667	35			
Variance	4.333333	14.33333	33.6			
Total						
Count	9	9				
Sum	167	183				
Average	18.55556	20.33333				
Variance	94.27778	231				
ANOVA						
Source of Variation	*SS*	*df*	*MS*	*F*	*P-value*	*F crit*
Sample	2403.111	2	1201.556	207.9615	4.86E-10	3.885294
Columns	14.22222	1	14.22222	2.461538	0.142644	4.747225
Interaction	129.7778	2	64.88889	11.23077	0.001783	3.885294
Within	69.33333	12	5.777778			
Total	2616.444	17				

Figure 150. The results of a two-way ANOVA carried out by the *Analysis ToolPak*

The final table in the results shows us the standard ANOVA results table. The Source of Variation column is not labelled with the treatment names but we can work it out easily; the Sample row relates to the rows (grazing in our example) and the Columns row relates to the columns (the different sites). We also have an interaction row.

The limitation is that we must have equal numbers of observations in each of the treatments; if this condition is not met then the analysis routine will simply not function. We cannot undertake more complex analyses than two-way ANOVA and of course our data must be set out in the on-the-ground layout rather than in strict biological-recording layout.

10.3 Kruskal–Wallis test

When you wish to look at differences between more than two samples but your data are not normally distributed (i.e. are skewed) then analysis of variance (Section 10.2) is not appropriate and we should use the Kruskal–Wallis test instead. This is often referred to as a non-parametric analysis of variance.

The Kruskal–Wallis test examines for differences between multiple samples and is analogous to one-way ANOVA. In common with most non-parametric tests the original data are replaced with the ranks of the observations. Look at the data in Table 69. Here we see data for counts of a freshwater invertebrate at three sites.

Table 69. Counts of hoglouse at three sites in a stream in Devon

Upper	Mid	Lower
3	4	11
4	3	12
5	7	9
9	9	10
8	11	11
10		
9		

Before we start we ought to determine that the data are not normally distributed. The graph, Figure 151, is a box–whisker plot that displays the median and quartiles of the data samples.

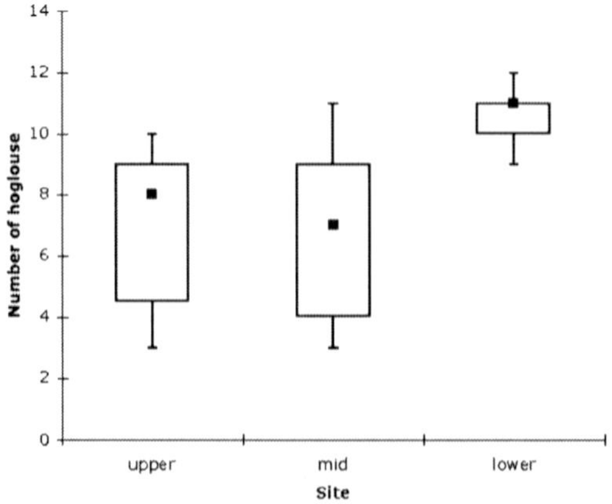

Figure 151. Box–whisker plot showing counts of hoglouse at three sites in a stream in Devon. Points show median, boxes are inter-quartiles and whiskers the range

We can see readily that the data are not normally distributed from this plot without needing to draw a frequency distribution. All three samples are non-symmetrical; we can see this especially from the lower sample where the median coincides with the quartile.

The first thing to do is to rank all the data; we lump all the values together for this purpose. In other words we do not rank each sample separately. If there are big differences between samples then the ranks of one sample will tend to be smaller (or bigger) than the others. The more separate the samples, the more separate the ranks. If there are no real differences then the ranks will be all muddled up.

Once we have our data ranked we can start to calculate the Kruskal–Wallis statistic (called *H*) using the formula given in Figure 152.

$$H = \frac{12}{N(N+1)} \sum_{i=1}^{k} \frac{R_i^2}{n_i} - 3(N+1)$$

Figure 152. The formula used to determine the Kruskal–Wallis statistic

The Kruskal–Wallis formula looks pretty horrendous but actually it is not that bad. The numbers 12 and 3 are constants. Uppercase *N* is the total number of observations (look back at the data in Table 69 and see that there are 17). The R refers to the ranks of the observations in each sample and *n* is the number of observations per sample.

We start by working out the ranks. The lowest value gets the lowest rank. Where there are tied values we determine an average rank; then the next value gets the next available rank. In Table 70 we see the data converted to ranks. The two lowest values were 3; these occupy ranks 1 and 2 and so get a tied/average rank of 1.5. The next available rank is therefore 3; however, there are two 4 values and these need to take up ranks 3 and 4, getting an average rank of 3.5.

Table 70. Hoglouse at three sites in a stream in Devon. Original data converted to ranks

Upper	Mid	Lower
1.5	3.5	15
3.5	1.5	17
5	6	9.5
9.5	9.5	12.5
7	15	15
12.5		
9.5		

Next we add up the ranks for each sample. In this case we end up with three rank totals as we have three samples. Now we square each one and divide by the number of observations in each sample. These calculations are shown in Table 71. Now this gives us the R^2/n values from the formula. We add these together (that is the sigma part) giving 1540.29.

Table 71. Hoglouse at three sites in a stream in Devon. Sums of ranks

	Upper	Mid	Lower
Sum	48.5	35.5	69
Sum^2	2352.25	1260.25	4761.00
Sum^2/n	336.04	252.05	952.2

The final step is to combine the rest of the formula. We have the sigma part so now we multiply that by $12/N(N + 1)$ to give $1540.29 \times 12/(17 \times 18) = 1540.29 \times 0.04 = 60.4$. Last of all we subtract the $3(N + 1)$ at the end (remember that we do all the multiplication and division bits before the adding and subtracting bits). This gives us $60.4 - (3 \times 18) = 60.4 - 54 = 6.404$.

We now have a result, *H*. In order to see if our result is significant we need to compare our calculated value of *H* to a critical value. The Kruskal–Wallis statistic is a funny beast! If we have at least five observations in each sample, then *H* is very close to Chi-squared. We can compare our value of *H* to the Chi-squared value for appropriate degrees of freedom (Table 55). The degrees of freedom are the number of samples – 1. In our example we have three samples so we have two degrees of freedom.

Where the sample sizes are smaller, the *H* statistic behaves in a rather more peculiar fashion. Fortunately for us, statisticians have worked out exact *H* values for a range of sample sizes, therefore we are able to look at tables of values that have been calculated for us. There are two tables, one for equal sample sizes and another for unequal.

Our calculated value was $H = 6.404$, which is greater than the Chi-squared critical value (of 5.991) and thus the result is significant at the 5% level.

10.3.1 Post-hoc testing in Kruskal–Wallis

Once we have our result, we know if the samples are significantly different or not; however, we do not know if all the samples are different from each other or if some are actually not different from each other (implying that others are). If we look back to our graph (Figure 151) we can see that the two sites on the left are overlapping quite a lot and it seems unlikely that they could be significantly different. It looks like the site on the right (labelled lower) might be different from the other two.

As discussed previously (Section 10.2.1) we should not resort to looking at each pair in turn (using a *U*-test for example, since we have non-parametric data) as the very act of performing multiple tests alters the probability of our getting a result. As before we run a post-hoc test. In the case of Kruskal–Wallis, there is *non-standard* approach and many statistical programs will not even provide a routine to calculate a post-hoc test for the Kruskal–Wallis analysis. We can, however, run a modified version of the *U*-test where the calculation is based loosely on Tukey's approach for his HSD test, and this is suitable for post-hoc testing. What we actually do is to run a standard *U*-test but calculate a special critical value for the post-hoc analysis.

Unlike the situation we had with ANOVA (see Section 10.2.1 on post-hoc testing) we don't have an error term. As a result we need a way to take into account the within groups variability. The formula for the post-hoc test is shown in Figure 153.

$$U = \frac{n^2}{2} + Qn\sqrt{\frac{(2n+1)}{24}}$$

Figure 153. Post-hoc testing for the Kruskal–Wallis test

There are only two variables in the formula, n and Q. The variable Q is derived from Student's t-statistic and is called the Studentised range (values are shown in Table 72). The value of n is the common sample size and is calculated using the formula shown in Figure 154; it is essentially the harmonic mean of the sample sizes.

$$n = \frac{2}{(\frac{1}{n_1} + \frac{1}{n_2})}$$

Figure 154. Calculating the harmonic mean of two samples

To start with we work out the sample sizes and calculate the common sample size using the formula above. We then look up the value of Q, which depends only upon the number of samples. Even though we are only comparing two samples at a time, we use the value for the total number of samples we have. In this case we have three samples (the three different sites) so our value of $Q = 3.314$ (at the 5% significance level).

Table 72. Values of Q, the studentised range

	Significance	
Number of groups	5%	1%
2	2.772	3.643
3	3.314	4.120
4	3.633	4.403
5	3.858	4.603
6	4.030	4.757

The value of U we have calculated is the critical value (at 5%, although we could easily calculate the 1% value as well), determined for our combination of sample groups and sizes. What we do now is to work out a regular U-test for each pair of samples we are interested in. Unlike a regular U-test however, we select the larger of the two U-values as our test statistic. Now finally we get to compare our calculated value to the critical value (that is the one we worked out previously). If our pairwise U-value is equal or greater than the critical value, then the post-hoc test is significant (at that level of significance).

If we do this for our hoglouse data (Table 69) we find that the only difference is between the Upper and Lower sites (see Table 73). Looking back at Figure 151 we might not have guessed this and would probably have assumed that the Mid site was also different from the Lower. This is of course why we undertake the statistical analysis in the first place.

Table 73. Post-hoc test results for Kruskal–Wallis analysis of hoglouse data. Main vales shown are calculated *U*-values and significance

	Upper	Mid	Lower
Upper	-	18	32.5
Mid	n.s.	-	21.5
Lower	5.00%	n.s.	-

Table 73 shows the main results of the post-hoc test. We have shown the calculated *U*-values in the upper-right of the table and the significance in the lower-left.

Post-hoc tests in general are rather conservative and tend to err on the side of caution. The post-hoc procedure outlined above for the Kruskal–Wallis test is a bit long-winded but is really your only option with non-parametric data and small sample sizes. The procedure outlined here probably should be carried out only when the sample sizes are fairly equal. As the sample sizes become more and more unequal, the accuracy of the formula diminishes and we cannot rely on the calculated probabilities being correct.

10.3.2 Using Excel for Kruskal–Wallis

Excel does not have any built-in formulae for working out Kruskal–Wallis tests. It is of course reasonably easy to carry out the procedures, although getting the ranks to work out correctly is not that simple in Excel (see Figure 101).

10.3.3 Using R for Kruskal–Wallis testing

R has the routines for carrying out Kruskal–Wallis testing via the *kruskal.test()* command. We can perform the Kruskal–Wallis test in several ways. If we have our data as separate samples we can run the command like so:

```
> kruskal.test(list(item1, item2, item3...))
```

An example of this is shown below, where we use three samples:

```
> upper
[1]  3  4  5  9  8 10  9
> lower
[1] 11 12  9 10 11
> mid
[1]  4  3  7  9 11
> kruskal.test(list(upper, lower, mid))
```

```
  Kruskal-Wallis rank sum test

data:  list(upper, lower, mid)
Kruskal-Wallis chi-squared = 6.5396, df = 2, p-value = 0.03801
```

The three samples are the same data we had earlier in Table 69 and refer to counts of hoglouse in different parts of a stream. The *list* part of the command is a way of temporarily joining the data items together before running the analysis.

We get a simple output; R reminds us of the data and gives us the test statistic (the Kruskal–Wallis test results in a distribution similar to Chi-squared), the degrees of freedom and the final *p*-value. Here we see that $p = 0.03801$ and since this is <0.05 we can reject the null hypothesis and accept the alternative that the samples are different. In other words, the abundance of the invertebrate is different across the sites.

It is more likely that we will have our data in a spreadsheet file and we could save this as a CSV and then use R to read the file. In the following example we have taken the same hoglouse data as above and created a single data item *hl*.

```
> hl = read.csv(file.choose())
```

Notice how R creates the data as a rectangular matrix. Any gaps are filled with NA.

We now run the Kruskal–Wallis test on the whole data item. Since this is comprised of three samples, the command compares all those it finds. The results are of course exactly the same as in the previous example:

```
> hl
  upper lower mid
1     3    11   4
2     4    12   3
3     5     9   7
4     9    10   9
5     8    11  11
6    10    NA  NA
7     9    NA  NA
> kruskal.test(hl)

  Kruskal-Wallis rank sum test

data:  hl
Kruskal-Wallis chi-squared = 6.5396, df = 2, p-value = 0.03801
```

In general it is better to collect and record data in separate columns so that you end up with one column for each item and each row containing a complete record. In the case of our freshwater hoglouse, we might record our data with two columns. The first would be the count of hoglouse and the second would be the site where they were collected from. In the following example we see the data written in this fashion:

```
> hog
   count  site
1      3 upper
2      4 upper
3      5 upper
4      9 upper
5      8 upper
6     10 upper
7      9 upper
8     11 lower
9     12 lower
10     9 lower
11    10 lower
12    11 lower
15     4   mid
16     3   mid
17     7   mid
18     9   mid
19    11   mid
> kruskal.test(count ~ site, data = hog)

  Kruskal-Wallis rank sum test

data:  count by site
Kruskal-Wallis chi-squared = 6.5396, df = 2, p-value = 0.03801
```

Now we have the first column as our response variable and the second column as our predictor variable. This time we run the Kruskal–Wallis test using the formula type of command. Here we put the response factor on the left of the tilde (~) and the predictor factor to the right. The output is of course the same as in the previous tests because we are using the same data.

Post-hoc testing and Kruskal–Wallis

There is no built-in command to work out post-hoc comparisons for the Kruskal–Wallis test. What we must do is work out the pairwise U-values we are interested in and then use the post-hoc formula we met earlier.

We would start by running a U-test on the pair of samples we wished to examine. Then we need to check which is the largest of the U-values. Remember that:

$$U_a + U_b = n_a \times n_b$$

So, if we know the two sample sizes we can work out the other U-value once we have our result. We can now also determine the harmonic mean of the pair of samples using the formula as before. We can now re-arrange the post-hoc formula to make Q the subject.

Once we have solved the equation for *Q* we have two options. We can work out the critical value we need to exceed for any particular level of significance or we can determine the *p*-value directly. R provides links to the studentised range via two commands:

```
> ptukey(Q-value, nmeans, df = Inf)
> qtukey(CI, nmeans, df = Inf)
```

In the first case we can use the *ptukey()* command to determine the *p*-value if we have solved the post-hoc equation and found *Q*. We also need the total number of samples in the original analysis (nmeans). Since we are always comparing two samples, we set our degrees of freedom to *Inf*, in other words infinity. The result we get is a confidence interval so we need to use 1 – CI to get our *p*-value so:

```
> 1 - ptukey(Q-value, nmeans, df = Inf)
```

For our hoglouse data we had three samples originally so we set nmeans = 3.

We can also work out the value of *Q* that we have to exceed for any given level of significance using the *qtukey()* command. Here we need to insert the confidence interval (rather than the *p*-value) and in general we'd stick to 0.95, which equates to a *p*-value of 0.05.

This procedure is a bit complicated. To help out a bit, here are a few lines of command that you can copy to help work out the post-hoc test. The first works out the harmonic mean and the second takes the harmonic mean, the *U*-value and the number of groups and determines the *p*-value.

```
> hm = 2 / ((1 / n1) + (1 / n2))

> 1-(ptukey((uval-(hm^2/2)) / (hm * sqrt(((2 * hm) + 1) /
24)), groups, df = Inf))
```

In the first command you need to insert the length of each sample for the *n1* and *n2* parts.

In the second command you need to replace the uval with the largest *U*-statistic for your chosen pair of samples (you will have to run the *wilcox.test()*). The groups part is the total number of samples in the original Kruskal–Wallis test.

In the following example we show the process from start to finish. We begin with the data as a reminder. These data are called hl. First we run a *U*-test on the data. Since the two samples we want are inside the hl item, we add the $ to get at them. We get a statistic of *W* = 2.5 (R calls the *U*-test statistic *W*).

```
> hl
  upper lower mid
1     3    11   4
2     4    12   3
3     5     9   7
4     9    10   9
```

```
5      8     11   11
6     10     NA   NA
7      9     NA   NA
> wilcox.test(hl$upper, hl$lower)

  Wilcoxon rank sum test with continuity correction

data:  hl$upper and hl$lower
W = 2.5, p-value = 0.01732
alternative hypothesis: true location shift is not equal to 0

Warning message:
In wilcox.test.default(hl$upper, hl$lower) :
  cannot compute exact p-value with ties
> 7 * 5 - 2.5
[1] 32.5
> hm = 2 / ((1 / 7) + (1 / 5))
> hm
[1] 5.833333
> 1 - (ptukey((32.5 - (hm^2 / 2)) / (hm * sqrt(((2 * hm) +
1) / 24)), 3, df = Inf))
[1] 0.02641968
```

To check if this is the largest of the values we use the simple formula to work out that 32.5 is the other. If we had typed in the samples in the reverse order we would have ended up with 32.5. Next we work out the harmonic mean of the two samples. We can view the information (the mean is 5.833) but this is not essential.

Finally we use the command to generate the *p*-value for the pairwise comparison. We insert 32.5 in the uval position and replace the groups part with 3 as we had 3 samples in the original Kruskal–Wallis test. The result of $p = 0.0264$ tells us that there is a significant difference between these two samples.

If this sounds a bit complicated, it is! There is no sensible alternative for this kind of analysis. It does however illustrate one of the strengths of R. Once you have worked out a complicated formula you can copy it to a simple text file and keep it for another time.

10.3.4 Non-parametric tests for two-way ANOVA

If you have more than two predictor variables and your data are not normally distributed, your options are limited! The best way around this is to design your experiment carefully so that you are likely to get parametric data. If this fails then you may be able to do some cunning maths to alter your data and make it more normal. For example you might take the log of your values or transform your data using some other mathematical process (look back to Section 4.6 on transforming data).

11. Tests for linking several factors

11.1 Multiple regression

In Section 8.1 we looked at correlation, which describes the relationship between factors when we suspect a link (as opposed to looking for differences). Examples might include water speed and stonefly abundance, light levels and abundance of bluebell. We can draw a graph of the relationship and it would naturally be a scatter plot (see Section 6.3.1, we'll cover graphs in more detail in Section 12.4). The relationship might be positive or negative (or there might be no relationship at all). The strength of the relationship was determined using Spearman's rank correlation coefficient (r_s).

Spearman's rank works if your data are parametric or not; however, the coefficient says nothing about the link other than how strong it is. In regression however, we can use the properties of the normal distribution curve to tell us more. The simplest regression is worked out using Pearson product moment (Section 8.2, often just called r). Using this we can tell the strength of the link (correlation) as well as something about its nature.

For example the equation that describes a straight line on a scatter graph is:

$$y = mx + c$$

The y is the dependent variable (the vertical axis, also called the response factor) and x is the independent variable (the horizontal axis, also called the predictor factor). In our examples above, y would be stonefly or bluebell abundance whilst x would be speed or light. Now m is the slope of the line, a steep slope tells us that the abundance changes rapidly with the independent factor (speed or light) whilst a low slope tells us that the abundance changes slowly. This is **not** related to the correlation coefficient so it is perfectly possible to have a weak link that alters rapidly or a strong link that changes slowly. The correlation coefficient (the link) tells us how close to the best-fit line the data points are.

When we draw graphs of our correlations we naturally want to emphasise the link between the two factors. For Spearman's rank tests we should **never** try to show a line of best-fit. For the Pearson correlation we **may** produce a line of best-fit if we are looking to determine how quickly one factor changes the other.

So far, so good. Now consider this example: we are looking at butterfly abundance at a number of sites and are trying to work out which environmental factor is most important in determining how many butterflies there are. We might have things like: % cover of food-plant, % cover of trees, max temperature, abundance of crab spiders, number of tourists per hectare (you may be able to think of other factors).

The simplest thing to do is to take each of these potential predictor variables and run a correlation for each one against our response factor (the number of butterflies). We can then simply determine which factor has the highest correlation coefficient. This one would be the most important one in determining butterfly abundance.

As in our example of ANOVA above (Section 10.2.2), the factors may interact with one another in unexpected ways. In order to examine this we use multiple regression. In mathematical terms, we can write multiple regression in a similar way to the equation for a straight line so:

$$y = m_1x_1 + m_2x_2 + m_3x_3 + m_4x_4 + c$$

Here four factors are shown. In multiple regression, the slopes ($m_1 \ldots m_4$) are not exactly the same as in the simple example. We can better think of them as expressing how our predictor factor (independent variable) is affecting our response factor (dependent variable).

It can be hard to visualise what is going on. In the simplest case we have our response variable and a single predictor variable. We can draw a scatter graph and use the $y = mx + c$ formula to help us determine the line of best fit. The graph in Figure 155 shows a simple linear situation.

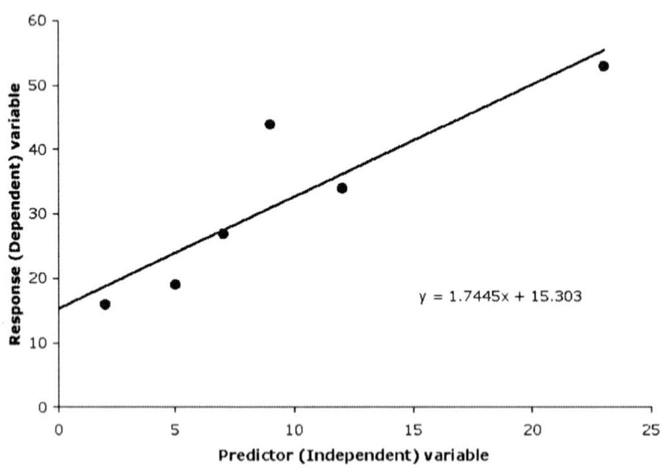

Figure 155. The simplest form of linear regression with a single predictor variable. The line of best fit passes through the mean of *x* and mean of *y* and its slope is adjusted so it minimises the distance (vertically) to the points

We can see the scatter of points and the line of best fit. This line is fitted by the equation (shown on the figure) and this passes through the mean of both variables. The slope is such that the cumulative distance to the points (measured vertically) is minimised (these distances are called residuals). When we add a second predictor variable the situation becomes a little harder. We could draw a 3D graph using the third axis to represent our new variable. Instead of a straight line we now attempt to fit a flat plane (like a sheet of paper or plank of wood) through the 3D cloud of points so that the cumulative distance to those points is minimised. Figure 156 shows a plane of best fit in a 3D setting.

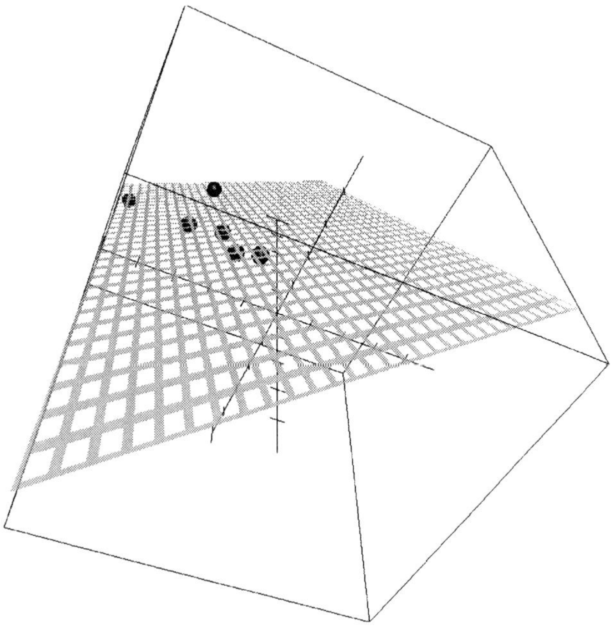

Figure 156. A 3D representation of a plane of best fit. The equation for the line is $y = 2.39x_1 - 0.83x_2 + 23.04$ calculated by multiple regression

The graph in Figure 156 was created using the same data as for Figure 155 but now has a second predictor variable added. This naturally alters the equation as we now have an added factor to take into account (the actual equation is given in the caption). We can just about visualise this 3D situation but if we add more predictor variables we don't have enough planes to work with and it is impossible to draw! Mathematically, however, the situation is just the same as before – trying to fit a plane to all the points to minimise the distance from it to those points.

In simple regression we end up with a correlation coefficient. With multiple regression we end up with several slopes, which are usually called coefficients. We can examine the statistical significance of the various factors and it is usual to remove the non-significant factors and re-run the maths using only those factors that are significant. At the end we finish up with an overall coefficient (usually called R^2 and written as a proportion, e.g. 0.64), which tells us what proportion of the variability is explained by our factors. If you like, it is a way of seeing how good your equation is. If you find that you get several significant terms (another word for the independent factors) but that R^2 is fairly low, then you might think that there are other factors out there that you haven't thought of that might explain the situation better!

We now need to think about how to go about calculating the regression for some data. Table 74 shows the data that were used to draw the two previous graphs (Figures 155 and 156). The first column shows the abundance of a butterfly species, this is the dependent (response) variable. The next two columns show the abundance of the larval food plant and the availability of nectar sources (two examples of predictor variables).

Table 74. Abundance of butterflies, larval food-plant and nectar plants for a meadow in Shropshire

Count	Food	Nectar
16	2	9
19	5	13
27	7	17
44	9	15
34	12	21
53	23	26

We are going to need to determine the sums of squares for all the columns so we start by determining the mean of each column. Then we'll determine the correlation between all the columns.

Start by evaluating the correlation formula for each pair of columns (recall Section 8.2). The formula is shown again in Figure 157.

$$r = \frac{\sum (x-\bar{x})(y-\bar{y})}{\sqrt{\sum (x-\bar{x})^2 \sum (y-\bar{y})^2}}$$

Figure 157. The correlation coefficient

We need the mean of each column and we then subtract it from each observation. The top line of the formula shows us that we need to multiply each difference by the corresponding difference in the column we are working on (we call one column x and the other column y). The bottom of the formula looks at the sums of squares, which we met when looking at ANOVA in Section 10.2. We multiply the sums of squares for each of the two columns together and take the square root.

In our case we end up with three values: one for the correlation between count and food, one for count and nectar, and one for the correlation between the two predictor variables, food and nectar. Because of the multiple predictor variables, we need to come up with a system for naming the various components. Generally a subscript is used, e.g. r_{y1}, r_{y2}, r_{12}, which gives us the correlation between y and the first predictor variable, between y and the second predictor and finally the correlation between the first and second predictors.

Table 75 shows the correlation coefficients and a reminder of which variables they relate to.

Table 75. Correlation coefficients for butterflies, larval food-plant and nectar plants for a meadow in Shropshire

Coef label	Coef value	Variables
r_{y1}	0.8915	Count vs. food
r_{y2}	0.8148	Count vs. nectar
r_{12}	0.9045	Food vs. nectar

We can see several things already: the coefficients are quite high and all are positive, so as food increases so does count for example. There is a strong link between the two predictor variables. We can also see that count and food have a stronger link than count and nectar. This implies that larval food-plant availability is more important than nectar-plant availability; however, we are getting a little ahead of ourselves.

We will also need to determine the value of these correlation coefficients squared and the standard deviation of each factor (i.e. column). We will not go into all the details of the calculations but we will look at the results in detail, as this will help us interpret the situation. We will also look at how to get Excel to calculate multiple regression for us and also use the R program.

When we have two predictor variables we can work out a beta coefficient using the correlations between the columns we already determined. The formula for that is shown in Figure 158.

$$b'_{y12} = \frac{(r_{1y} - r_{2y} r_{12})}{(1 - r_{12}^2)}$$

Figure 158. Calculating a beta coefficient using correlation coefficients

We can see that we need all three correlation coefficients (the subscripts tell us which variables the coefficients relate to) to determine this one value, which we can think of as being analogous to the slope (for our first predictor variable). There is a similar formula used to determine the other beta coefficients. The beta coefficients (there will be two in this case, one for each predictor variable) are standardised against one another by this formula. Effectively they assume the same units. Now in multiple regression we really want to produce a formula that describes the relationship between our dependent variable and the various predictor variables. We need coefficients that assume the units of the original measurement. We do this by using the standard deviations and the formula shown in Figure 159.

$$b_{y12} = b'_{y12} \frac{s_y}{s_{x1}}$$

Figure 159. Calculating regression coefficients from the beta coefficients

The final coefficient (which we will use to describe the relationship) is determined by taking the beta coefficient and the standard deviation of the two factors we are comparing.

Our final list of values is shown in Table 76.

Table 76. Regression coefficients for butterflies, larval food-plant and nectar plants for a meadow in Shropshire

	Coefficients	Beta coefficients
Intercept	23.0468	
Food	2.3889	1.2209
Nectar	–0.8301	–0.3463

We can use the coefficients column to construct the formula that describes the complete relationship between butterfly abundance and the two predictor factors in the following manner:

$$y = 2.38x_1 - 0.830x_2 + 23.05$$

Better still would be to substitute the names of the variables to give:

$$Count = 2.38\ Food - 0.830\ Nectar + 23.05$$

This tells us much about the regression but it is not the final story. The beta coefficients are useful; they are standardised against one another and we can readily see that food is more important than nectar. We have details about the relationship but we do not know if the factors are statistically significant. Nor do we know if the overall relationship is itself significant and how much of the variability (of the dependent factor) is explained by our model.

When we conduct a multiple regression we can also determine an overall regression coefficient, called R^2, which is a measure of how good our model is. In our case we get an overall R^2 of 0.806. This means that our two predictor variables explain about 80% of the variability in y, the dependent factor (in our case butterfly abundance).

We can also determine the statistical significance of the various elements. We can tell if each individual factor is significant and if our overall model is significant. Before we get to that stage it would seem prudent to look at how we can calculate all these results.

11.1.1 Using Excel to undertake multiple regression

It is not surprising to learn that multiple regression is not easy to calculate on your own, many of the calculation steps are long, involved and tedious (although not in themselves mathematically difficult). Most calculations are done using computer software. You can undertake regression fairly easily using your spreadsheet (e.g. Excel).

We met the LINEST function earlier (Section 8.3.1) when we looked at correlation between two variables. We can extend the range of the function to include several predictor variables and so perform multiple regression. Recall how the function works:

LINEST(*y_values*, *x_values*, 1, 1)

The numbers 1 and 1 tell the program to give us full statistics (rather than just a brief summary) and to produce a regular intercept (0 forces the intercept through the origin). Recall also that the function is an array function and we need to highlight the range of cells where the answers are going to be displayed and to press CTRL+SHIFT+ENTER (in Windows) instead of just ENTER (on a Mac cmd+ENTER).

In Figure 160 we can see the LINEST command used on the butterfly data that we met earlier in Table 74.

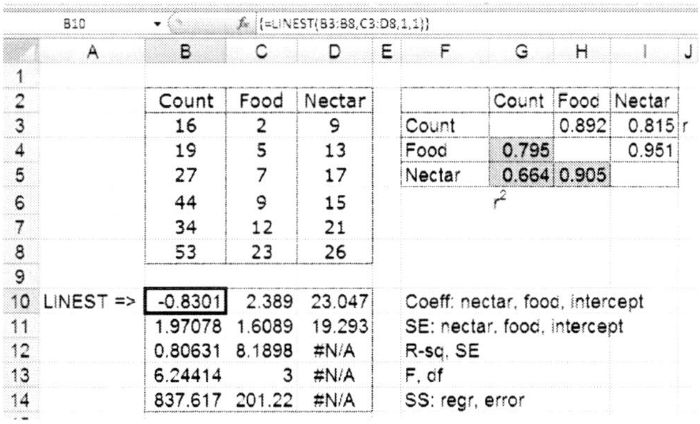

B10 {=LINEST(B3:B8,C3:D8,1,1)}

	A	B	C	D	E	F	G	H	I	J
1										
2		Count	Food	Nectar			Count	Food	Nectar	
3		16	2	9		Count		0.892	0.815	r
4		19	5	13		Food	0.795		0.951	
5		27	7	17		Nectar	0.664	0.905		
6		44	9	15			r^2			
7		34	12	21						
8		53	23	26						
9										
10	LINEST =>	-0.8301	2.389	23.047		Coeff: nectar, food, intercept				
11		1.97078	1.6089	19.293		SE: nectar, food, intercept				
12		0.80631	8.1898	#N/A		R-sq, SE				
13		6.24414	3	#N/A		F, df				
14		837.617	201.22	#N/A		SS: regr, error				

Figure 160. Using the LINEST function for multiple regression

The LINEST function requires us to enter the range for the response factor and we can see from Figure 160 that we use the count data from cells B3:B8. Previously we entered the predictor variable as a single column but here we have two (food and nectar) so we enter them both (cells C3:D8). We do need our columns to be in a continuous block, so if necessary we must move or copy them. We always get five rows of answers so we need to highlight five rows. We will get three columns of results in this case, because we have three columns of data; if we have more columns then we would highlight the appropriate range of cells.

In Figure 160 we have labelled the output of the LINEST function since the spreadsheet will not give us any labels. The first row shows the coefficients and these are shown in reverse order. The first coefficient is the last of the predictor factors. The last one is always the intercept. The next row shows the standard errors of the coefficients and they are in the same order as the coefficients. The third row shows us the overall fit of the model, i.e. the R^2 value and its standard error. We then see the F-value and the degrees of freedom. The last row shows the sums of squares.

To the right of the data we have also worked out the individual correlation between each pair of columns using the CORREL function. Also shown are the r^2 values (simply the result of the CORREL function squared). They are perhaps not strictly necessary but they can be informative.

The LINEST function gives us a variety of statistics but it does not compute the statistical significance of the overall regression nor of the individual factors. We must do this for ourselves. Previously we looked at the significance of the *F*-value and because we only had one predictor variable, it was not really necessary to do any more. Here though, we have two predictor values and should examine then separately. In Figure 161 we see how to do this.

B17 =TDIST(ABS(B16),$C$13,2)

	A	B	C	D	E	F
9						
10	LINEST =>	-0.8301	2.389	23.047		Coeff: nec
11		1.97078	1.6089	19.293		SE: necta
12		0.80631	8.1898	#N/A		R-sq, SE
13		6.24414	3	#N/A		F, df
14		837.617	201.22	#N/A		SS: regr,
15						
16	t	-0.4212	1.4849			Coef÷SE
17	P-val	0.702	0.234			TDIST

Figure 161. Determining the significance of individual factors in a multiple regression

We can work out a *t*-statistic for each of the components in our regression model. This is done from the LINEST results by taking the coefficient and dividing by the standard error. In Figure 161 we see this carried out for the two predictor factors. Because coefficients may be negative values we may end up with a negative *t*-value (as we have here). We now use the TDIST function to determine the *p*-value for each *t*-statistic.

TDIST(*t_value, degrees_of_freedom*, 2)

The degrees of freedom are found in the LINEST results in the fourth row (cell C13 in this case). The 2 tells the spreadsheet to look at both ends of the normal distribution (this is called a two-tailed test).

However, the TDIST formula expects a positive value for *t* and cannot cope with a negative value. We use the ABS function to force the *t*-value to be positive (ABS stands for absolute value). We could have done this when calculating the *t*-value in the previous line but the negative *t*-value reminds us that the statistic relates to a negative coefficient. Here we can see that the *p*-values for both of the predictor factors are > 0.05 meaning that neither are statistically significant.

The LINEST results give us an overall *F*-value but do not tell us if the regression model is statistically significant. We need to use the FDIST function that we saw before.

FDIST(*F_value, df_num, df_den*)

The function requires an *F*-value and two degrees of freedom, one for the denominator and one for the numerator. The *F*-statistic is derived from the ratio of two variances, which is why we talk about numerator and denominator (like a fraction). In Figure 162 we can see the elements of the command extracted and used to determine the *p*-value.

F	6.24414			
df (num)	2			No. predictor vars (all cols -1)
df (den)	3			
P-val	0.08525			FDIST
b'	-0.3463	1.2209		b' = b (s_x/s_y)

Figure 162. Determining overall significance of a regression model from the *F*-statistic and calculating the beta coefficients

The *F* and the *df(den)* values are given in the LINEST results. We need to work out the degrees of freedom for the numerator. This is the total number of columns – 1; that is the number of predictor factors. This works out to be 2 for our current situation.

The final *p*-value shows us that the result (0.085) is not statistically significant; there is an 8% chance that this result could have occurred by random chance.

The program does not calculate the beta coefficients for us, which is a shame but simple to do for ourselves. First of all we need to determine the standard deviation for each column. We can use the built-in formula STDEV for this. Now we use the standard deviations and the value of the regression coefficients to determine the beta coefficients using the formula shown in Figure 163.

$$b' = b\frac{s_x}{s_y}$$

Figure 163. Calculating a beta coefficient

In the formula *b* is the regular coefficient. The two standard deviations (s_x and s_y) relate to the *x* and *y* values, i.e. the predictor and response variables. In Figure 164 we can see the summary statistics for each of the columns of data. We can also see the calculations for the separate beta coefficients. We do not really need to work out all these separate summary statistics, as we only require the standard deviation to calculate the beta coefficients.

B23 =B10*D27/B27

	A	B	C	D	E	F	G	H	I
18									
19	F	6.24414							
20	df (num)	2				No. predictor vars (all cols -1)			
21	df (den)	3							
22	P-val	0.08525				FDIST			
23	b'	-0.3463	1.2209			b' = b (s_x/s_y)			
24									
25	#data	6	6	6					
26	mean	32.1667	9.6667	16.833					
27	sd	14.4141	7.3666	6.0139					
28	var	207.767	54.267	36.167					
29	SS	1038.83	271.33	180.83		Variance x (n-1)			

Figure 164. Calculating beta coefficients for a multiple regression

In this case we have found that the individual factors were not statistically significant and that the overall model is also not statistically significant. The correlation coefficients were quite high but we have only six replicates and the dataset is too small for us to be certain that this result is not due to random chance.

When we have a situation where we have several predictor variables our first step should be to determine the correlation coefficients singly (compared to the dependent variable). This will tell us which is the most important factor and our starting point. We could then run a multiple regression by adding the next most important factor to the analysis. If our result is not significant then we should stop. If, however, our result is significant we could continue to add factors and re-run multiple regression. In this way we can build-up a model for each situation.

Some computer programs will perform what is called stepwise analysis; that is they start with a single factor and keep adding the next best option until the best model is attained. The next best option is usually not the one with the next highest correlation, so it requires extra work! We will look at this as an option when we look at how to calculate multiple regression using the R program. We should draw a graph too at some point, ideally before we carry out the analyses. We'll look at graphs in more detail in Section 12.4.

Using the *Analysis ToolPak* for multiple regression

We can use the *Analysis ToolPak* to carry out multiple regression. If we have this add-in installed (see Section 1.9.1), we can run it by selecting the *Data Analysis* button on the *Data* menu on the ribbon.

In the following example (Figure 165) we see three columns of data: the first corresponds to our y-data (i.e. the response variable) and the other two correspond to the predictor variables (i.e. x-data).

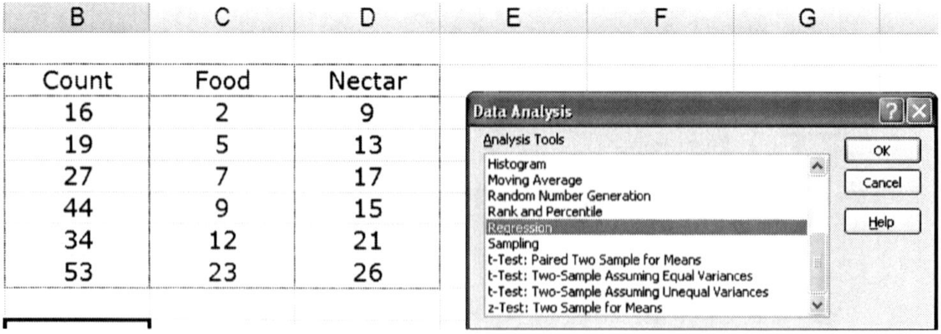

Figure 165. Selecting Regression using the *Data Analysis* menu

We select the *Regression* option from the *Data Analysis* selection window and can then select our data (Figure 166). Note that our predictor variables are in a block, which allows us to select them more simply.

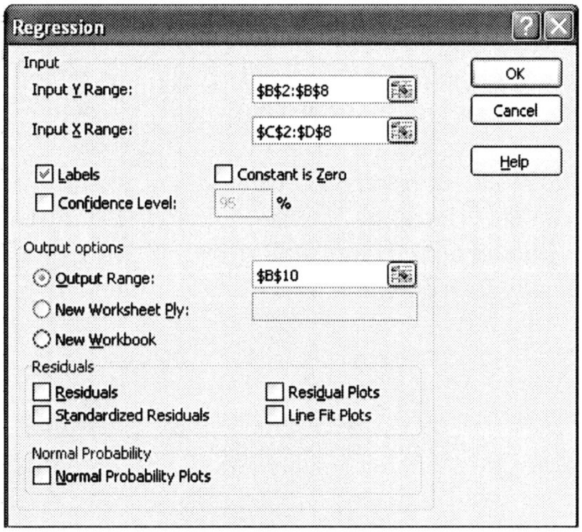

Figure 166. In multiple regression, we select a single response variable but may have several predictor variables

Once we have selected the data and the options we require, we carry out the analysis by clicking the *OK* button. The results appear where we selected the output range to be and are comprised of three main tables (Figure 167).

SUMMARY OUTPUT

Regression Statistics	
Multiple R	0.89794499
R Square	0.8063052
Adjusted R S	0.67717533
Standard Errc	8.18976227
Observations	6

ANOVA

	df	*SS*	*MS*	*F*	*Significance F*
Regression	2	837.616715	418.808358	6.24414169	0.08524658
Residual	3	201.216618	67.072206		
Total	5	1038.83333			

	Coefficients	*Standard Error*	*t Stat*	*P-value*	*Lower 95%*	*Upper 95%*	*Lower 95.0%*	*Upper 95.0%*
Intercept	23.0468094	19.2932009	1.19455603	0.31809451	-38.3527666	84.4463854	-38.3527666	84.4463854
Food	2.38898769	1.60888441	1.48487217	0.23424662	-2.73120056	7.50917595	-2.73120056	7.50917595
Nectar	-0.83012022	1.97077559	-0.42121499	0.7019556	-7.10200773	5.44176728	-7.10200773	5.44176728

Figure 167. The multiple regression results from the *Analysis ToolPak* are quite comprehensive

The top table shows us some overall statistics, including the R^2 value and the number of observations. The middle table shows us the overall significance of our regression model. The final table shows us the coefficients for the various variables as well as their significance. We see for example that neither of the two predictor variables are statistically significant.

At the lower part of the data selection window (Figure 166), we can see that there are some additional options. We can produce some plots and tables of residuals. We can think of the residuals as being the difference between a particular point of data and the idealised regression, which we get when we work out the line of best fit. Ideally we would like our residuals to be normally distributed and we can use the tables of results to check for this. We can also check the *Normal Probability Plot* option to produce a graph that presents the data plotted against the ideal probabilities; essentially this allows us to see if our data are normally distributed. If the points lie more or less on a straight line then we have normality.

11.1.2 Using R to perform multiple regression

R will perform multiple regression quite easily and the basic command to do this is *lm()*.

Multiple regression and the *lm()* command

The *lm()* command uses a model syntax similar to that we met earlier when looking at ANOVA (Section 10.2.4). We place our response factor on one side of the equation and list the predictor variables on the other. We start by getting our data into R and giving it some sensible name. In the following example we see the entire process carried out on the butterfly data that we met earlier:

```
> bf
  count food nectar
1    16    2      9
2    19    5     13
3    27    7     17
4    44    9     15
5    34   12     21
6    53   23     26
> bf.lm = lm(count ~ food + nectar, data = bf)
> summary(bf.lm)

Call:
lm(formula = count ~ food + nectar, data = bf)

Residuals:
      1       2       3       4       5       6
-4.3537 -5.2002  1.3423 11.9041 -0.2821 -3.4104

Coefficients:
            Estimate Std. Error t value Pr(>|t|)
(Intercept)  23.0468    19.2932   1.195    0.318
food          2.3890     1.6089   1.485    0.234
nectar       -0.8301     1.9708  -0.421    0.702

Residual standard error: 8.19 on 3 degrees of freedom
Multiple R-squared: 0.8063,     Adjusted R-squared: 0.6772
F-statistic: 6.244 on 2 and 3 DF,  p-value: 0.08525
```

Here we have our data and these have been read in from a CSV file and called *bf*. We start by creating a new item to hold the result of our regression analysis (the linear model).

```
> bf.lm = lm(count ~ food + nectar, data = bf)
```

We make a new object and call it *bf.lm*. Notice how we simply list the factors additively; we do not use the * sign like we did in the ANOVA. In ANOVA we need to tell the program explicitly to look for interactions; in linear modelling the process of fitting the points to the best-fit automatically takes the other factors into account. We can use the *summary()* command to see the result afterwards.

```
> summary(bf.lm)
```

The main part(s) that we are interested in are the coefficients and their significance and the overall model significance. We see a block of results headed coefficients and can see values for these (our m_n values). We see also the value for *c*, the intercept. The end of each row shows us the significance and we can see here that neither of the factors are statistically significant. R provides us with asterisks as a quick code of significance in a similar fashion to the output of the ANOVA we saw previously (Section 10.2). The final block details the overall significance of our model and shows us to what extent we have explained variability in the response factor. Here we can see that neither of the predictor factors was statistically significant and indeed the overall model, whilst explaining about 80% of the variability, is not statistically significant either.

R does not work out the beta coefficients for us and if we wish to determine what they are, we need to calculate them ourselves. In order to do that we need to work out the standard deviations for each of the columns of data and use the formula we met previously. In words, this is the regression coefficient multiplied by the standard deviation of the predictor variable divided by the standard deviation of the response variable.

First of all we need a way to get at the individual coefficients from the linear model we created in R. The coefficients are stored inside the main result, which we called *bf.lm*. To extract them, we need to add the $ sign and the word *coefficients*. In practice we only need to type *co* after the $ sign like so:

```
> bf.lm$co

(Intercept)         food       nectar
 23.0468094    2.3889877  -0.8301202
```

The *co* part is just enough for R to work out that we want the coefficients; we cannot use just a *c* as there is an item *call*, which shows the model we used. To get a single coefficient we can add another bit at the end in square brackets:

```
> bf.lm$co[2]

    food
2.388988
```

Here we extract the second coefficient, which happens to correspond to the one for the food variable.

Now we need to create the standard deviations and determine the beta coefficients. We could create objects for each of the standard deviations like the following example for the food variable:

```
> food.s = sd(bf$food)
```

On the other hand, we can type the formula to calculate the beta coefficients in one go as shown below:

```
> food.b = bf.lm$co[2] * sd(bf$food) / sd(bf$count)
> food.b
    food
1.220935
> nectar.b = bf.lm$co[3] * sd(bf$nectar) / sd(bf$count)
> nectar.b
    nectar
-0.3463437
```

When we have only a few predictor variables, we can explore the situation by typing in various combinations of factors. Once we get to the best situation we can stop and report the most effective model for our data. As we get more and more variables, it becomes a lot more time consuming to try out every combination of factors. It would be better if we could find a way to find the best starting point and then add the next best factors one at a time. This approach is called stepwise analysis.

Stepwise regression

A good starting point is to select the single predictor variable that has the best correlation with the response factor we are considering. In the following example we see a summary of some data:

```
> summary(mf)
     Length          Speed          Algae           NO3              BOD
 Min.   :13.00   Min.   : 9.0   Min.   :25.0   Min.   :1.050   Min.   : 55.0
 1st Qu.:18.00   1st Qu.:12.0   1st Qu.:40.0   1st Qu.:1.750   1st Qu.:110.0
 Median :20.00   Median :16.0   Median :65.0   Median :1.950   Median :145.0
 Mean   :19.64   Mean   :15.8   Mean   :58.4   Mean   :2.046   Mean   :146.0
 3rd Qu.:21.00   3rd Qu.:20.0   3rd Qu.:75.0   3rd Qu.:2.350   3rd Qu.:180.0
 Max.   :25.00   Max.   :26.0   Max.   :85.0   Max.   :2.950   Max.   :235.0
```

There is a single response factor, the length of a freshwater invertebrate (a species of mayfly). There are four predictor variables, the water speed, percentage cover of algae, nitrate concentration and biological oxygen demand.

We have read in the data from a CSV file and called it *mf* for brevity. Here we use the *summary()* command to see some details. We can also use:

```
names(mf)
[1] "Length" "Speed"   "Algae"   "NO3"    "BOD"
```

The *names()* command shows us the column names (the variables), which is useful as a reminder of what we are dealing with. In our data files it is a good idea to place the response factor first and the predictor factors in subsequent columns. We need to determine the best starting point for our multiple regression. One way is to use the *cor()* command. We met this previously (Section 8.4.1) when looking at correlation between two variables. If our data comprises of several columns of data, R will attempt to correlate all the variables with one another. Below we see the result of a correlation for all the variables in our mf data item:

```
> cor(mf)
           Length       Speed      Algae         NO3        BOD
Length  1.0000000 -0.34322968  0.7650757  0.45476093 -0.8055507
Speed  -0.3432297  1.00000000 -0.1134416  0.02257931  0.1983412
Algae   0.7650757 -0.11344163  1.0000000  0.37706463 -0.8365705
NO3     0.4547609  0.02257931  0.3770646  1.00000000 -0.3751308
BOD    -0.8055507  0.19834122 -0.8365705 -0.37513077  1.0000000
```

The command produces a rectangular matrix, with each pair of variables being represented twice. In this case we have length as the first variable and this is our response factor. We can look down the first column (or along the first row) and see which of the four predictor variables produces the highest correlation coefficient. We can see that BOD has the hi ghest coefficient (–0.806) with Algae (0.765) coming next.

This means that when we start our multiple regression, *BOD* should be the first of the predictor factors included in our model. The *Algae* factor has the second highest correlation coefficient but this may not be the next best factor to add after the *BOD* variable. Let us look at an alternative way to proceed.

Rather than create a linear model with the best predictor factor, we will create a blank model, i.e. one with nothing except an intercept:

```
> mf.lm = lm(Length ~ 1, data = mf)
```

Now we have created a blank model that contains nothing except an intercept. R can now look at the available predictor factors and give us some information about which one will produce the best fit. This command is *add1()* and it works like so:

```
> add1(object, scope)
```

The object part is the linear model we wish to add new terms to. The scope part is the list of data where the potential new terms actually are to be found. We use the command repeatedly to help us select the next best factor to add to our growing model. Below we see the starting point for our *mf* data. We have made our starting blank model and called it *mf.lm*, next we run the *add1()* command:

```
> add1(mf.lm, scope = mf)
Single term additions

Model:
Length ~ 1
       Df Sum of Sq     RSS    AIC
<none>               227.760 57.235
Speed   1    26.832 200.928 56.102
Algae   1   133.317  94.443 37.228
NO3     1    47.102 180.658 53.443
BOD     1   147.796  79.964 33.067
```

We tell R that the object is the *mf.lm* article that we just created (our blank model) and the scope is the *mf* data. We see that we get several columns of results. We are mainly interested in the final column labelled AIC.

Our best factor to add to our regression model is the one with the lowest AIC. This is the Akaike information criterion (named after a Japanese statistician) and is a measure of how good our model is with this factor in place. We do not need to know much about AIC as it is determined from the residual sums of squares (RSS) like so:

$$AIC = 2k + n[\text{Ln}(RSS)]$$

The residuals are worked out from a line (or plane) of best fit. Essentially they are a measure of how far the points lie from the line of best fit. In the AIC equation, k is the number of terms in the regression model and n is the number of replicates (observations).

From our point of view we only need to know that we are looking for a low AIC and this will also go with a low RSS and high sum of squares. The term that has these characteristics will be the best to add to our regression model. In our *mf* case, the freshwater invertebrate, we see that *BOD* has the lowest AIC and this corresponds to the highest correlation that we saw previously. We can now add the *BOD* term to our regression model.

We can save ourselves a bit of typing by using the up arrow to bring up the previous R command that we used to create the blank model. We can edit this by replacing the 1 with the *BOD* term. In the example below we see this new model and the result summarised:

```
> mf.lm = lm(Length ~ BOD, data = mf)
> summary(mf.lm)

Call:
lm(formula = Length ~ BOD, data = mf)

Residuals:
   Min     1Q Median     3Q    Max
-3.453 -1.073  0.307  1.105  3.343

Coefficients:
             Estimate Std. Error t value Pr(>|t|)
(Intercept) 27.697314   1.290822   21.46  < 2e-16 ***
```

```
BOD          -0.055202   0.008467   -6.52 1.18e-06 ***
---
Signif. codes:  0 '***' 0.001 '**' 0.01 '*' 0.05 '.' 0.1 ' ' 1

Residual standard error: 1.865 on 23 degrees of freedom
Multiple R-squared: 0.6489,    Adjusted R-squared: 0.6336
F-statistic: 42.51 on 1 and 23 DF,  p-value: 1.185e-06
```

We can see that the *BOD* is statistically significant and that our overall model is also significant. We now have three predictor factors remaining unused and we need to look to see if any of these will add significantly to our regression. We will run the *add1()* command again but once again we can save typing and recall this using the up arrow. We get a new list of potential terms:

```
> add1(mf.lm, scope = mf)
Single term additions

Model:
Length ~ BOD
       Df Sum of Sq    RSS    AIC
<none>              79.964 33.067
Speed   1    7.9794 71.984 32.439
Algae   1    6.3081 73.656 33.013
NO3     1    6.1703 73.794 33.060
```

The *add1()* command recognises that we have used some of the factors and shows us only those terms that remain to be selected. We see that the speed factor has the lowest AIC; so we should add this to our model. Note that the *Speed* variable was not the one with the next highest correlation (see the correlation matrix we produced earlier). We can use the up arrow to bring up the R command we used to make the last model and edit it to add the *Speed* term. We can then use the up arrow again to bring up the *summary()* command and look at the result.

We see the new regression model below; note that we use the + symbol to add the terms together. This is unlike ANOVA where we usually use the * symbol to take into account the interaction:

```
> mf.lm = lm(Length ~ BOD + Speed, data = mf)
> summary(mf.lm)

Call:
lm(formula = Length ~ BOD + Speed, data = mf)

Residuals:
    Min      1Q  Median      3Q     Max
-3.1700 -0.5450 -0.1598  0.8095  2.9245

Coefficients:
            Estimate Std. Error t value Pr(>|t|)
(Intercept) 29.30393    1.62068  18.081 1.08e-14 ***
BOD         -0.05261    0.00838  -6.278 2.56e-06 ***
```

```
Speed          -0.12566    0.08047  -1.562     0.133
---
Signif. codes:  0 '***' 0.001 '**' 0.01 '*' 0.05 '.' 0.1 ' ' 1

Residual standard error: 1.809 on 22 degrees of freedom
Multiple R-squared: 0.6839,     Adjusted R-squared: 0.6552
F-statistic:  23.8 on 2 and 22 DF,  p-value: 3.143e-06
```

Our new and improved model is still statistically significant but the *speed* terms we added is not in itself significant. We ought to remove it from our final model; we can use the up arrow to call up the previous *lm()* command and delete the "+ Speed" part.

This brings us to the final point. We now have the best regression model from the factors available. We only retain terms that are themselves statistically significant. You might have thought that the *Algae* variable would have a place in our model; it did have a high correlation. If we look back to the correlation matrix we created earlier, we can see that *BOD* and *Algae* are highly correlated with one another (–0.837). So, adding the *Algae* term to the *BOD* already present is not giving us **new** information.

In the following example we see the *BOD* and *Algae* terms used to create a regression model. When we use the *summary()* command to view the result we can see that the *BOD* term is significant but that *Algae* is not:

```
Call:
lm(formula = Length ~ BOD + Algae, data = mf)

Residuals:
    Min      1Q  Median      3Q     Max
-3.1246 -0.9384 -0.2342  1.2049  3.2908

Coefficients:
            Estimate Std. Error t value Pr(>|t|)
(Intercept) 22.34681    4.09862   5.452 1.78e-05 ***
BOD         -0.03779    0.01517  -2.492   0.0207 *
Algae        0.04809    0.03504   1.373   0.1837
---
Signif. codes:  0 '***' 0.001 '**' 0.01 '*' 0.05 '.' 0.1 ' ' 1

Residual standard error: 1.83 on 22 degrees of freedom
Multiple R-squared: 0.6766,     Adjusted R-squared: 0.6472
F-statistic: 23.01 on 2 and 22 DF,  p-value: 4.046e-06
```

So, now we have a way to step along from a bare model to an ever more complex regression. We keep only those terms that are significant and stop when we start to get non-significant terms.

11.2 Curved-linear regression

We first encountered the possibility that a correlation could be based on a trend that was not more or less straight in Section 8.5, where we looked at polynomial and logarithmic equations. Any mathematical equation that could be employed can be pressed into service and used to create a form of correlation. Here we will focus on the polynomial and logarithmic as the methods can be applied easily to other equations.

11.2.1 Polynomial regression

As we saw previously, the equation that describes a polynomial is:

$$y = ax + bx^2 + c$$

This tends to produce an inverted U-shape when plotted as a regular scatter graph. We can see how the equation is analogous to the multiple linear regression formula that we used earlier:

$$y = m_1x_1 + m_2x_2 + c$$

In Table 77 we see some data that follow a polynomial trend. We have measured the abundance of a plant (bluebell, *Hyacinthoides non-scripta*) in woodlands in southern Britain. The light levels were also measured. Rather than an exact unit we have a relative unit for light (a digital camera can work out the shutter speed and aperture and this can be used to create a unit for light).

Table 77. Data on abundance of bluebell in response to light levels. The data form a polynomial correlation

Light levels (arbitrary units)	Abundance (plants per m²)
2	2
4	3
6	8
8	13
10	16
16	23
22	26
28	25
30	20
36	17
48	6

What we see is that at low light levels we have few plants. As the light increases we see more plants. At higher levels of light the abundance decreases once more. The abundance data is our response variable and the light level is the predictor variable; however, we need to express the light in terms of light^2 in addition to light. In R this can be done on-the-fly but in Excel we need to modify the data; we will describe the use of R first.

Polynomial regression in R

We can run our regression analysis in much the same manner to regular linear regression using the *lm()* command, which is a powerful and flexible tool. We start with our data and they will generally be in a spreadsheet, although with a small dataset we could type them into R directly. The first task is to save the original spreadsheet as a CSV file. We may then read the file into R using the *read.csv()* command. We give our data some meaningful (but short) name. Then we create a model for our regression and ensure that we hold it in a container object so that we can see the results afterwards.

The steps in the process are thus:

1. Save original spreadsheet data as CSV file.
2. Read CSV file into R and give it a sensible name.
3. Look at a graph to see what the data look like.
4. Create the regression and give it a memorable name.
5. View the result of the regression.
6. Create final graphical summary.

In the example below we see the process of creating the regression. We have named our data *bbel*, which is pretty short and also reminds us that the data refer to bluebell abundance.

```
> bbel
  abund light
1     2     2
2     3     4
3     8     6
4    13     8
5    16    10
6    23    16
7    26    22
8    25    28
9    20    30
10   17    36
11    6    48
> bbel.lm = lm(abund ~ light + I(light^2), data = bbel)
> summary(bbel.lm)

Call:
lm(formula = abund ~ light + I(light^2), data = bbel)

Residuals:
   Min     1Q Median     3Q    Max
-3.538 -1.748  0.909  1.690  2.357

Coefficients:
             Estimate Std. Error t value Pr(>|t|)
(Intercept) -2.004846   1.735268  -1.155    0.281
light        2.060100   0.187506  10.987 4.19e-06 ***
I(light^2)  -0.040290   0.003893 -10.348 6.57e-06 ***
---
Signif. codes:  0 '***' 0.001 '**' 0.01 '*' 0.05 '.' 0.1 ' ' 1
```

```
Residual standard error: 2.422 on 8 degrees of freedom
Multiple R-squared: 0.9382,    Adjusted R-squared: 0.9227
F-statistic: 60.68 on 2 and 8 DF,  p-value: 1.463e-05
```

Since we called the data *bbel* it seems logical to call our linear model *bbel.lm* but anything would do. The main command is as follows:

```
> bbel.lm = lm(abund ~ light + I(light^2), data = bbel)
```

The abund part is our response variable and light is our predictor. We use the *I()* part to tell R that we want to include light^2 as a factor. We have to use this because ^2 means something special in a formula of this type (it is used in complicated ANOVA to describe complex interactions). The *I()* part insulates the bit in the brackets and whatever is inside is evaluated as a regular mathematical expression.

We can look at the main results with a simple *summary()* command. We interpret these in exactly the same way that we did previously. The first column shows the coefficients (although it is labelled Estimate) and we can see the three values that we need to create our polynomial equation. In this case we get:

$$abund = 2.060\ light - 0.040\ light^2 - 2.005$$

We also see that both light and light^2 are statistically significant. The bottom part shows details for the overall model and we see for example the R^2 value, around 0.93, which shows that this model explains quite a lot of the variability. We also see the overall *p*-value, which is highly significant.

We ought to look at this in a graphical manner. R provides a huge range of graphical commands; we will look at the graphing capabilities in some detail in Section 12.4.1. For the moment we will merely describe the commands we need to type to produce a sensible output.

The main plotting command in R is *plot()* and we can use this in a similar fashion to our *lm()* command to make a scatter graph of abundance against light. We can add titles to each axis from within the *plot()* command or, as here, add them later using the *title()* command.

It is also helpful to add the line of best fit, which is one based on the formula we just created. Here are the lines of text we type in to make our graph (shown below as Figure 168):

```
> plot(abund ~ light, data = bbel, xlab = '', ylab = '')
> title(xlab = 'Light (arbitrary units)')
> title(ylab = 'Bluebell abundance per sq metre')
> lines(spline(bbel$light, fitted(bbel.lm)), col = 'blue')
```

The first line creates a plot. We want to graph the original data so we use the abund and light variables in a simple formula. Essentially this reads, "take abund and plot it against light, and take the data from within the bbel data". We could add the titles but for the moment we make them blank using a pair of quotes (two lots of single quotes will suffice).

We then use the *title()* command to add a label to the x-axis. This is followed by a command to add a title to the y-axis. We could have combined these but then the line would be really long and would not be so easily read (by us).

The final part needs a bit more explaining. It is a combination of two commands. The *lines()* command adds lines to a graph based on some formula or co-ordinates. The lines are generally short sections of straight line. The *spline()* command takes short bits of straight lines and works out how to make them look nicer.

We are looking for x and y co-ordinates for our lines and we start with the light values (which are on the x-axis). We'll have to use the $ sign to tell R where to find the values. Next we will need the y-values and we can get these from the model we just created. The *lm()* command works out what values of y (in this case abund) correspond to each value of x (in this case light); however, it hides the values and does not clutter up the screen with them. We can get to them in a couple of ways but these fitted values are so important that there is a special command that reads them, *fitted()*.

Finally we make the line blue so it stands out better. The final result is shown in Figure 168.

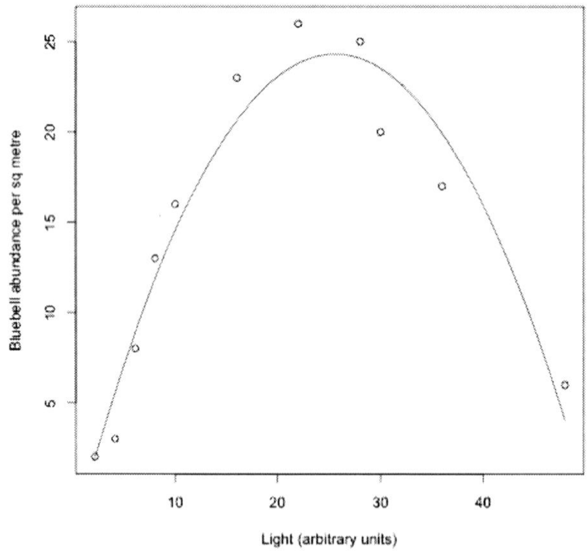

Figure 168. Polynomial graph with line of best fit

This gives us a pretty good impression of the situation and helps us to visualise the data in a way that is simply not possible from a plain text summary.

Polynomial regression in Excel

We can use Excel to determine our regression fairly easily using the LINEST function that we met earlier. First of all we need to add a new column for the x^2 data, then we simply carry out the regression as we did earlier.

We need to arrange our columns of data so that we can carry out the regression. A most sensible arrangement is to have the response variable first, then we have our predictor variable and the extra column where we square these predictor values. This arrangement is shown in Figure 169 along with the results of the LINEST function.

A	B	C	D	E	F	G	H	I	J
	abund	light	light^2			light^2	light	intercept	
	2	2	4		coeff	-0.04029	2.0601	-2.0048	
	3	4	16		SE	0.00389	0.1875	1.7353	
	8	6	36		R Sq	0.9382	2.4218	#N/A	SE
	13	8	64		F	60.68	8	#N/A	df
	16	10	100		SS	711.81	46.92	#N/A	
	23	16	256						
	26	22	484		t value	-10.348	10.987	-1.155	
	25	28	784		P-value	6.57E-06	4.19E-06	0.281	
	20	30	900						
	17	36	1296		F	60.68084659			
	6	48	2304		df(num)	2			
					df(den)	8			
					P-value	1.46E-05			

Figure 169. Using LINEST in Excel to carry out a polynomial regression

Here we have annotated the results, which Excel does not do automatically. Remember that this is an array formula so we need to highlight a block of cells three wide and five high, to hold the output. The result columns are shown backwards! The first one is the second of our response variables, followed by the first response variable. The last column shows the intercept.

We will not repeat what we said regarding the calculations of the t-values, p-values and so on as they are exactly the same as before (Section 11.1.1). We do however want to make a graph of the output. We will cover graphs in more detail in Section 12.4 so what follows is a brief overview.

We will need to create a scatter plot using the x and y variables (we will show a detailed example in Section 12.4.3). This might cause us a slight problem as Excel expects the data to be set out with the x-column first and the y-column second. We had them the other way around to make it easier to run the regression calculations.

The simplest way to get around this is to ensure that you do not highlight any data at all and then create the graph. In some spreadsheets you will be asked for the data series, in others (e.g. Excel 2007) you will not and a blank graph will be created. This is fine as we may then right-click in the graph and select our data. Then we can ensure that we get the x and y variables the correct way around.

Once we have our graph data we can edit the axis titles and so on to get the look we require. We will deal with this sort of thing in Section 12.4 but for now we'll look at how to add the line of best fit. We start by right-clicking a data point, this brings up a menu and we select *Add Trendline...* from the options (Figure 170).

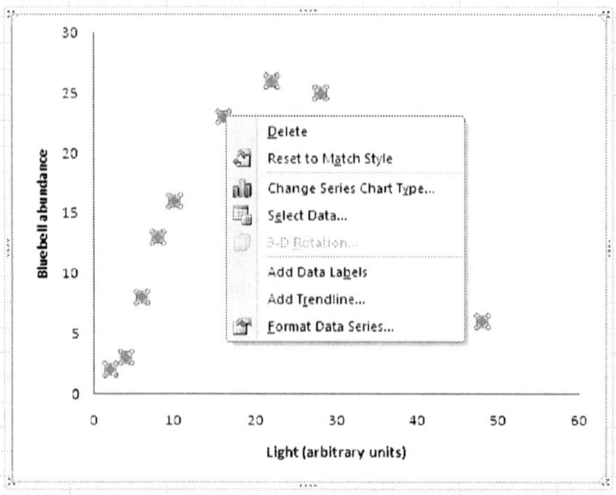

Figure 170. Adding a trend line to an Excel 2007 graph

After saying that we require a trendline, we are presented with further options. You can see from Figure 171 that there are several types we may select. Here we require a polynomial line.

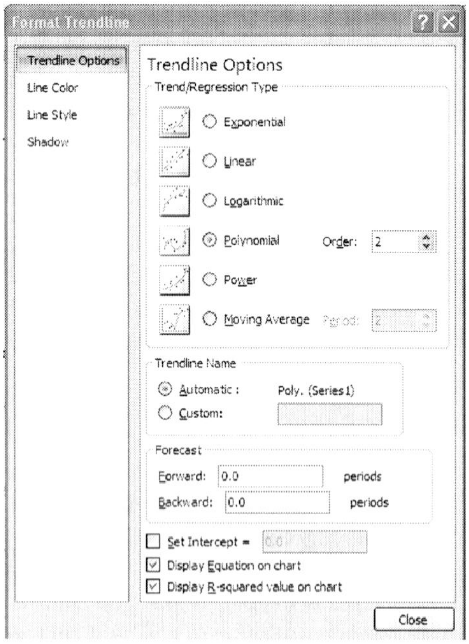

Figure 171. Trendline options in Excel 2007

Once we have selected the polynomial we also tell Excel to use second order; in other words use the x^2. Selecting 3 would include x^3 as well as x^2. This kind of line goes up, down and back up again! In our example we have chosen to display the equation and the R^2 value on the graph (they appear as a text box). After we click "Close", the trendline is added and the resultant graph looks like Figure 172.

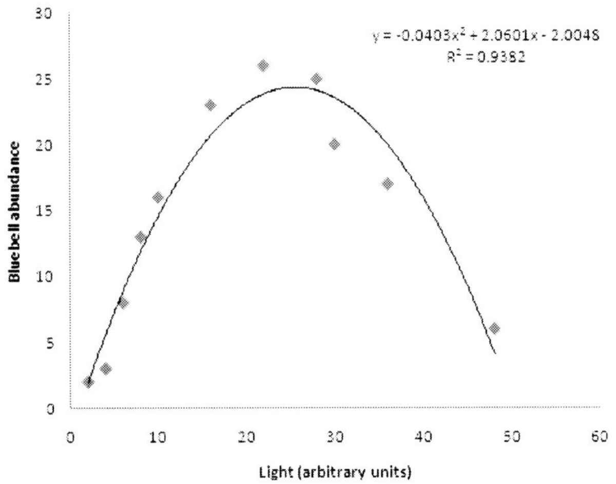

Figure 172. Final data graph with polynomial regression line, equation and R^2 value displayed

Using the trendline with the equation and R^2 value is a useful way to explore your data before running the LINEST function.

Using the *Analysis ToolPak* for polynomial regression

We can use the *Analysis ToolPak* (Section 1.9.1) to carry out polynomial regression. In order to use it we set out our data in a similar fashion to Figure 169, where we have our y-values in the first column followed by the predictor values for light and light^2. We then run *Data Analysis* on the *Data* menu and highlight the appropriate data. Our results (Figure 173) follow the same format as those we saw when looking at regular multiple regression (Section 11.1.1).

SUMMARY OUTPUT

Regression Statistics	
Multiple R	0.96858551
R Square	0.93815789
Adjusted R Square	0.92269736
Standard Error	2.42180963
Observations	11

ANOVA

	df	*SS*	*MS*	*F*	*Significance F*
Regression	2	711.8059776	355.903	60.680847	1.4626E-05
Residual	8	46.92129511	5.865162		
Total	10	758.7272727			

	Coefficients	*Standard Error*	*t Stat*	*P-value*	*Lower 95%*	*Upper 95%*
Intercept	-2.00484592	1.735267535	-1.155353	0.281285	-6.00638003	1.99668819
light	2.06010027	0.187506339	10.98683	4.187E-06	1.62770988	2.49249067
light2	-0.04028952	0.003893326	-10.34835	6.57E-06	-0.04926755	-0.0313115

Figure 173. The results from a polynomial regression using the *Analysis ToolPak*

We see the summary in the top table including the R^2 value, then we see the overall model significance as an ANOVA table. Finally we see the regression table and the coefficients and their significance.

11.2.2 Logarithmic regression

As we saw previously, the equation that produces a logarithmic correlation is:

$$y = a\ \mathrm{Ln}(x) + c$$

This tends to form a curve that begins quite steeply and then flattens off. In Table 78 we see some data that follows this form. We have the growth rate of a plant subjected to different levels of nutrients. Initially the more nutrients we add, the faster the growth rate. At higher concentrations however, there is a flattening off of the increase (in other words we get diminishing returns).

Table 78. Data of plant growth rate in response to nutrient concentration. The data show logarithmic

Nutrient concentration	Growth rate
2	2
4	9
6	11
8	12
10	13
16	14
22	17
28	19
30	17
36	18
48	20

We should draw a basic scatter graph of this before we do anything else, as this will help us to visualise what is happening. Figure 123 back in Section 8.5 shows the situation. We could carry out a simple correlation as we only need the response variable (growth) and a single predictor variable, Ln(nutrients). In R we can create the logarithm on-the-fly but in Excel we need to add a separate column for the Ln(nutrient) data.

Logarithmic regression in R

We can create the logarithmic values as we go along and do not need to alter the original data. To start with we must get our data into R. The easiest is to make a CSV file and use the *read.csv()* command. In the following example we see the data (which we called *pg*) and the process of creating the regression:

```
> pg
   growth nutrient
1       2        2
```

```
2        9        4
3       11        6
4       12        8
5       13       10
6       14       16
7       17       22
8       19       28
9       17       30
10      18       36
11      20       48
> pg.lm = lm(growth ~ log(nutrient), data = pg)
> summary(pg.lm)

Call:
lm(formula = growth ~ log(nutrient), data = pg)

Residuals:
    Min      1Q  Median      3Q     Max
-2.2274 -0.9039  0.5400  0.9344  1.3097

Coefficients:
              Estimate Std. Error t value Pr(>|t|)
(Intercept)     0.6914     1.0596   0.652     0.53
log(nutrient)   5.1014     0.3858  13.223 3.36e-07 ***
---
Signif. codes:  0 '***' 0.001 '**' 0.01 '*' 0.05 '.' 0.1 ' ' 1

Residual standard error: 1.229 on 9 degrees of freedom
Multiple R-squared: 0.951,      Adjusted R-squared: 0.9456
F-statistic: 174.8 on 1 and 9 DF,  p-value: 3.356e-07
```

Because we want to examine the logarithm of the predictor variable we simply put this on the right of the tilde (the ~) in the formula. We do not need to use the *I()* format this time (although it would work fine); the *I()* set-up is only needed if we use regular mathematical operators (like +, –, *, ^ or /) as they could be used as part of the experimental design formula itself.

We can see the results by using the *summary()* command. Here we can see that the relationship can be summarised by:

$$growth = 5.10\ \mathrm{Ln}(nutrient) + 0.69$$

Remember that in R all logarithms are the natural log unless you say otherwise, there is no separate symbol like Ln. In the summary formula above we write Ln to make it clear and avoid ambiguity; it would also be possible to write Log_e.

We should draw a graph of the situation, probably a basic scatter plot, before we ran the regression. We can use similar commands to those we used for the polynomial regression:

```
> plot(growth ~ nutrient, data = pg, xlab = '', ylab = '')
> title(xlab = 'Nutrients', ylab = 'Plant Growth')
> lines(spline(pg$nutrient, fitted(pg.lm)), col = 'blue')
```

The resulting graph is shown in Figure 174. The raw data have been plotted and the line represents the logarithmic curve.

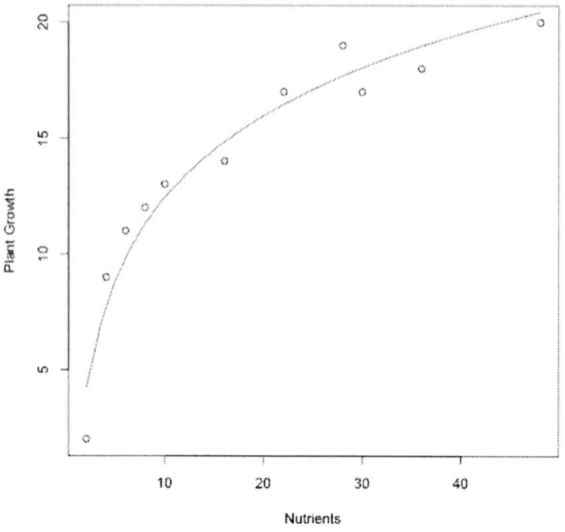

Figure 174. Logarithmic graph with line of best fit

As before we use the *fitted()* command to get the idealised *y*-values from the regression. The *lines()* command draws the main curve but we add the *spline()* part to interpolate and produce a nice curve (rather than a series of short straight lines).

When we have a single variable on the right of the regression formula, we can choose to draw the graph in a slightly different way. Rather than plot growth against nutrient we could plot growth against Log(nutrient). If we do that we see an interesting pattern (Figure 175).

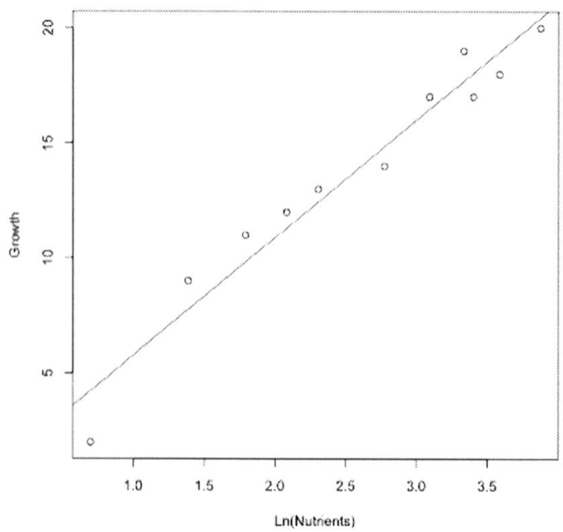

Figure 175. Logarithmic graph with line of best fit. Drawn with *x*-axis on logarithmic scale

The relationship resembles a straight line one. We added a line of best fit to emphasise the pattern. The lines we typed are:

```
> plot(growth ~ log(nutrient), data = pg, xlab =
'Ln(Nutrients)', ylab = 'Growth')
> abline(pg.lm, col= 'blue')
```

The *plot()* command is pretty much the same as we have seen before. This time however, we use *abline()* to add the regression line of best fit. The *abline()* command allows us to draw straight lines and we usually specify the intercept and slope. Here we can refer to the regression directly and R is smart enough to determine that we want these items.

Logarithmic regression in Excel

We can conduct a logarithmic regression easily enough using Excel but as in our polynomial example we need to create a new column to hold the logarithmic data. Once we have done this, we can use the LINEST function to perform the regression. In Figure 176 we see the process.

A	B	C	D	E	F	G	H	I
	growth	nutrient	Ln(nutrient)			Ln(nutrient)	Intercept	
	2	2	0.6931		Coef	5.1014	0.6914	
	9	4	1.3863		SE	0.3858	1.0596	
	11	6	1.7918		R Sq	0.951	1.229	SE
	12	8	2.0794		F	174.84	9	df
	13	10	2.3026		SS	264.04	13.592	SE
	14	16	2.7726					
	17	22	3.0910		T-value	13.223	0.652	
	19	28	3.3322		P-value	3.36E-07	0.530	
	17	30	3.4012		df(num)	1		
	18	36	3.5835		df(den	9		
	20	48	3.8712		F	174.84		
					P-value	3.36E-07		

Figure 176. Using LINEST for logarithmic regression in Excel

In column D we calculate the natural logarithm of the predictor variable (nutrient) using the LN function. We then use the LINEST function, remembering to set the y-values as column B (growth) and the x-values as column D (natural log of nutrient). In Figure 176 we have annotated the resulting output. As in the previous examples we have to calculate the values for the t-statistic and the p-values ourselves (this was covered in Section 11.1.1).

We can produce a scatter graph of the data and add a logarithmic trendline much like we did earlier with the polynomial regression. The resulting graph for these data is shown in Figure 177.

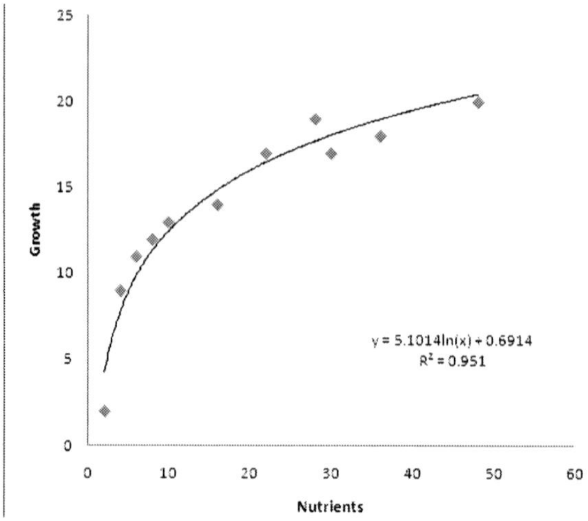

Figure 177. Excel can produce a logarithmic line of best fit

We saw previously that we can also plot the graph using a logarithmic axis. In Excel this is achieved directly from the existing graph. First of all we right-click the *x*-axis and select the Format Axis option. We then are presented with a new menu (Figure 178).

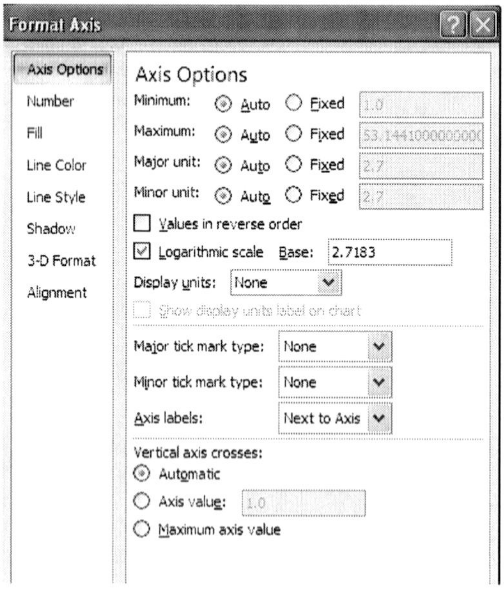

Figure 178. Setting axis options to produce a logarithmic scale

It is simple enough to click the logarithmic scale box. In this instance we have typed the value of 2.7183 into the box that says "Base". This needs a tiny bit of explanation. A regular logarithm is to base 10 so $\log_{10}(100) = 2$ whilst $\log_2(8) = 3$. We can turn our examples around and say $10^2 = 100$ and $2^3 = 8$. The base of the natural logarithm is called *e*. The actual value is approximately 2.7183. Since we used the natural log for our regression we ought to use this base for our axis. We could of course have used a regular base 10 logarithm for our regression!

Once we have set the log axis we get a plot that looks like Figure 179.

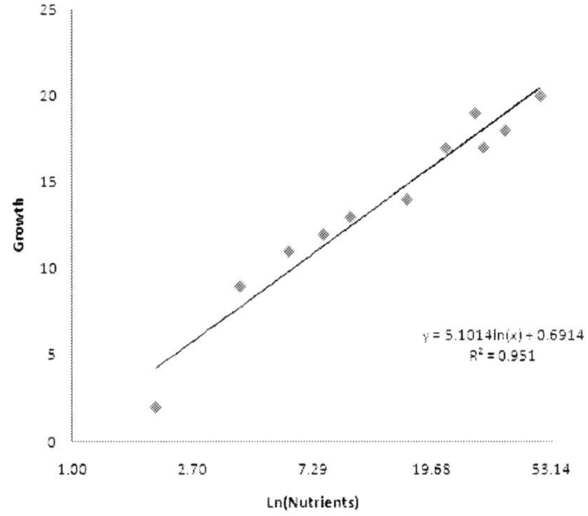

Figure 179. The regression plotted with a logarithmic scale

Here we have made a few alterations, e.g. renamed the *x*-axis title and tidied up the numbers (the spreadsheet has a habit of adding lots of decimal places).

Using the *Analysis ToolPak* for logarithmic regression

We can use the *Analysis ToolPak* (Section 1.9.1) to carry out logarithmic regression. We simply select the log values as our *x*-values in the *Data Analysis*. We follow the same processes we met before when looking at regular multiple regression (Section 11.1.1) and polynomial regression (Section 11.2.1).

11.2.3 Logistic regression

Logistic regression is largely concerned with odds, for example "what are the odds of finding a newt in a pond when there is 50% coverage by macrophytes?" We use logistic regression for cases where we have one variable that is nominal and another that is ordinal (or interval). A nominal variable is one where there is a restricted set of categories. The most extreme case would be presence or absence. Taking our amphibian example, we either find a newt or there are no newts. Our variable can only assume two forms, generally written as 1 or 0. This is also called binomial data.

Let's follow the amphibian example further. We examine a series of ponds and look for the presence of newts (in this case the great-crested newt *Triturus cristatus*). We also measure a series of environmental factors including the coverage of large plants (macrophytes). We could run a differences test (a *t*-test or a *U*-test) and compare the average macrophyte coverage with and without newts. This would be perfectly fine but, as we suggested above, we could use a different analysis and get a more sensitive answer. Using logistic regression, we could determine the chances of finding a newt with the coverage of macrophytes was at a particular level.

Another example of when logistic regression would be useful is where the experimenter sets the levels of the predictor variable, for example, many reptiles have temperature-dependent sex determination. You might incubate reptile eggs at various temperatures and see what sex results from these differing conditions. If you used a *t*-test to compare average temperatures for male and female offspring you would be making an incorrect assumption. Your test would be assuming that temperature depended upon sex of the offspring. Clearly this is the other way around; the sex of the offspring is dependent upon the temperature. It might be possible to take each incubation temperature and determine the sex ratio for each. You might then use a regular linear regression. Unfortunately this also makes an assumption and each of the proportions would be given equal weight in the analysis. If you had equal numbers of eggs in each temperature band then you would get away with this but if not then your analysis is flawed. Logistic regression looks at each observation separately and so automatically takes unequal sample sizes into account.

In regular regression we find the equation that best predicts the value of y for each value of x. In logistic regression we do not measure the values of y directly but take the probability of getting a particular result. Think of it this way: we visit many ponds and find newts in 25% of them. The probability of finding a newt is 0.25 and of course the range of probability goes from 0 to 1. This limits our regression so we look at the odds like so:

$$y/(1-y)$$

For our 0.25 example, we get 0.25/(1 – 0.25) = 0.25/0.75 = 1/3. In other words we have a 3:1 against chance of finding a newt.

We now take the natural logarithm of the odds and use this in our regression. We end up with an equation that looks like this:

$$\mathrm{Ln}[y/(1-y)] = a + bx$$

In this equation a is the intercept and b is the slope or coefficient. Don't forget that y is the probability. Once we have solved the equation we can get back our original proportions, y, using the following:

$$y = e^{a+bx}/(1 + e^{a+bx})$$

Or the following, which is the equivalent:

$$y = 1/(1 + e^{-(a+bx)})$$

The usual way to solve the equation is to use a maximum likelihood method, which alters the various parameters until the most likely solution is found. This is a fairly intensive method and it comes as no surprise to learn that it is not straightforward to do without a computer!

Working out logistic regression using a spreadsheet is not trivial and the iterative altering of the parameters to obtain the best solution requires special additional routines to be loaded into Excel. It may be done using the Solver add-in for Excel but once the equation is completed we still need to examine the significance of the model.

Logistic regression using R is fairly straightforward and the routines to carry it out are built-in to the basic program. We will therefore focus on conducting logistic regression using R.

Logistic regression using R

The command that conducts logistic regression is called *glm()*, which is short for generalised linear modelling. It is an impressively powerful command, which in its basic form will carry out regular linear regression. We will look at two main examples. We will start with the newt example we mentioned previously. Here we have presence and absence data. We will also look at data on relative frequency of alleles in an amphipod crustacean in relation to latitude.

We will begin by looking at the newt example. In Table 79 we see some of the data concerned. A series of ponds were investigated; the researcher spent time looking for signs that newts were resident. If signs were found (e.g. eggs found or actual adults spotted) then the ponds was scored a "1". If, after a certain time, no signs were found then the pond was scored a "0". In other words, we have a 1 or 0 representing the presence or absence of the amphibian. At the same time a variety of habitat factors were recorded.

Table 79. Partial dataset showing presence or absence of great-crested newts and macrophyte cover

Newt pres/abs	Macrophyte % cover
0	20
1	30
1	0
0	5
0	5
1	5
0	0
0	30
0	20
1	10
0	35

Here we will focus on just one factor, the percentage cover or macrophytes. These data are shown in the table alongside the presence/absence data. Overall, 200 ponds were surveyed. In this kind of survey it is good to get approximately equal numbers of presence and absence observations, here there were 72 presence observations.

We could compare the macrophyte cover when newts were present with macrophyte cover when newts were absent; to do that we would first examine the distribution of the present and absent samples. We would find in fact that they are not normally distributed and a *U*-test would be suitable.

To visualise the data we could draw a box–whisker plot like the one shown in Figure 180. We can see fairly clearly that when newts are present the cover of macrophytes is higher.

This is fine but it sort of implies that newts have some effect on macrophyte coverage so it is not really what we should do. A better solution is to carry out a logistic regression, then we would be able to predict the probability of finding a newt for any level of macrophyte coverage.

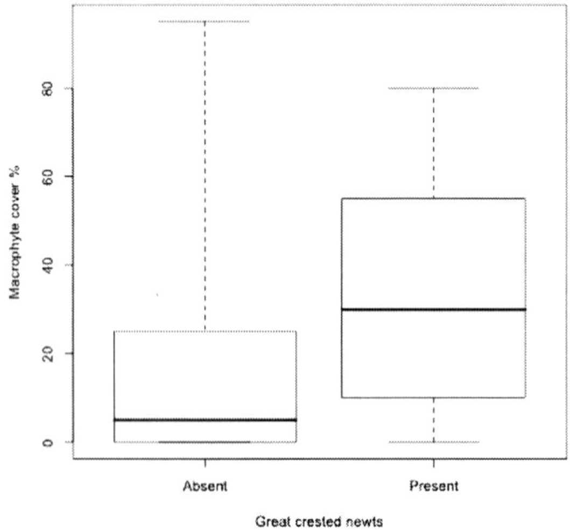

Figure 180. Great-crested newts and macrophyte cover for ponds in Buckinghamshire

First of all we ensure that our data are read into R using the *read.csv()* command. Here we call the data newt as a short but memorable name. Next we use the *glm()* command to execute a regression. We need to tell R that the data are binomial. The command we actually type here looks like this:

```
> newt.glm = glm(gcn ~ macro, data = newt, family = binomial)
```

We always make an object to hold the result, here we make it brief, *newt.glm* in this case seems most appropriate. As with the formulae we have seen previously, we have the response variable first, then the predictor on the right of the tilde. Here our response is called *gcn* and this is either 1 or 0. The predictor is called *macro,* which is a bit shorter than macrophyte. We remind R that these variables are actually contained within the newt data. Finally we add family = binomial at the end to tell R that we have that kind of data and therefore wish to carry out a logistic regression.

To get the actual results we simply use the *summary()* command. The results we obtain for the newt example are shown below. The output looks very similar to the other regressions that we have seen. We see the various parameters and their coefficients, along with the significance:

```
> newt.glm = glm(gcn ~ macro, data = newt, family = binomial)
> summary(newt.glm)

Call:
glm(formula = gcn ~ macro, family = binomial, data = newt)

Deviance Residuals:
    Min       1Q    Median       3Q       Max
-1.6042  -0.8209  -0.7464   1.2860   1.6817

Coefficients:
             Estimate Std. Error z value Pr(>|z|)
(Intercept) -1.135512   0.217387  -5.223 1.76e-07 ***
macro        0.022095   0.005901   3.744 0.000181 ***
---
Signif. codes:  0 '***' 0.001 '**' 0.01 '*' 0.05 '.' 0.1 ' ' 1

(Dispersion parameter for binomial family taken to be 1)

    Null deviance: 261.37  on 199  degrees of freedom
Residual deviance: 246.55  on 198  degrees of freedom
AIC: 250.55

Number of Fisher Scoring iterations: 4
```

Here we see that the cover of macrophytes (*macro*) is significant. What about R^2? This is not quite appropriate in logistic regression; however, we can determine a statistic that is broadly equivalent. This is often called D^2 as it relates to the deviance parameters. If we look at the bottom of the result we can see two lines headed Null deviance and Residual deviance. We can use the ratio of these values to determine D^2 and get something that broadly represents how good our model is.

To do this we need to evaluate the following:

1 – (Residual Deviance/Null deviance)

If we do that for our model, we get:

```
>  1 - (newt.glm$dev / newt.glm$null)
[1] 0.05669383
```

Note how we used the results of the model directly: the null deviance and deviance values are already stored for us (if we know how to get them). Here we obtain a value of 0.0567, which is not that good. It tells us that there are probably a lot of other factors determining the presence or absence of newts.

We should draw some sort of graph to illustrate our results. There would not be a lot of point in plotting presence and absence against macrophyte cover because our response variable is either 0 or 1. What we can do is to show the predictive model where we have the probability of finding a newt against the macrophyte cover.

To make a fairly good plot we need the following four lines of code:

```
> plot(newt$macro, fitted(newt.glm), type = 'n', xlab =
'Macrophyte cover %', ylab = 'Probability')
> lines(spline(newt$macro, fitted(newt.glm)), col =
'blue', lwd = 2)
> abline(v = mean(newt$macro), lty = 2)
> abline(h = mean(fitted(newt.glm)), lty = 2)
```

We start with the basic graphing tool in R, *plot()*. The basic form takes *x* and *y* co-ordinates. We want our *x*-values to be the cover of macrophytes so this is the first part. The *y*-values need to be the probabilities as calculated by our regression. Fortunately R has a special command to get those, *fitted()*. Since we are not plotting real values but predicted model values, we will create a blank graph to begin with! This is what *type* = '*n*' does. It creates everything except draw in the points; we will add a line in a moment. The final couple of parts simply add labels to the axes.

The next command starts with *lines()* and this allows us to add trendlines to plots like we saw earlier. The *spline()* part interpolates and makes the line curved. We will make the line blue and a bit wider than usual as well (*lwd* = 2).

The *abline()* command we met previously and it allows us to add straight lines to existing graphs. Here we use two separate commands to draw two lines: one vertically at a position representing the mean of the *x*-values, the other at the mean probability (going horizontal of course). We'll keep them a standard colour but make them dashed (*lty* = 2).

The final logistic model plot looks like Figure 181. The blue regression line is a subtle S-shape.

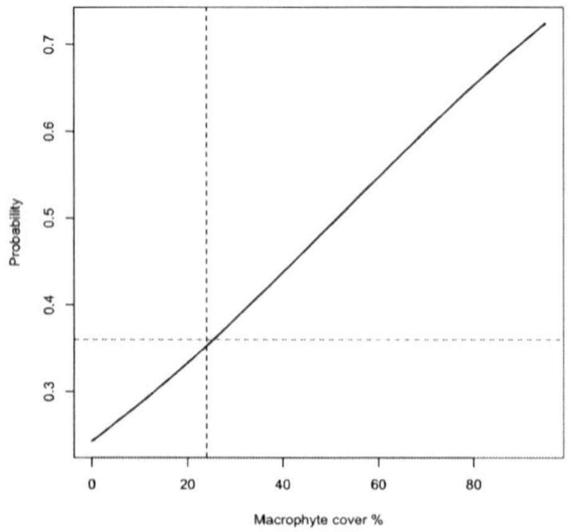

Figure 181. Fitted probability model for newts and macrophyte cover at ponds in Buckinghamshire

From the figure we can estimate the likelihood of finding a newt for any percentage of macrophyte cover.

We will return to our newt example later when we will look at adding extra factors to make a multiple logistic regression. For the next example we will look at data from a study of the Californian beach hopper. We see the data in Table 80 where we have incidence of two versions of an allele in relation to latitude.

Table 80. Allele frequencies at the mannose-6-phosphate isomerase (Mpi) locus in the amphipod crustacean *Megalorchestia californiana*, Californian beach hopper. Data from McDonald, J.H. 1985 (Heredity 54: 359–366)

Latitude	Mpi90	Mpi100	p.Mpi100
48.1	47	139	0.74731
45.2	177	241	0.57656
44	1087	1183	0.52115
43.7	187	175	0.48343
43.5	397	671	0.62828
37.8	40	14	0.25926
36.6	39	17	0.30357
34.3	30	0	0.00000

The first column is self-explanatory; it is the latitude where the sampling took place. The next two columns show the numbers of beach hoppers, which had one version of the allele or the other. There are two versions so we can work out the proportion that had the Mpi100 allele; this value is in the final column. This is done simply enough; for the first row this is 139/(139 + 47) = 0.747.

The temptation would be to attempt a regular linear regression using the proportion of the Mpi100 allele and latitude. This would give us an answer that would seem quite convincing, but which would be misleading. As we mentioned before, the proportions here are the result of quite different sample sizes, varying by three orders of magnitude. The simple linear regression would give equal weight to each proportion and this would be quite wrong. As an exercise you could try running a *lm()* and see how this differs from the logistic regression we will undertake now.

It would be tempting to use the proportion of the Mpi100 allele as the response variable in our analysis. We can make it work but the result is not correct! What we do is to use the two alleles we have counts for to define our success or failure. We could assign either allele to be a success but here it looks like the Mpi100 allele increases in relation to latitude. For some reason the graphs look nicer if the line goes up to the right rather than the other way around! So, we will use the Mpi100 allele as our success. What we are doing is allowing R to see how many observations there were for each latitude and to take sample size into account. We can do this in one of two ways: we can create a separate success/failure matrix or do it in on-the-fly in the *glm()* command.

In the example below we see the success/failure matrix being created as a separate item:

```
> cbh
  latitude Mpi90 Mpi100  p.Mpi100
1     48.1    47    139 0.7473118
2     45.2   177    241 0.5765550
3     44.0  1087   1183 0.5211454
4     43.7   187    175 0.4834254
5     43.5   397    671 0.6282772
6     37.8    40     14 0.2592593
7     36.6    39     17 0.3035714
8     34.3    30      0 0.0000000
> cbh.sf = cbind(cbh$Mpi100, cbh$Mpi90)
> cbh.sf
     [,1] [,2]
[1,]  139   47
[2,]  241  177
[3,] 1183 1087
[4,]  175  187
[5,]  671  397
[6,]   14   40
[7,]   17   39
[8,]    0   30
```

Our allele data have been read via the *read.csv()* command and named *cbh*. We use the *cbind()* command to bind together columns, which we list. Remember that we wanted to focus on the Mpi100 allele so we name that first (it is regarded as a success). We will give the success matrix a name, *cbh.sf*.

Now we can run our logistic regression using the success matrix as our response variable. R will see the two columns and convert them into appropriate probabilities during the analysis. We run this analysis in a similar fashion to the newt example. For our allele data the results look like this:

```
> cbh.glm = glm(cbh.sf ~ latitude, data = cbh, family = binomial)
> summary(cbh.glm)

Call:
glm(formula = cbh.sf ~ latitude, family = binomial, data = cbh)

Deviance Residuals:
    Min       1Q   Median       3Q      Max
-3.4457  -2.3765  -0.8526   0.8579   6.4205

Coefficients:
            Estimate Std. Error z value Pr(>|z|)
(Intercept) -7.64686    0.92487  -8.268   <2e-16 ***
latitude     0.17864    0.02104   8.490   <2e-16 ***
---
```

```
Signif. codes:  0 '***' 0.001 '**' 0.01 '*' 0.05 '.' 0.1 ' ' 1
(Dispersion parameter for binomial family taken to be 1)

    Null deviance: 153.633  on 7  degrees of freedom
Residual deviance:  70.333  on 6  degrees of freedom
AIC: 116.59

Number of Fisher Scoring iterations: 4
```

We can see from the results that the intercept is –7.64686 and that the slope is 0.17864. We could solve the equation for any latitude like so:

$$y = e^{-7.64686 + 0.17864\ Latitude}/(1 + e^{-7.64686+0.17864\ Latitude})$$

We could have achieved the same result by creating the success matrix directly in the *glm()* command like so:

```
> cbh.glm = glm(cbind(cbh$Mpi100,cbh$Mpi90) ~ latitude,
data = cbh, family = binomial)
```

Which method you choose is up to you!

We need to examine the D^2 statistic. We can do this in exactly the same manner as before. If we determine D^2 we find it works out to 0.5422, which is quite a lot better than for our newt model.

We ought to plot some sort of graph. In the previous example our response variable was either 0 or 1 and we plotted the final fitted model only. Here we do have allele ratios so we can make a slightly different graph.

We will start by plotting the probability of encountering the Mpi100 allele against latitude. We already had the *p.Mpi100* variable in our data but if we didn't, it would be simple enough to make it from the existing allele frequencies like so:

```
> p.Mpi100 = cbh$Mpi100 / (cbh$Mpi100 + cbh$Mpi90)
```

Because we have repeated observations at each latitude we can calculate a confidence interval for each of the observed p.Mpi100 values. We met this measure in Section 4.5.2. Usually we use the standard deviation to determine the confidence interval but we cannot calculate standard deviation here so we will use the value of the proportion instead:

$$CI_{95} = 1.96 * p.Mpi100/\sqrt{sum\ of\ alleles}$$

This should give us a fair indication of variability with the larger sample sizes generally having a smaller confidence interval.

The text we need to type to create our graph is as follows:

```
> plot(cbh$latitude,cbh$p.Mpi100, xlab = 'Latitude', ylab
= 'Mp100 proportion', ylim = c(0,0.9))
> lines(spline(cbh$latitude, fitted(cbh.glm)),col='blue',
lwd=2)
> CI = cbh$p.Mpi100/sqrt(cbh$Mpi90+cbh$Mpi100)*1.96
> segments(cbh$latitude, cbh$p.Mpi100+CI, cbh$latitude,
cbh$p.Mpi100-CI)
```

We start with a plot of latitude as the *x*-axis and p.Mpi100 as the *y*-axis. We use the *xlab()* and *ylab()* commands to add appropriate text labels. R will make the axis scales to the most sensible fit; however, it does not know in advance that we wish to add error bars so we need to estimate how much extra room we need to accommodate the tallest value plus bar. The *ylim()* command allows us to specify the minimum and maximum (we need to specify both) values for the *y*-axis. There is a corresponding *xlim()* command but we do not need that here. With a bit of luck and a trial or two we determined that 0.9 was a good maximum.

The *lines()* command in tandem with the *spline()* curve allows us to add the predicted model based on our logistic regression equation. Like before we'll make the line blue a bit wider than standard.

The next part is the calculation of the confidence intervals. This could have been done previously but it cannot be done after the creation of the error bars. These are added with another command, *segments()*. This allows straight lines to be drawn from one *x*, *y* co-ordinate to another *x*, *y* co-ordinate. The general form of the command is:

```
> segments(x1, y1, x2, y2)
```

Both the *x1* and *x2* will be our latitude data. The *y*-data will be the fitted values with the confidence interval added on for the top of the bar and the confidence interval subtracted for the bottom of the bar. It does not matter if we draw them from the top down or the bottom up! The final graph looks like Figure 182.

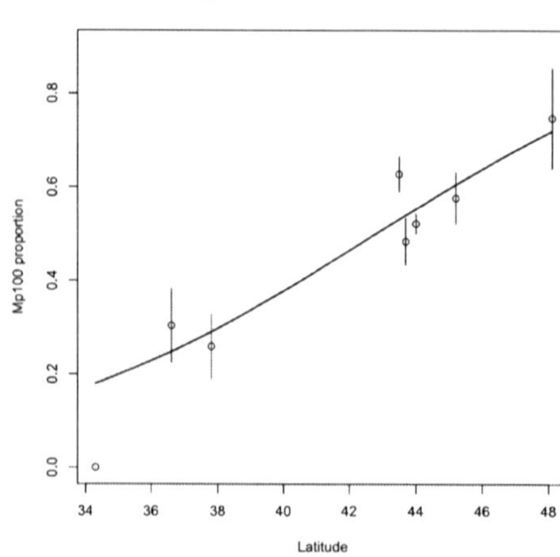

Figure 182. Proportion of Mp100 allele with latitude for the Californian beach hopper *Megalorchestia californiana*. Data from McDonald, J.H. 1985 (Heredity 54: 359–366)

We can see from the regression graph that most of the points lie near the line, which has a slight S-shape. If we had run a regular regression this would have had a large influence on the fit, and our regression line would have been quite different.

Multiple logistic regression

We can carry out logistic regression where we have one response variable and multiple predictor variables much as we can with regular multiple regression. We may simply add as many factors as we like to our regression model! Previously we saw how to undertake a stepwise regression (Section 11.1.2). We can do pretty much the same thing here by looking at our data and reminding ourselves what factors we have available. Then we create a blank model:

```
> newt.glm = glm(gcn ~ 1, data = newt, family = binomial)
```

Here we a 1 on the right of the tilde, which says, "take the *gcn* factor as the response variable, create an intercept only and do not add any terms. Look in the newt data for the variables and base this regression on the binomial".

Now we can look at the factors we have available to see which would be the best to begin building our model. We use the *add1()* command like before.

```
> add1(object, scope)
```

The object is our regression model and the scope is the place where all the factors are to be found. In the following example we see the beginning of this process. We have our blank model and R has looked through our list of possibilities and given us an output similar to before (but with different values of course):

```
> add1(newt.glm, scope = newt)
Single term additions

Model:
gcn ~ 1
            Df Deviance    AIC
<none>          261.37 263.37
area         1   258.88 262.88
dry          1   261.31 265.31
water        1   255.07 259.07
shade        1   248.52 252.52
bird         1   261.03 265.03
fish         1   257.16 261.16
other.ponds  1   252.05 256.05
land         1   260.80 264.80
macro        1   246.55 250.55
```

Here we see the AIC values as well as a column headed Deviance. We are looking for low AIC values. We can see than the macro term has the lowest AIC so this would be a good starting point. This was the factor we used when we examined the simple logistic regression previously. We now use the up arrow to recall the previous R commands until we get to the one where we created our blank model. We can now edit this and replace the 1 with our new term, macro. This will create a simple regression model like we had before. We can check the model using the *summary()* command before running the *add1()* command again. Each time we add the term with the lowest AIC value.

Once we get to a situation where the last term is not a significant one, we must step back by removing it. This then leaves us with the best model available; here the final model is shown below:

```
> newt.glm = glm(gcn ~ macro + other.ponds + shade + fish
+ water, data = newt, family = binomial)
> summary(newt.glm)

Call:
glm(formula = gcn ~ macro + other.ponds + shade + fish +
water, family = binomial, data = newt)

Deviance Residuals:
    Min       1Q   Median       3Q      Max
-1.8668  -0.8542  -0.4975   0.9801   2.3311

Coefficients:
             Estimate Std. Error z value Pr(>|z|)
(Intercept) -4.130592   0.965606  -4.278 1.89e-05 ***
macro        0.016121   0.006518   2.473  0.01339 *
other.ponds  0.238078   0.079631   2.990  0.00279 **
shade       -0.015162   0.005372  -2.822  0.00477 **
fish         0.562670   0.178477   3.153  0.00162 **
water        0.446487   0.224915   1.985  0.04713 *
---
Signif. codes:  0 '***' 0.001 '**' 0.01 '*' 0.05 '.' 0.1 ' ' 1

(Dispersion parameter for binomial family taken to be 1)

    Null deviance: 261.37  on 199  degrees of freedom
Residual deviance: 218.52  on 194  degrees of freedom
AIC: 230.52

Number of Fisher Scoring iterations: 4
```

We can work out the fit of the model like before and obtain a D^2 value. If we calculate that from the values we see something like the following:

```
 > 1-(newt.glm$dev/newt.glm$null)
[1] 0.1639223
```

Here we have a value of 0.162. This is a lot better than we had with the single term model although it is not very good.

We ought to draw a graph of our model but this is something of a problem as we have many factors. What we can do is show how good our overall model is. As part of the regression calculations, R determines the fit of the model and we can access this using the *fitted()* command. The values that result are in order of the values of the samples. We want them to be in ascending order so we simply sort them using *sort()*. To plot our graph we type commands similar to the following:

```
> plot(sort(fitted(newt.glm)), type= 'l',col= 'blue',lwd=
2, xlab= 'Sorted sample number', ylab = 'Probability')
> abline(h = mean(fitted(newt.glm)), lty= 2)
> abline(v= length(newt.glm$fit)/2, lty= 2)
```

The main values are the fitted regression values: we sort them in order and therefore the x-axis is simply the samples sorted in ascending order of probability. We make the graph show us a line rather than points; this is the *type = 'l'* part. We also make the line blue a bit wider than standard. The axes are labelled using the *xlab()* and *ylab()* commands.

It is helpful to have some guides on the plot and we add a horizontal line to correspond to the mean probability. The *abline()* command adds straight lines and we set the line to a dashed line using *lty* = 2. The vertical line corresponds to the mean of the sample number, simply half-way along and because we had 200 samples, this is at 100. We can use the *length()* command to get the number of samples and divide by 2. The final graph looks like Figure 183.

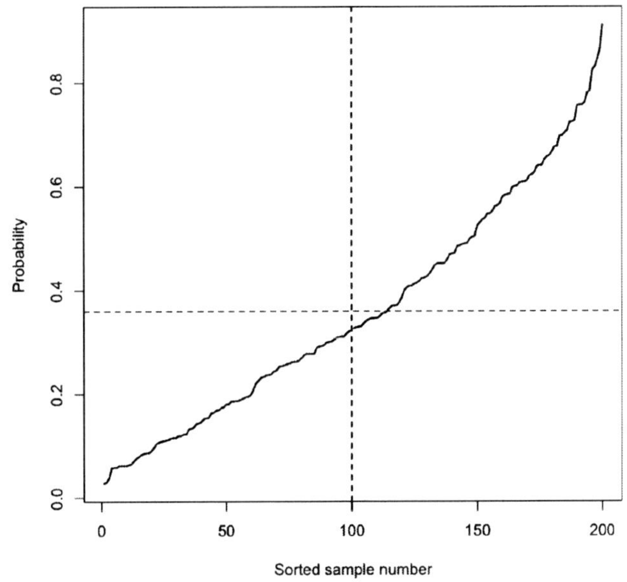

Figure 183. The success of the multiple logistic regression model

In an ideal model we would expect the line to be a broad S-shape with the middle straight section at 45° and passing through the centre of the plot at a mean probability of 0.5. A very flat or very steep line would indicate a poor fit. Here our model is not too bad (although the D^2 value is low); it goes from close to 0 to about 0.9 and this is another good indication. A poor model would not cover the entire range of probabilities.

12. Reporting results

There is undoubtedly some personal satisfaction in planning a project, collecting the data and exploring the results; however, if you do not convey these results to the wider world then the work might as well not exist. The presentation of your work is an important stage in the scientific process. It helps you to move forwards and to determine "what next?" as well as adding to the body of scientific knowledge and helping other researchers in the future. Your work may be presented to the general public, the scientific community, a client or simply your tutor. There are various elements in presenting results and various ways to do the actual presenting.

12.1 Presenting findings

There are three main ways to publish your findings. In general there is no point doing some scientific study if you aren't going to tell anyone what you have done. The three main ways to get your message out into the world are:

- Publishing a written article (e.g. paper, report, internet, newsletter)
- Giving a talk at a conference or meeting
- Presenting a poster at a conference or meeting

This is the final stage in the loop that started with the planning process. You might even have an idea of where your work will be presented right at the planning stage. As you move through the process to recording and analysis you end up with the business of reporting.

12.2 Publishing

Whenever you undertake a piece of scientific work you should be thinking about where the results will be published. There are many options ranging from a scientific journal to a press release. Here are a few ideas:

- Thesis or dissertation
- Assignment
- Report for a client
- Scientific journal
- Local press

- Wildlife magazine
- Book
- Website
- Newsletter

The standards and styles needed for these different places may vary but many of the elements will be the same. If you know where you are going to publish, it will help you to structure your work appropriately. You will need to know about any guidelines for authors and find out what you need to do to get your work published in the place you want.

12.3 Reporting results of statistical analyses

It is important to present your findings in a manner that can be understood by your peers. If you are presenting results to the general public then you may have to alter the presentation to suit your audience, but we still keep to the conventions used by scientists in displaying results.

Usually your raw data are not helpful and big lists of things generally get in the way of understanding. Of course your aim may be to produce a massive site list and if this is the case then carry on and show your big list. Even large lists can be summarised in some way in the main text and the appendix can be used to store the whole list (out of the way).

In the main, we need three pieces of information to summarise numeric data:

- Measure of central tendency (the middle) – e.g. an average
- A measure of dispersion (spread) – e.g. standard deviation or range
- The sample size (how much data was collected) – e.g. sample size or degrees of freedom

Say what your results were in a few words in the text and then refer to a graphic that summarises your data. Graphs are fundamentally important in getting your message over. A good graph can convey a large amount of information in minimal space and with minimal effort (on the part of the reader at least). For example you may have looked at the abundance of early purple orchids on reserves in the southeast UK. You found *mean abundance was 3.5 plants per* m^2 *(*$s = 1.7$*,* $n = 93$*, Figure 2)*. From this, the reader now knows that you had 93 quadrats and can see what sort of average you are giving (mean implies the data are normally distributed), the standard deviation is quite high compared to the mean so we can also get an idea of the variability. The reader also knows to look at Figure 2 where they can see a nice picture (your graph).

Statistical tests need to be reported in a similar fashion, say a few words in the text and then refer to a graph (or possibly a table). There are conventions for reporting statistics results. You need to show several items:

- The calculated statistic (e.g. your value of t or U or r_s)
- The number of items in the sample or degrees of freedom
- The significance (or not)

In some cases the critical value(s) might also be expected but mainly it is only students who are expected to show these. Here is an example: you might have looked at the abundance of Duke of Burgundy butterflies at two sites. You start by summarising the means (assuming normal distribution, use median if not) and the spread as above, then follow up with the stats result and finally the reference to the graph. So *mean abundance was blah blah blah,* $t = 2.6$, $df = 32$, $p < 0.05$ *(Figure 3)*. In this case we show degrees of freedom, but with unequal sample sizes you might prefer to show the sample sizes like so $t_{20,14} = 2.6$, $p < 0.05$. You give the sample sizes as subscripts (in the same order you presented the means). The reader can now see that you have undertaken a *t*-test and can also see that the result is statistically significant.

If you are looking at a correlation then you need to include the sign of the coefficient; this is fundamentally important but is easily overlooked! With correlations of course your data are in pairs so the sample size refers to the number of pairs of data. The same goes for matched pairs tests (e.g. Wilcoxon test). With multiple regression and multi-way ANOVA, you may have several important pieces of statistical information to impart. Showing a table of the ANOVA or regression results is a good idea here but do also say a few words in the text and give the important salient points, e.g. for regression you can give the overall model significance and R^2 value. For ANOVA you may have to select the most appropriate line; this will likely be the main interaction, e.g. *two-way ANOVA showed the interaction between site and management to be highly significant,* $f_{23,2} = 7.34$, $p < 0.01$ *(Table 1).*

If your result is highly significant and p is really quite small you should resist the urge to quote the value directly (e.g. $p = 0.0000024$), it is seen as showing off! Instead you use an appropriate number from the standard range of options (i.e. 0.05 0.02 0.01 0.001).

If your result is **not** significant you may do one of three things:

- Quote the p-value exactly, e.g. $p = 0.37$
- Quote $p > 0.05$
- Just say *not significant* (or *n.s.*)

The option you select depends a bit on how close to being significant the result was and what emphasis you want to place on it. For example saying "$p > 0.05$" can be easily misread by your reader and you may want to gloss over it. Usually the best approach is to be direct, clear and honest, "not significant".

In tables and on figures, you often wish to show significance in a coded manner (to save space and clutter). The acceptable format is to use asterisks (often superscript) in the following manner: $p < 0.05$ *, $p < 0.01$ **, $p < 0.001$ ***. Note that the 2% level ($p < 0.02$) does not have a star rating. Although it is a fairly standard convention you might also include the legend in the caption saying how many stars equate to which level of significance.

12.4 Graphs

A graph should convey as much information as possible in the simplest manner. You need to present a summary (i.e. not raw data) and of course follow the same rules as above (Section 12.3) showing centrality and spread. You can produce very good graphs in Excel but sometimes the default settings need to be tweaked to make a really good figure (see Section 12.4.3).

Good graphs have the following attributes:

- Clear and not too cluttered. If you have too many bars or lines then consider making several graphs rather than one.
- No fancy fonts or effects. A basic 2D graph is perfectly sufficient and 3D bar charts with multi-coloured backgrounds are not essential to delivering the message (and may be misleading or hard to read).
- Labels. Axes need to be clearly labelled and include units where appropriate. Make sure the labels are not too small (or too large).
- A caption. The caption is the most important thing. In general you do not need to use the title facility in Excel (or any other graphical program) and it is better to use a caption generated by your word processor. Captions for figures go below the figure (table captions go above). The caption allows you to cite the figure or table from the text and so has a second, useful function.
- The plot area needs to be a good size in order to display the relevant information (e.g. mean, standard deviation).

Here are some basic graphs. The first one is a bar chart (Figure 184). This sort of graph is useful for showing differences between samples (two or more).

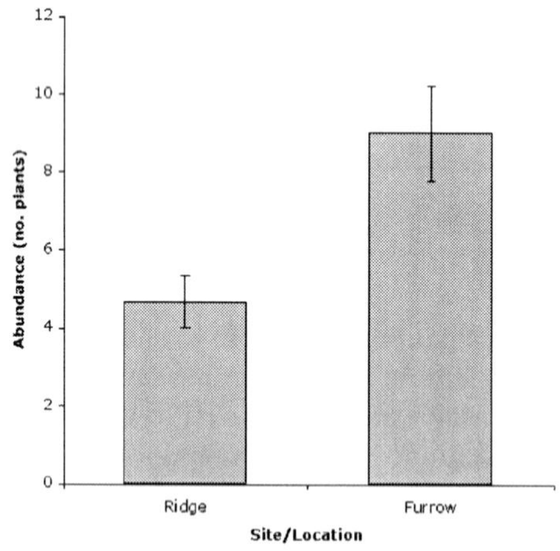

Figure 184. Example of a bar chart. Abundance of *Ranunculus repens* per m^2 in ridges and furrows of a medieval field in Buckinghamshire. Bars show mean and error bars represent standard error

Here our caption says pretty much all we need. The reader can get quite a lot out of this without even reading the text (a most important attribute). The axes are labelled appropriately. The main bars show the mean (so we already know the data are normally distributed) and the error bars are standard error (which is standard deviation divided by the square root of *n*, Section 4.5.1). Error bars can be added reasonably easily in Excel (see Section 12.4.3). We should probably include the units to the abundance axis (no. per 1 m^2).

The next graph (Figure 185) shows the same data but is a box–whisker plot. It conveys a little bit more information.

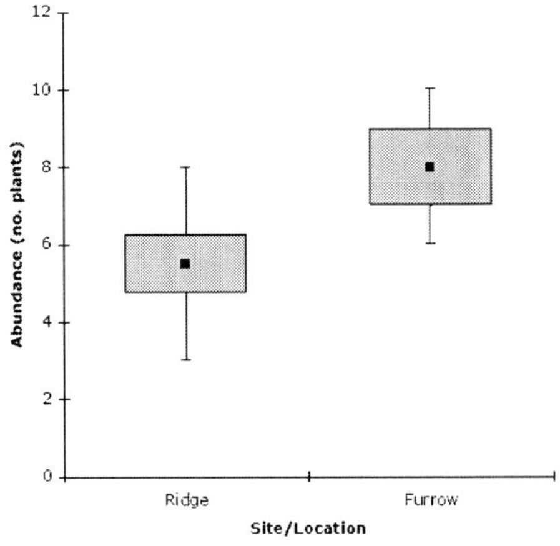

Figure 185. Example of a box–whisker plot. Abundance of *Ranunculus repens* per m^2 in ridges and furrows of a medieval field in Buckinghamshire. Points show mean, boxes show standard error and error bars represent standard deviation

The graph is labelled as before but this time we see two measures of dispersion. This sort of graph is especially useful for showing not normally distributed data, as the point, box–whiskers may be asymmetric. Here they are symmetrical and we can see at a glance that the data are parametric. This graph was produced in Excel. It is not so easy to produce this sort of graph in Excel but (evidently) it **is** possible (see Section 12.4.3).

Sometimes it is useful to convey the statistical result right on the graph and this can be done with a text annotation. The following graph (Figure 186) shows an example of this in a box–whisker plot.

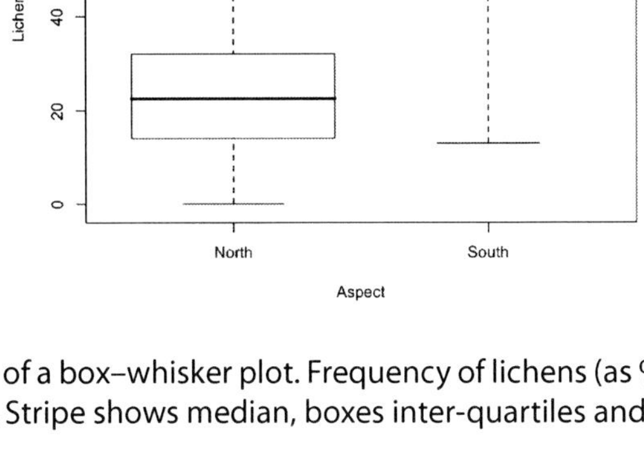

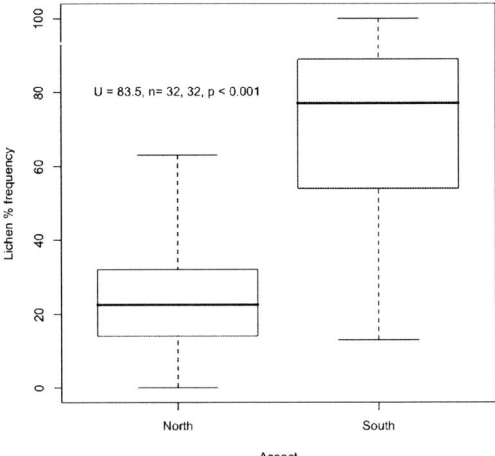

Figure 186. Example of a box–whisker plot. Frequency of lichens (as %) on north and south sides of trees in Kent. Stripe shows median, boxes inter-quartiles and whiskers show range

In this graph we can see immediately that the data are non-parametric, the plots are not symmetrical around the stripes. We could add a few words in the caption to describe the stats results but it is evident from reading the graph that we performed a *U*-test and had 32 replicates in each sample (there is no need to start your caption with *"graph to show..."*, that should be pretty obvious). The result is highly significant. We should perhaps have run a Wilcoxon matched pairs test but since the result is overwhelmingly significant this is not a problem.

When we are showing the result of a correlation, we present a more appropriate graph, a scatter plot. The following graph (Figure 187) shows the relationship between water speed and abundance of stonefly nymphs.

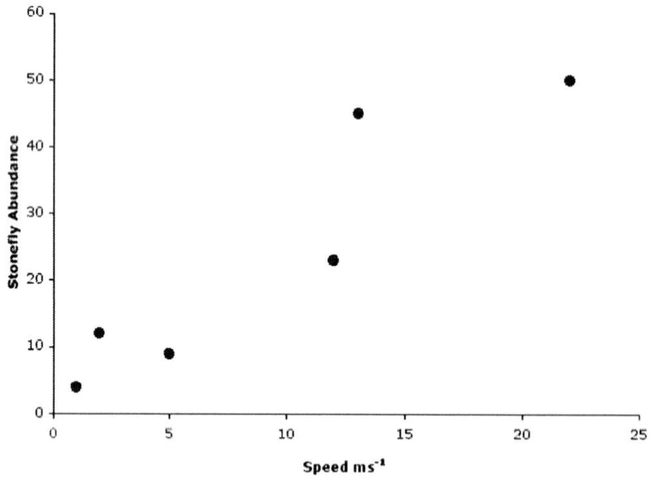

Figure 187. Example of a scatter plot. Abundance of stonefly and water speed in a stream in Shropshire (2 minute kick-samples), $r_s = 0.943$, $n = 6$, $p < 0.05$

Here we see a scatter plot with our dependent variable labelled as abundance and the independent axis as speed (with the units). This time the stats result is given in the caption. We have included *n* even though we can see the points on the graph; some may overlap and if there are a lot it would be hard to count. Let's be kind to the reader and put the information in the caption (and it is good practice to include it every time).

Remember that with basic correlation we do not attempt to produce a line of best fit! We are looking to determine the strength of the relationship and this may not be in the form of a straight line (although this looks pretty good). It is acceptable to include a best-fit line if you performed a Pearson correlation (i.e. a linear regression) because this assumes the relationship is a straight line. Excel can do this easily and you may also display the equation and/or the r^2 value on the graph as text (we did this when looking at curvilinear regression in Section 11.2). We can right-click on a data point and select *Trendline...* from the options.

There are times when you want to display proportional data. Pie charts are one way to do this as the total makes a complete circle. The following (Figure 188) is an example of a pie chart created in Excel.

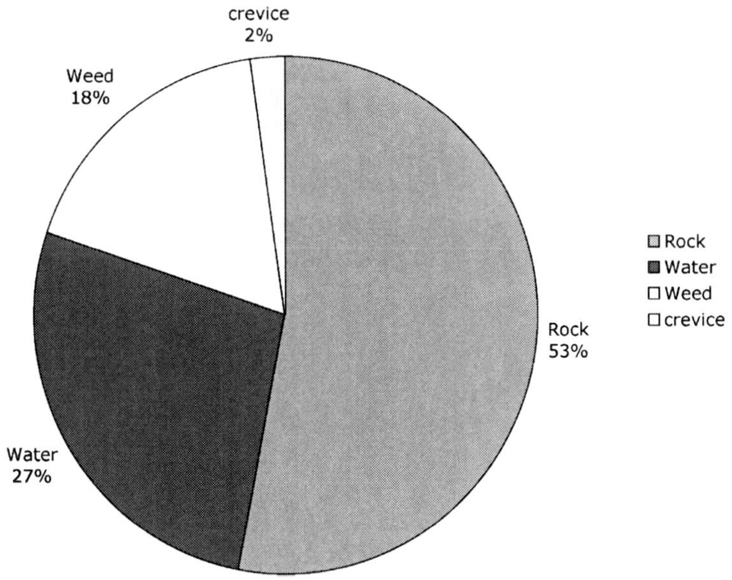

Figure 188. Proportion of topshells found in seashore habitats in Somerset. Here the pie chart (produced by Excel) is labelled with the category and percentage to make reading easier. The legend is not really necessary and could be omitted

The aim with a pie chart is to make it as readable as possible so we annotate the segments with the category as well as the percentage. Remember that this is the aim with all your graphs; to make the information as accessible as possible. You want to convey as much as you can with the minimum effort on the part of the reader; however, avoid clutter. Make every part of the graph useful and if something is not useful, then it is merely clutter so get rid of it! In our example we might consider losing the legend as we have annotated the slices with the habitats.

12.4.1 Introduction to graphing using R

R provides a wealth of commands that are concerned with producing graphs. There are two sorts of command: high-level and low-level plotting commands. The basic commands are high level; you make a new graph of some sort. The low-level commands add all sorts of extra things; you can alter the type of points displayed, change titles and so on.

We will start by presenting an overview of the main graph types before looking at some more in-depth examples in Section 12.5.

The *plot()* command

The *plot()* command is a very general one. Many of the statistical routines create objects that will produce some special sort of graphical output and by default *plot()* is the one they use. In a specific sense *plot()* is the one to use when you need to create a scatter plot, for example a correlation between two variables:

```
> plot(data4, data5)
```

Here we plot *data5* against *data4*. In other words we specify the *x* first and then the *y*.

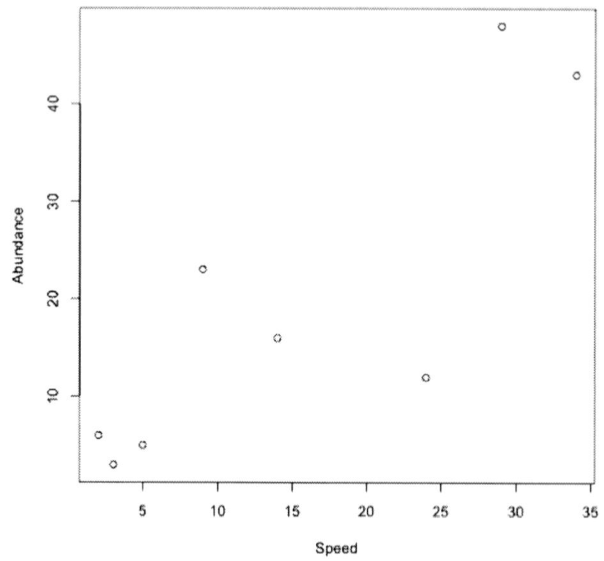

Figure 189. An example of a scatter plot. R labels the axes using the names of the variables as default

The plot produced above (Figure 189) is labelled using the names of the *x* and *y* variables. We can alter the labels and change the plotting characters using additional commands. So, the *plot()* command is the high-level command and produces the basic output. The low-level commands are the ones we use to alter the graph and tweak it to produce exactly what we want to present (more on this to come).

The *boxplot()* command

The *boxplot()* command allows us to summarise data using median, quartiles and range. It is also known as a box–whisker plot. This type of graph is especially useful for non-parametric data.

```
> boxplot(data5, data6)
```

Here we plot two samples, which appear on the *x*-axis in the order they are typed. We might specify three or more samples to display.

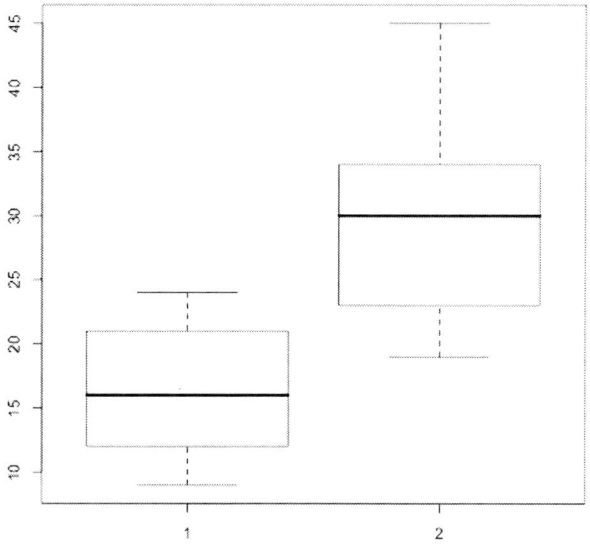

Figure 190. An example of a box–whisker plot. The stripe shows the median value, the box the inter-quartile range and the whiskers show the extremes

The plot above (Figure 190) is very basic and contains no labels. We can add labels and modify the plot using additional commands.

To create names for the samples we include a *names()* part to our *boxplot()* command:

```
> boxplot(data5, data6, names = c("Site a", "Site b"))
```

We can add axis titles using the *title()* command:

```
> title(xlab = 'Site', ylab = 'Counts of beetles')
```

This labels the *x* and *y* axes. If we use separate data items in this way then we have to specify the titles explicitly. Later we will see how to use the model syntax to create a boxplot. This will cause R to pick out the names for the samples and the *y*-axis title automatically. Of course our data have to be in the appropriate format first.

The *barplot()* command

The *barplot()* command is useful as it allows a standard bar chart to be produced.

```
> barplot(data2)
```

If, for example, we had values like the following:

```
[1]  9  2 17 43 18  4  9  3  6 19
```

we would end up with a bar plot like that below:

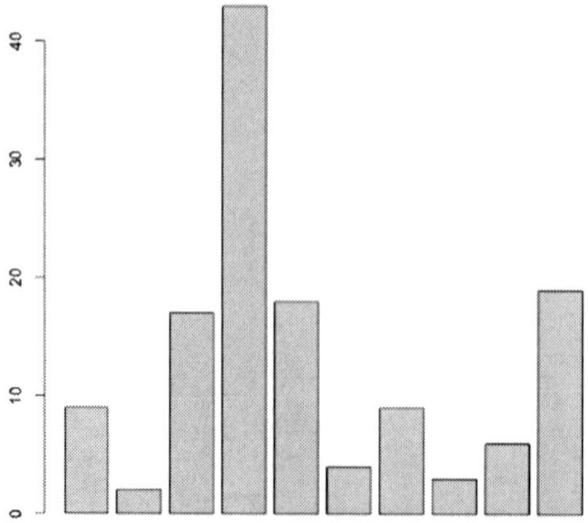

Figure 191. An example of a bar plot

In the simple bar plot above (Figure 191) we get a plain plot with no axis labels. We can add labels using additional commands to build up a more comprehensive plot, as we shall see shortly.

Pie charts, the *pie()* command

The *pie()* command allows a pie chart to be produced, an example might be as follows:

```
> pie(birds.chi$obs[,1])
```

This creates a pie chart of the *birds.chi* object (we ran a Chi-squared test on some bird data earlier) and we select the observed values (note how we do not need the full name). The square brackets tell R to look at the first column (if we had [1,] then R would examine the first row).

The pie chart looks like Figure 192 below:

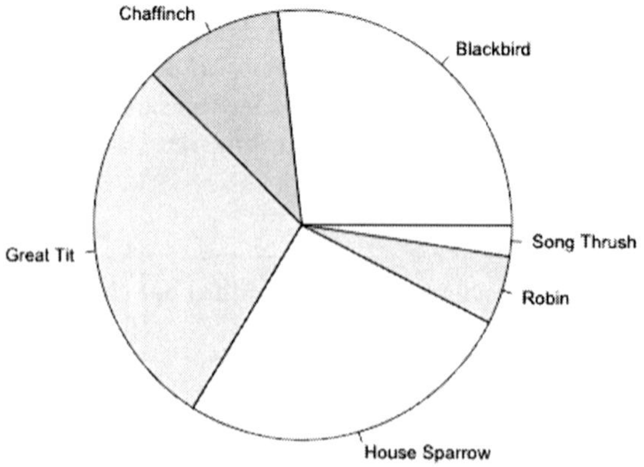

Figure 192. An example of a pie chart. R uses a default set of colours and adds labels by default

The full data look like the following:

```
                Garden Hedgerow Parkland Pasture Woodland
Blackbird           47       10       40       2        2
Chaffinch           19        3        5       0        2
Great Tit           50        0       10       7        0
House Sparrow       46       16        8       4        0
Robin                9        3        0       0        2
Song Thrush          4        0        6       0        0
```

Our pie chart (Figure 192) is drawn using the first column (Garden); to plot any other column we replace the [,1] with the appropriate column number. To plot a row, we switch the positions of the values in the square brackets, e.g. [1,] would plot the first row.

12.4.2 Additional graph commands in R

Many of the low-level plotting commands can be used with several of the high-level ones. To find out about the various plotting commands, use the help files in R; however, here are a few that are useful:

```
> plot(data1, data2, xlab= "X-Axis label")
> plot(data1, data2, ylab= "Y-Axis label")
> plot(data1, data2, main= "Main Title")
```

These may be joined together to form a longer command, e.g.

```
> plot(data1, data2, xlab= "X-label", ylab= "Y-label")
```

There are many other commands that are potentially useful as we will see shortly (we will look at some more examples in Sections 12.5 and 12.6).

The *title()* command

Although you may specify the title from the main *plot()* command, it is also possible to use the *title()* command as a separate entity.

```
> title(main= "A title")
```

This produces a title on an existing plot (which may be any graphics object). We can specify the main title or the individual axes like so:

```
main = "My title"
xlab = "X axis"
ylab = "Y axis"
```

The *text()* command to add to graphs

On occasion you may wish to add text to a graph, for example you might want to add the stats result. The *text()* command does this and is in the form:

```
> text(x, y, "your text")
```

You specify the *x* and *y* co-ordinates and the text you want in quotes.

12.4.3 Introduction to graphing using Excel

Excel can produce a wide variety of graphs. Many are subsets of the standard graph types we have mentioned and have fancy 3D effects that we do not really want; however, Excel is quite capable of producing good quality graphs if we are prepared to tinker with the basic settings a little. The general look and feel of Microsoft Office altered with the advent of their 2007 product. In previous versions you created a graph using the graph wizard and built up the options you required. In the new version, the program produces a finished graph based on the "most likely" options and you need to tinker with it afterwards.

Here we will illustrate using Excel 2007. If you have a previous version of Excel or have Open Office, the main principles will still apply but you may need a slightly different approach.

Preparing the data

The first step is always to prepare your data. If you need to chart mean values from samples then you will need to use standard formulae to create them. Excel expects labels to be adjacent to the data but it is not always necessary, as we shall see. Generally, Excel also expects the data to be in a continuous block of cells. For some graphs, the data are expected to be in a certain order too (for scatter plots for example), however we can usually get around that.

Once we have our data ready we can start to create our graph. We use the *Insert* menu. You may then select the type of chart you want to produce. If you have highlighted a range of cells then Excel will assume that these are the data you require and will create a basic chart accordingly. If you have not highlighted any data Excel, will look around the currently selected cell for data. If it finds anything "next door" it will select that. If it cannot find any data then it will create a blank outline and you may then select the data as you require. This is probably the best option as it provides better control over what data you are plotting.

Bar charts

Bar charts are useful to display values across a range of categories. In order to create one, we must first decide what data we wish to present. In most cases this will be some kind of average and some measure of variability. In Figure 193 we see some dataset out as three columns; each represents a sample.

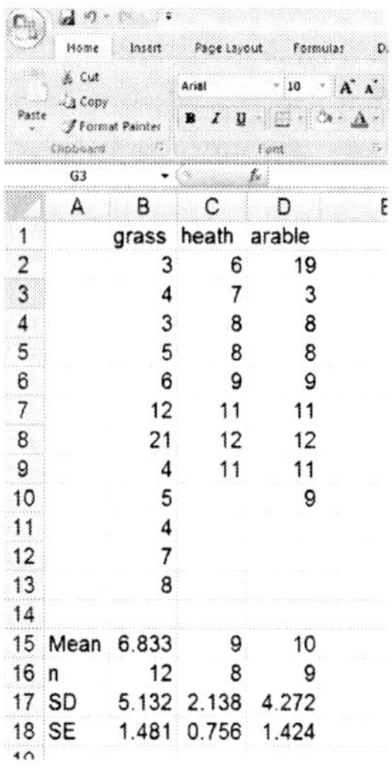

	A	B	C	D
1		grass	heath	arable
2		3	6	19
3		4	7	3
4		3	8	8
5		5	8	8
6		6	9	9
7		12	11	11
8		21	12	12
9		4	11	11
10		5		9
11		4		
12		7		
13		8		
14				
15	Mean	6.833	9	10
16	n	12	8	9
17	SD	5.132	2.138	4.272
18	SE	1.481	0.756	1.424

Figure 193. Summary data prepared and ready to be made into a bar chart using Excel 2007

The samples shown in Figure 193 are counts of butterflies from three different habitats. We must first determine which average we are going to use. Here we will use the mean although the grass sample is not really normally distributed. We use appropriate formulae to work out the mean, number of observations in each sample, standard deviation and standard error (Section 4.5.1). Now we are ready to create our graph.

We can select the graph types we need by going to the *Insert* menu. In previous versions of Excel this brought up the chart wizard; in Excel 2007 we see the various options in the ribbon. For our butterfly data we require a bar chart and select the *Column* option as shown in Figure 194.

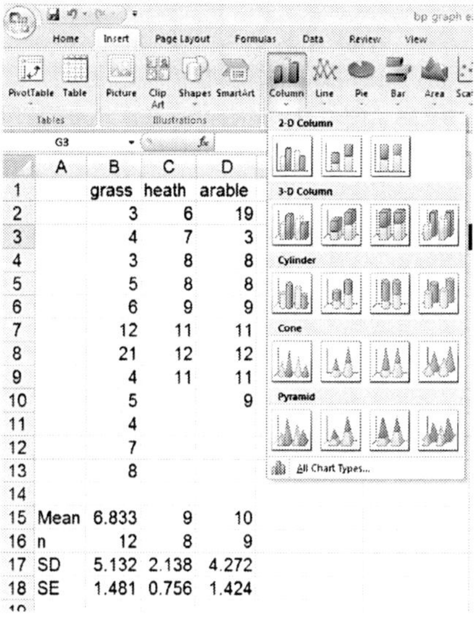

	A	B	C	D
1		grass	heath	arable
2		3	6	19
3		4	7	3
4		3	8	8
5		5	8	8
6		6	9	9
7		12	11	11
8		21	12	12
9		4	11	11
10		5		9
11		4		
12		7		
13		8		
14				
15	Mean	6.833	9	10
16	n	12	8	9
17	SD	5.132	2.138	4.272
18	SE	1.481	0.756	1.424

Figure 194. Selecting a chart type using the *Insert* menu in Excel 2007

The column chart presents us with a variety of options. Here we have a single item in each category (sample), the mean. Notice that Excel shows us a range of fancy layouts that we might select. Fancy 3D graphs rarely display the information as clearly as a basic plot so resist the temptation. You will notice that there is an option to produce a bar chart. The icon shows what this looks like; simply a column chart on its side.

If we had highlighted the mean data, then the graph would be created with standard layout for us; however, I suggest you start by ensuring the cursor is well away from any data. This will generate a blank graph (Figure 195). We can add the data we want to the graph and format it as we go along.

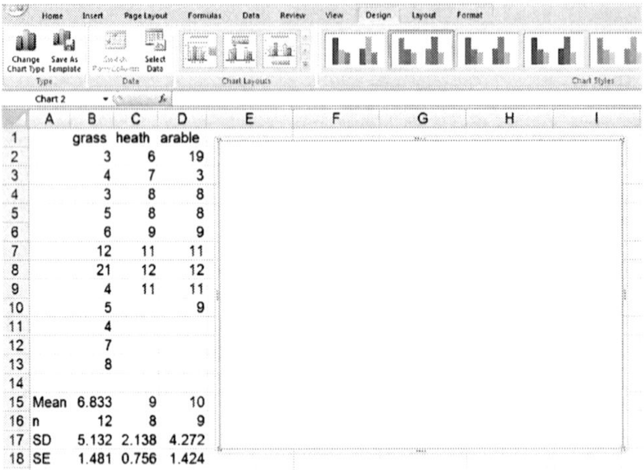

Figure 195. If no data are selected (or adjacent to the active cell) then a blank graph is created. Data may be added subsequently

Once we have our blank graph, the ribbon in Excel shows us some options specific to graphs. The one we will start with is labelled *Select Data*. Once we click this button, a menu window opens that allows us to add data series to the chart (Figure 196).

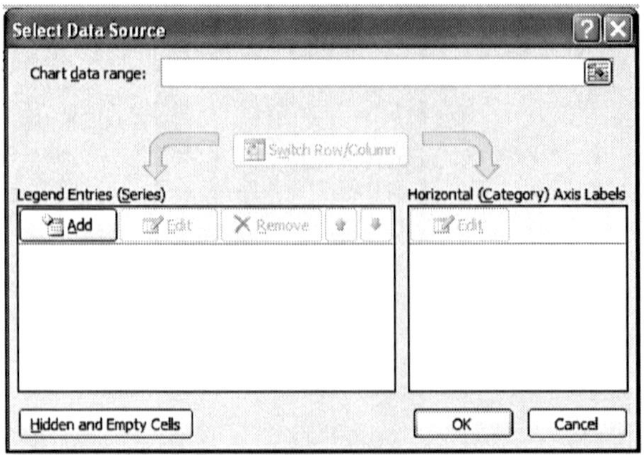

Figure 196. The *Select Data* button allows us to add data items to a blank graph in Excel

The first step to adding data is to add a data series. Once we click this button, we see a new menu allowing us to select the range of cells where the data lie (Figure 197). We can now select the range of values, in the "Series values" box.

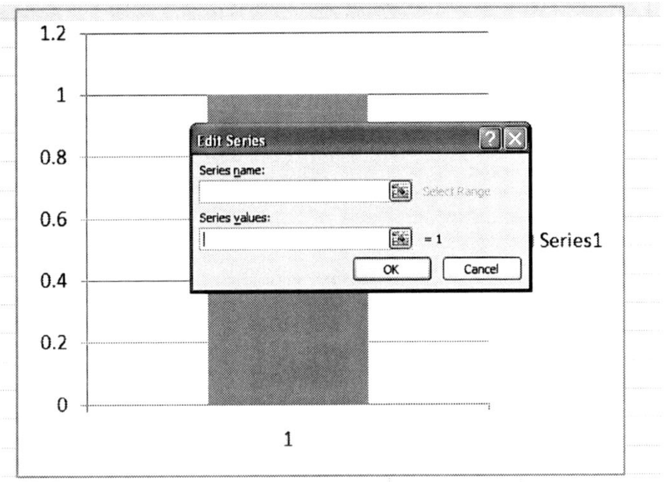

Figure 197. Adding a data series: we can select the range of cells for the values and the corresponding labels

Once the values are entered, we select a cell to act as the name for the series of data. We'll select the cell that says "Mean" in this case (Figure 198).

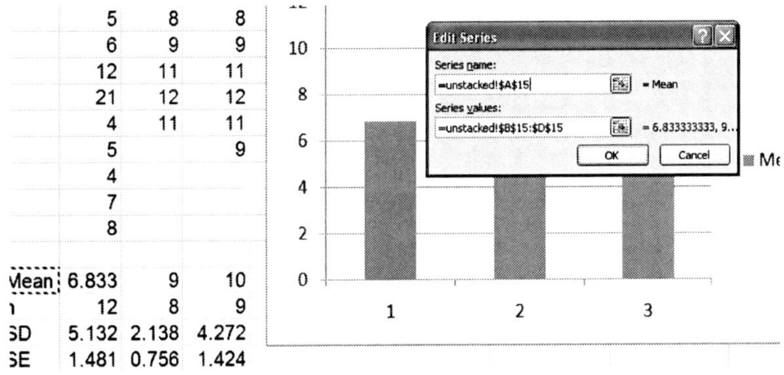

Figure 198. The data series needs a name. This permits several bars to be presented at the same time. In this case we have a single series representing the mean values

The data label relates only to the name of the data series, not to the categories. We may present more than one set of bars, perhaps the mean for the current year and the mean for a previous survey. Each series (set of means) would have its own name so you can tell which was which.

Once the series have been entered, we click OK and return to the previous menu window (Figure 199).

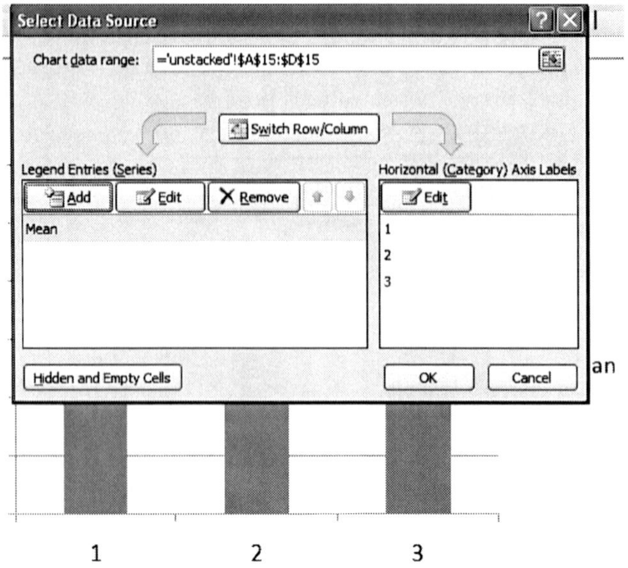

Figure 199. A data series is entered into a blank graph. Now we may add more series or edit the category labels

We can now add more data series or move on to edit the category labels (Figure 200). We simply click the *Edit* button on the right side where it says "Horizontal (Category) Axis Labels". Excel tries to put something in here and we see that it has filled out plain numbers 1–3 as default. We want to replace these with the names of the samples.

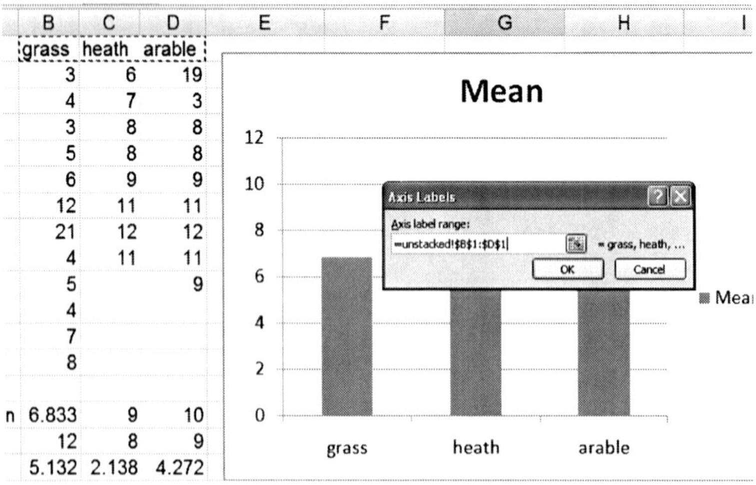

Figure 200. Once the data series are entered, we can select the data labels for the various categories (samples)

Now at last we can highlight the sample names as descriptive of the categories we wish displayed. Now we have a functional graph (Figure 201), but this is not quite complete.

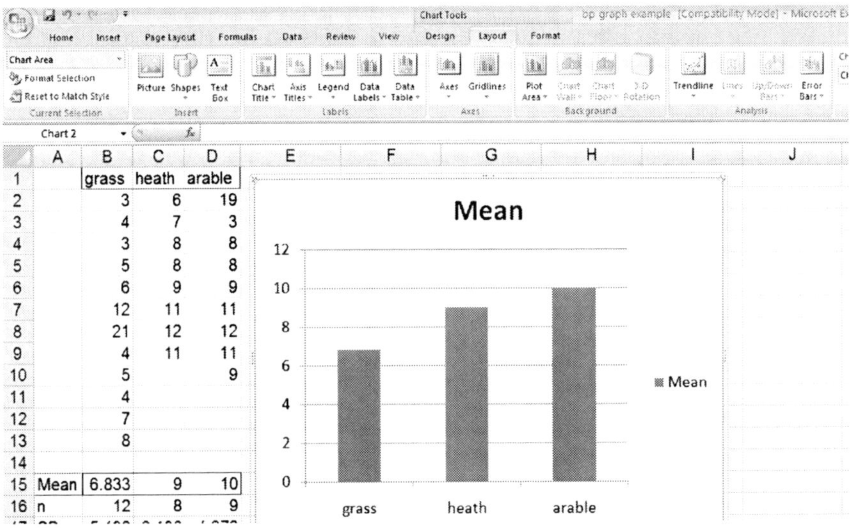

Figure 201. The default bar chart needs some tweaking before it is ready for final display

We need to use the *Layout* menu from the "Chart Tools" section. This gives us one way to edit, add and remove various features. For example, we do not require a legend on this graph, as there is only one data series; it is therefore obvious what the bars are. We can delete the legend by selecting it directly from the graph and pressing the delete key. We can also use the *Legend* icon on the ribbon. Some elements are also editable by selecting them directly from the graph and right-clicking (Figure 202).

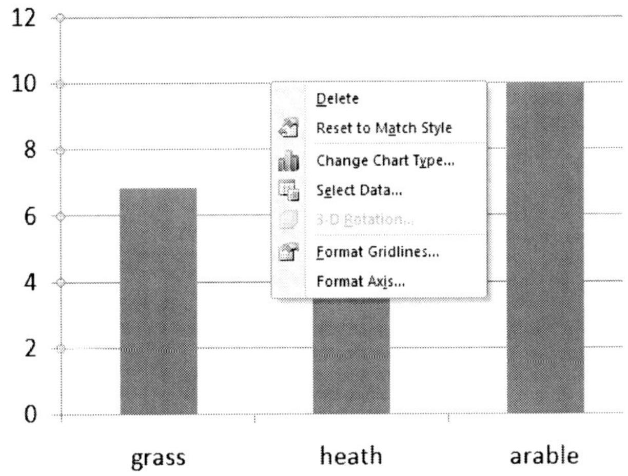

Figure 202. Some graph elements may be edited by right-clicking

For example, we may not wish the gridlines to be quite so bold and can right-click them to bring up a new menu. Here we select "Format Gridlines" and change the appearance of the lines (perhaps to a fainter dashed appearance).

We will also wish to add titles to the axes, and buttons on the ribbon allow us to select them; once activated we can edit the title text. We do not really need a graph title because we can add a caption using our word processor (in fact it is preferable); however, there may be occasions when an internal title is required and it can be added from the ribbon buttons. Our graph now looks something like Figure 203.

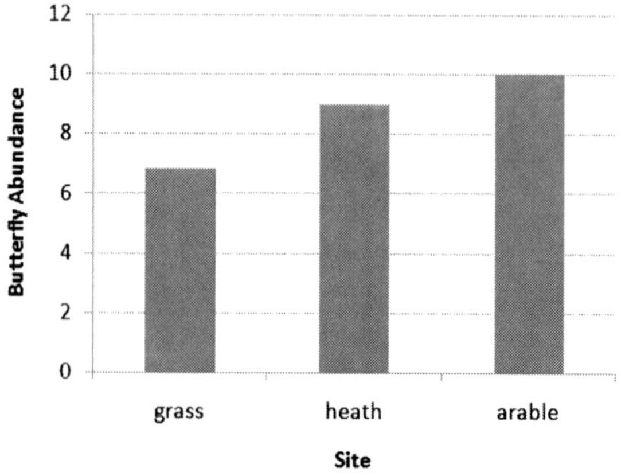

Figure 203. The final bar chart containing axis and category labels

The graph is now tolerably complete but we have data about the variability and it would be nice to add these data to our graph in some way using error bars.

Error bars on bar charts

In Figure 193 we used simple functions to create means and two measures of variability, standard deviation and standard error. We can use either of these to present as error bars, so adding variability information to our plot. If we click on the chart we can select the *Layout* menu item on the ribbon. One of the buttons is labelled *Error Bars* (Figure 204).

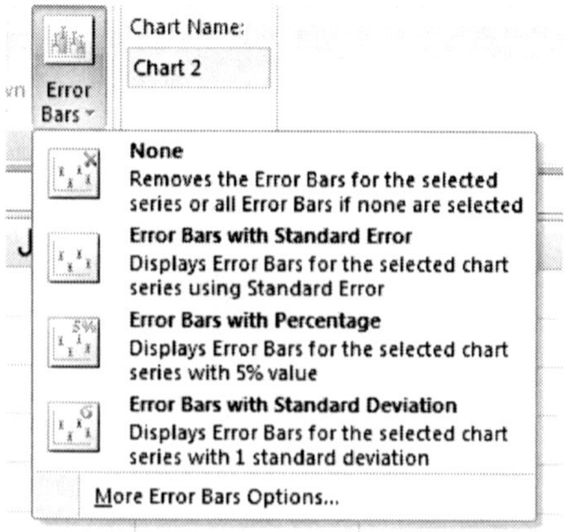

Figure 204. Error bars may be formatted and selected using the *Error Bars* button on the *Layout* menu in the *Chart Tools* ribbon

There are several options but we must select the "More error bars options" item. Although there are options for standard deviation and standard error, they do not work in this case because we are plotting a single value in each category. Excel cannot compute the correct values so we must show it where the correct values lie. We can alter the style of bar using the buttons available and we use the "Specify Value" button to select the range of cells where the values lie (Figure 205).

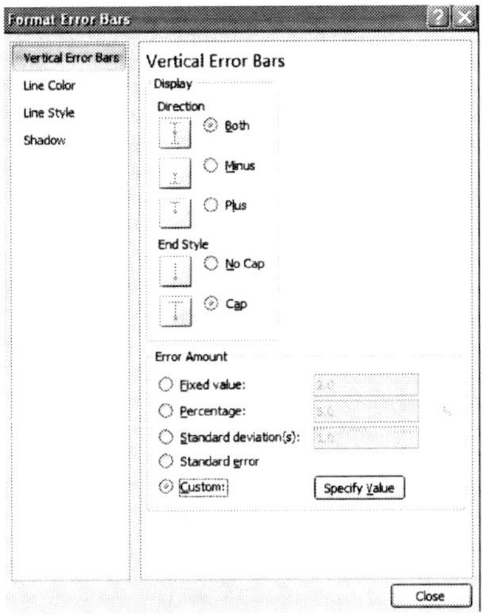

Figure 205. In most cases, we must specify where the values that form the error ranges lie

In this case, we will use the standard error (we could also use standard deviation). The values were calculated previously and we can use the same values for the up and down bars (Figure 206).

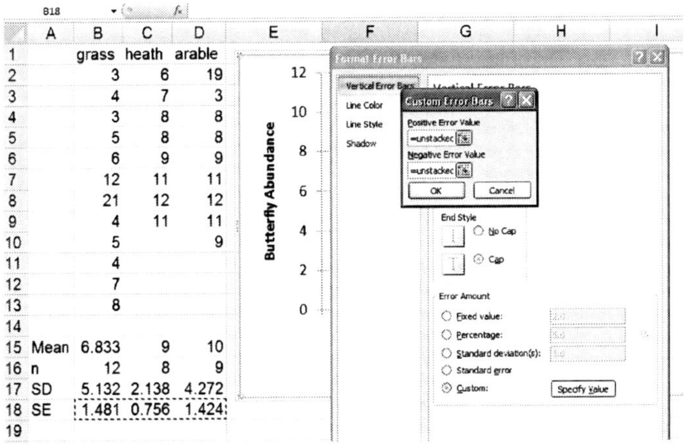

Figure 206. Selecting the range of values that will form the error bars

The error bars show us a measure of variability; generally the larger the sample size, the smaller the standard error (Section 4.5.1). We have no method of showing graphically the sample sizes without cluttering the graph. We could add text values to the bars but that might be confusing; better to add the information in the caption. The final graph looks like Figure 207.

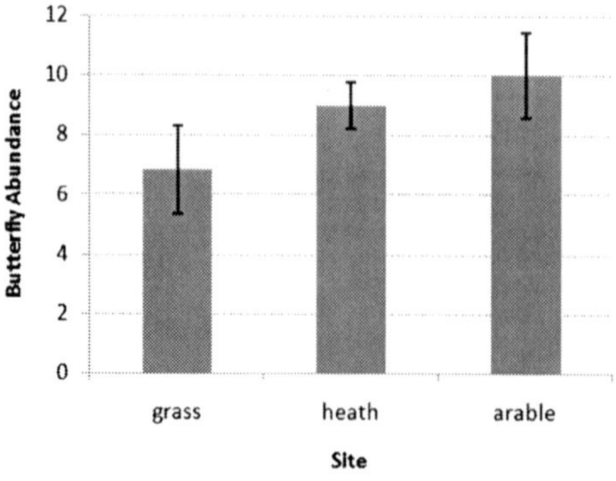

Figure 207. Total butterfly abundance at three habitats in Shropshire. Bars show mean values, error bars show standard error (n = 12, 8, 9)

Our final graph displays all the important information in the simplest manner. The caption says everything else that we need and cannot easily fit onto the graph and we can add more information (e.g. stats results) as needed.

Box plots

Box–whisker plots are probably the most useful type of graph we can produce as they display a great deal of information in a compact manner; however, it is not a standard option in Excel. We can produce box–whisker plots but we need to coerce a different sort of graph into displaying the information we require. We use the "Stock Graphs" option to do this (we find these under the "Other Charts" section). In Figure 208 we see some data. We have three samples; each relates to the abundance of butterflies determined by transect walks in the different habitats.

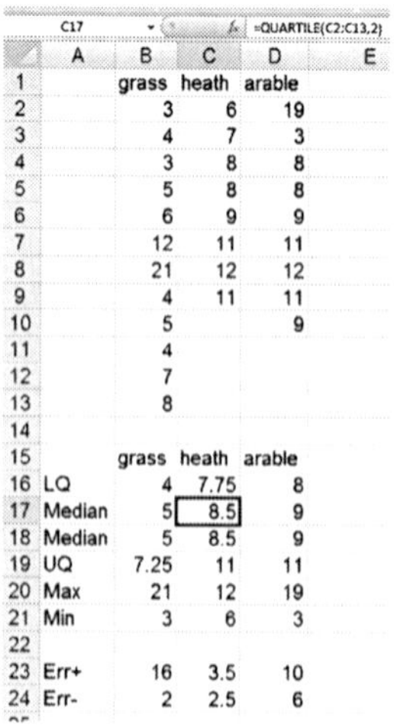
C17 =QUARTILE(C2:C13,2)

	A	B	C	D	E
1		grass	heath	arable	
2		3	6	19	
3		4	7	3	
4		3	8	8	
5		5	8	8	
6		6	9	9	
7		12	11	11	
8		21	12	12	
9		4	11	11	
10		5		9	
11		4			
12		7			
13		8			
14					
15		grass	heath	arable	
16	LQ	4	7.75	8	
17	Median	5	8.5	9	
18	Median	5	8.5	9	
19	UQ	7.25	11	11	
20	Max	21	12	19	
21	Min	3	6	3	
22					
23	Err+	16	3.5	10	
24	Err-	2	2.5	6	

Figure 208. Before creating a box–whisker plot, the summary data need to be prepared. Here we create median, quartiles, max. and min. values

Our final graph will need five values for each sample. We require the median and both inter-quartiles. These will form the box part of the graph. The whiskers will be generated using error bars. We need the maximum and minimum values and from them can determine how far above and below the median the whiskers should extend. We can use the QUARTILE function to determine the median and inter-quartile values as well as the maximum and minimum values (Section 4.1.8). Recall that the 0th quartile is the minimum value; the 2nd quartile is the median and so on. The error bars are simply calculated as the difference between the median and the maximum or minimum value; in other words, the up error bar has a size that will run from the median to the maximum (max – median); the down error bar has a size that will run from the median to the minimum (median – min).

Notice how we have created a duplicate row for the median. This is **not** a mistake. The type of graph we will use requires four values; the inter-quartiles are the outer values (making the box) and the two values in the middle are the same (median) value, as they will become the median point. Unlike other graph types we really do need to select the data before we start because Excel will simply refuse to create a blank graph! In Figure 209 we can see that we have highlighted the first four rows of data. We also created labels for the samples by copying the column headers. This makes it easier to follow the steps we need to create the final plot.

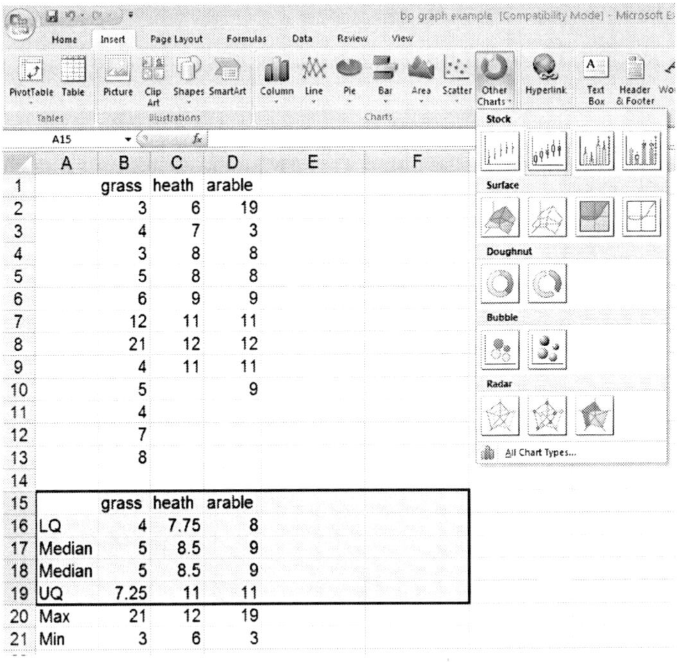

Figure 209. A stock chart is pressed into service to create a box–whisker plot. Note that we must select at least 5 columns of data, even if they are empty, for the graph to be created

We will use one of the stock graphs and these are found in the *Insert* menu under the *Other Charts* button. The top row shows Stock charts, we require the second option. We may create charts using any number of samples (each in its own column) but we need to highlight **at least five** columns (if this seems silly, it is, but it does **not** work with fewer columns). In our example (Figure 209) we see that we have highlighted the three samples as well as two empty columns. We include the cells containing the labels for the rows and columns too. We highlight four actual rows of data, the two median rows as well as their flanking inter-quartile values. Once we have selected the graph, it will be created (Figure 210). It is not complete of course and will need some editing.

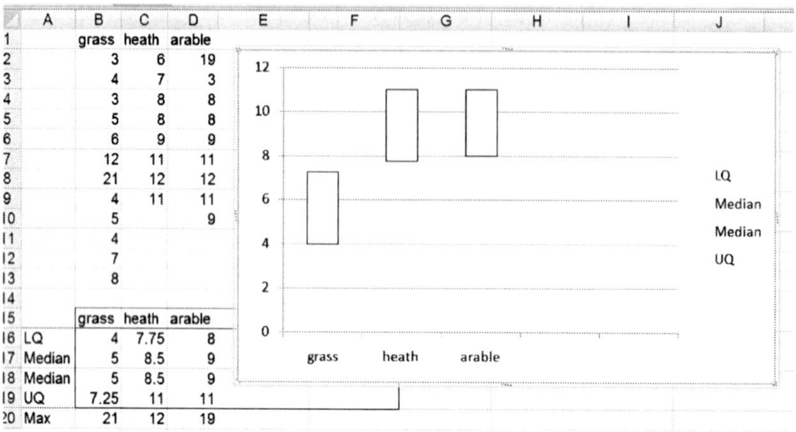

Figure 210. The initial stock graph will require some further editing

We can see that the samples are labelled for us and the legend shows the four values that are plotted. The stock graph was designed to show stock values and this chart shows the opening value, the highest value the stock achieved, the lowest value and the final closing value. The opening and closing values form the box. The high and low values usually form a stem. We set them to be the same so we can use them as the median value. At present they are not displayed, but we will fix that shortly. We will use error bars to add whiskers later. For the moment we need to remove the blank columns and so we use the *Select Data* button to bring up a dialog box (Figure 211).

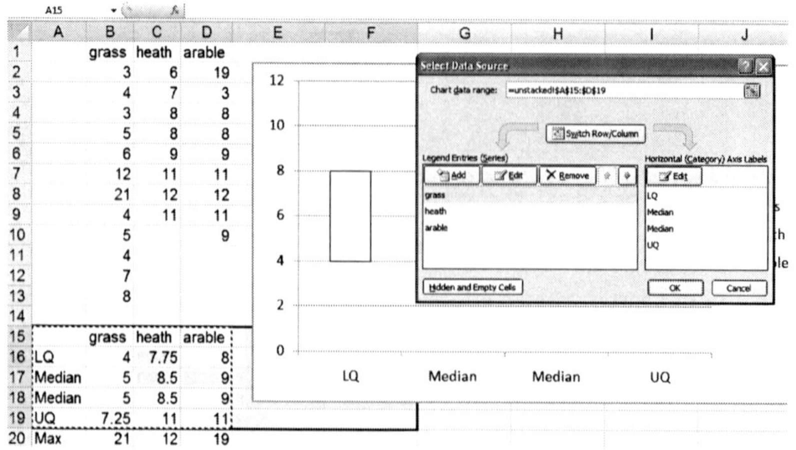

Figure 211. Editing the data series in a stock chart to form the basis for a box–whisker plot

We start by clicking the *Edit* button under Series. We can now highlight the three samples and the four rows of data plus the adjacent labels (Figure 211). The graph now changes but still looks wrong; it has now switched the rows and columns! Fortunately there is a button to allow us to switch them around (Figure 212).

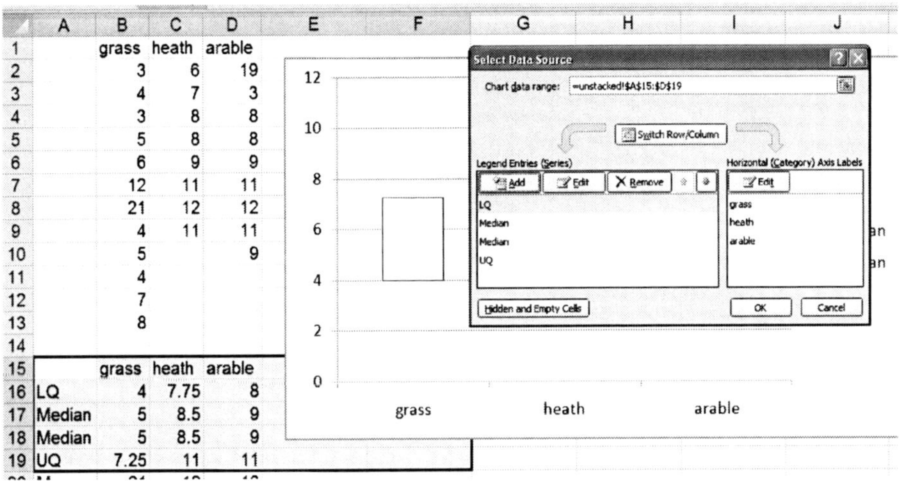

Figure 212. Editing the data series may cause the rows and columns to be transposed. We can switch them easily using the Switch Row/Column option

Once we click the *Switch...* button, we see the graph change to look a bit better. We now have only three samples displayed and can see the beginnings of boxes. These boxes represent the inter-quartile range of each sample. We now want to bring up points displaying the median values. To achieve this, we go to the *Layout* menu (Figure 213).

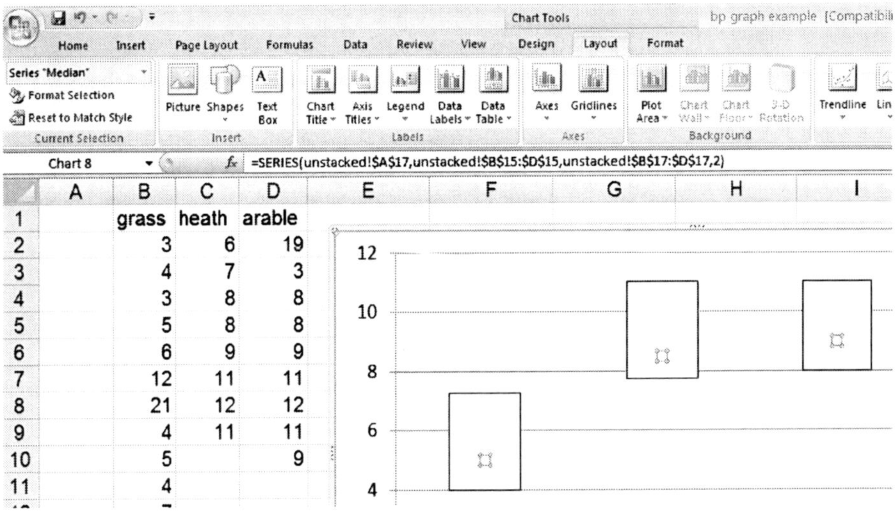

Figure 213. Highlighting the median values so that the points may be displayed as part of a box–whisker plot

On the left, we see a drop-down box where we can select various graph elements. Here we choose one of the median series then click the *Format Selection* button underneath. We see the median values highlighted briefly (Figure 213) before a new menu window opens (Figure 214).

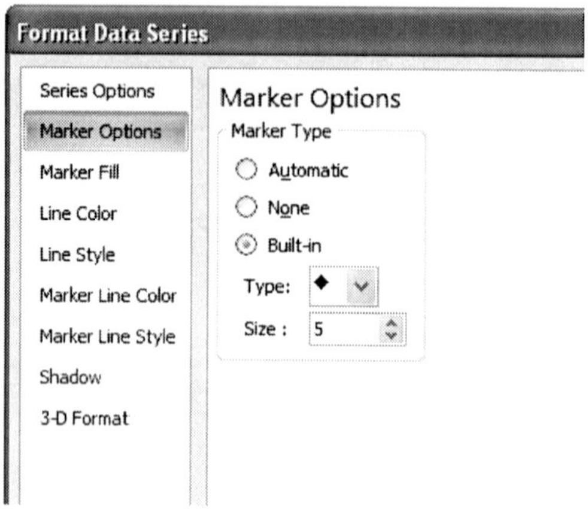

Figure 214. Options for formatting data points to display median values in a box–whisker plot

There are several things we can alter. Here we change the character to a diamond, change it to a solid fill colour (black) and set the marker line to none. Once we are finished, we see that the median values are now displayed (Figure 215). Our boxes are now complete but there are no whiskers.

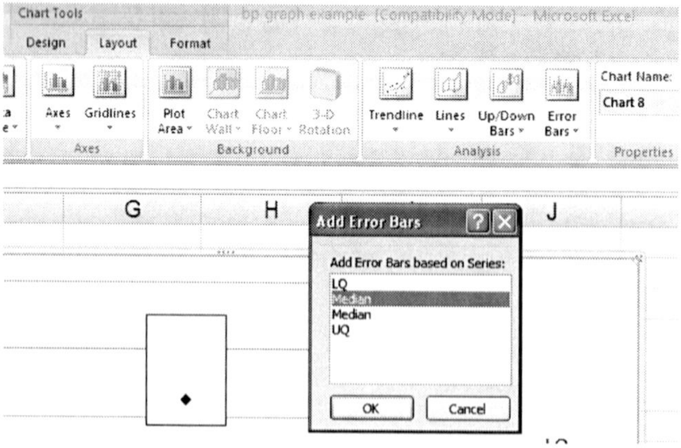

Figure 215. Median values are now displayed in the box–whisker plot but the whiskers need to be added using the Error Bars options

We use the *Error Bars* button to add the whiskers. In this case we select one of the data series to add the error bars to. It makes sense to use the same median series that we displayed earlier. We need to produce bars that extend up from the top of the box and down from the bottom. We need to select the Custom option and click the *Specify Value* button (Figure 216).

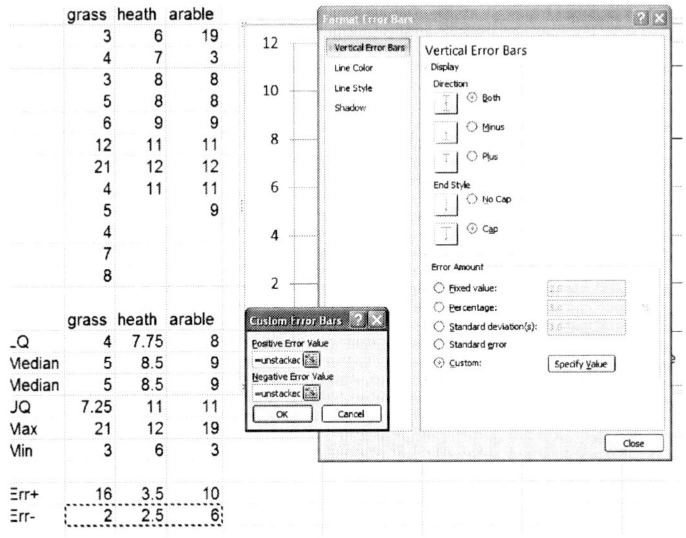

Figure 216. Selecting the data to produce error bars that form the whiskers on a box–whisker plot

The whiskers will actually sprout from the median points but because we have a solid colour for the boxes, we will see them only when they appear from behind the boxes. We select the upper and lower values from the cells we created earlier. Once we are finished, our final graph is nearly ready (Figure 217).

We no longer need the legend, which may be simply deleted. We should add axis titles using the *Layout* menu. The gridlines can be kept, deleted or modified. In this case we choose to edit them to produce dashed lines.

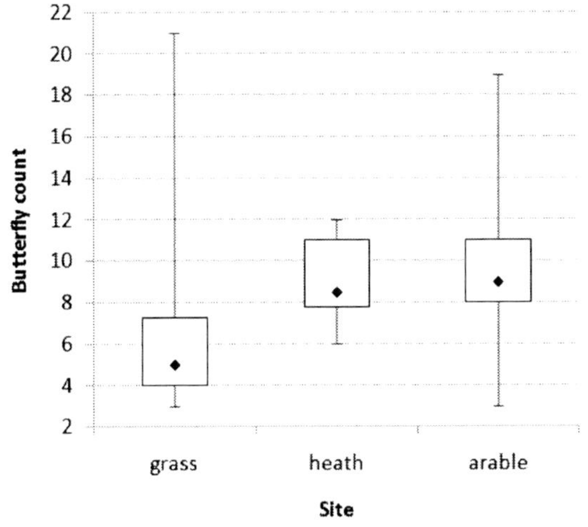

Figure 217. Final box–whisker plot. Butterfly abundance at three habitats in Dorset. Points represent median, boxes are inter-quartile and error bars are range (n = 12, 8, 9)

We should check that the error bars look correct; we want them to extend to the maximum and minimum values and it is easy to make a mistake calculating them. If we do need to modify them, the graph should update itself to reflect the altered values. Our final tidying up involves resizing the chart to make it display at its best. We might rescale the y-axis for example. Notice how Figure 217 does not start at 0.

Producing this box–whisker plot took quite a few steps. It is easy to get lazy and just accept the defaults and go with the flow; however, your graphs represent the culmination of all your work and are the most important thing you can produce! It is worth spending some time to get the graphs exactly right for the job; do not be tempted by shortcuts.

Re-arranging data into samples

The data we used for the bar chart and the box–whisker plot was already set out in separate sample columns (Figure 193). This makes it easy to create the summary statistics that we require but it is probably not the best layout for many statistical analyses. In most cases we will set out our data into columns where each column represents a different variable (see Section 2.2).

In Figure 218 we see some Excel data laid out in this fashion. The first column contains the abundance data and the second column contains the site data. For simple and quite small datasets, the previous multi-sample set-up (like Figure 193) is manageable but if we had a second predictor variable then this multi-column layout becomes impractical.

B	C
count	site
3	grass
4	grass
3	grass
5	grass
6	grass
12	grass
21	grass
4	grass
5	grass
4	grass
7	grass
8	grass
6	heath
7	heath
8	heath
8	heath
9	heath
11	heath
12	heath
11	heath
19	arable
3	arable
8	arable
8	arable
9	arable
11	arable
12	arable
11	arable
9	arable

Figure 218. Data in Excel set out in recording format. Each column represents a single variable. The alternative is for each column to be a single sample

When we have data in the recording format, that is one column per variable, we sometimes need to re-arrange the data to produce summary statistics. We can use a pivot table to re-arrange the data as we saw in Section 3.2.7. In Figure 219 we see the data re-arranged from the recording format into multi-column layout.

count	site	obs				
3	grass	1				
4	grass	2				
3	grass	3				
5	grass	4				
6	grass	5				
12	grass	6				
21	grass	7				
4	grass	8	**Sum of count**			
5	grass	9		**arable**	**grass**	**heath**
4	grass	10	1	19	3	6
7	grass	11	2	3	4	7
8	grass	12	3	8	3	8
6	heath	1	4	8	5	8
7	heath	2	5	9	6	9
8	heath	3	6	11	12	11
8	heath	4	7	12	21	12
9	heath	5	8	11	4	11
11	heath	6	9	9	5	
12	heath	7	10		4	
11	heath	8	11		7	
19	arable	1	12		8	

Figure 219. Data converted from recording format to multi-column layout using a pivot table

We need to do a bit of editing in order to get what we need. We first ensure that the data are sorted in order (see Section 3.2.1) so that each site forms a block. Then we add a new column of data. Here we label it *obs*, which can be thought of as short for observation. We might have used replicate as a label instead. When we create the pivot table, we use the obs column as our row data and set the site data as the column data. The actual values we want to see are the count data. In Figure 219 we have placed the pivot table next to the original data. In most cases you will want to place the pivot table into a new worksheet. You can use the pivot table directly and work out means etc. or you can copy to the clipboard and use *Paste Special* (see Section 3.2.3) to transfer only the data. This is probably preferable since you may wish to do something else with the pivot table later.

Scatter plots

We use a scatter plot to show the relationship between two variables. In Figure 220 we see some data in a spreadsheet arranged in two columns. The first is the abundance data (of mayfly nymphs); this is the response (dependent, y-axis) variable. The other column shows the water flow and this is the predictor (independent, x-axis) variable. If we had highlighted these data, Excel would produce a chart for us when we click the Scatter Chart option on the *Insert* menu; however, Excel expects the data to be arranged with the x-axis first and the y-axis second. We prefer to write our data with the response variable first and the predictor variable second (this makes a lot more sense, particularly when we have several predictor variables). It is probably a good idea to start with a blank graph by ensuring that the active cell is not adjacent to any data, then we can select the data we require and ensure that we get the *x* and *y* variables the correct way around. If we do create a graph with the axes reversed we can easily swap them. In any event, once we have our starting graph we click the *Select Data* button.

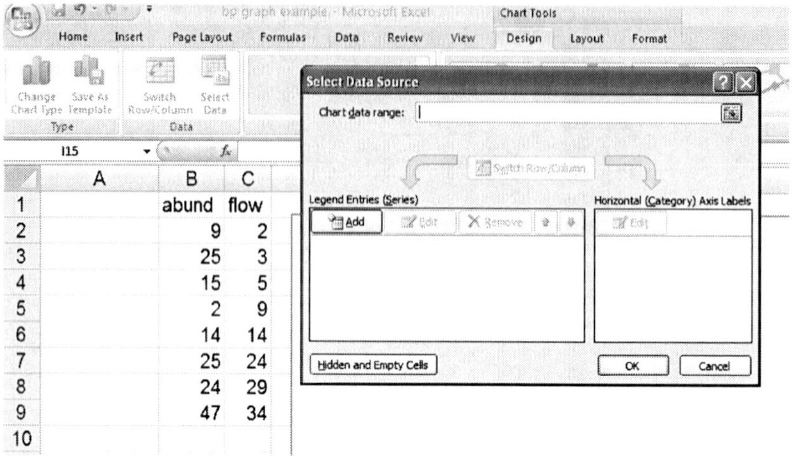

Figure 220. Creating a scatter plot. We start with a blank chart and then select the *x* and *y* data to create our plot

If we wish to swap the *x* and *y* axes, we click the *Switch Row/Column* button. In our example we start with a blank graph so we need to add a data series by clicking the button to the left (Figure 221).

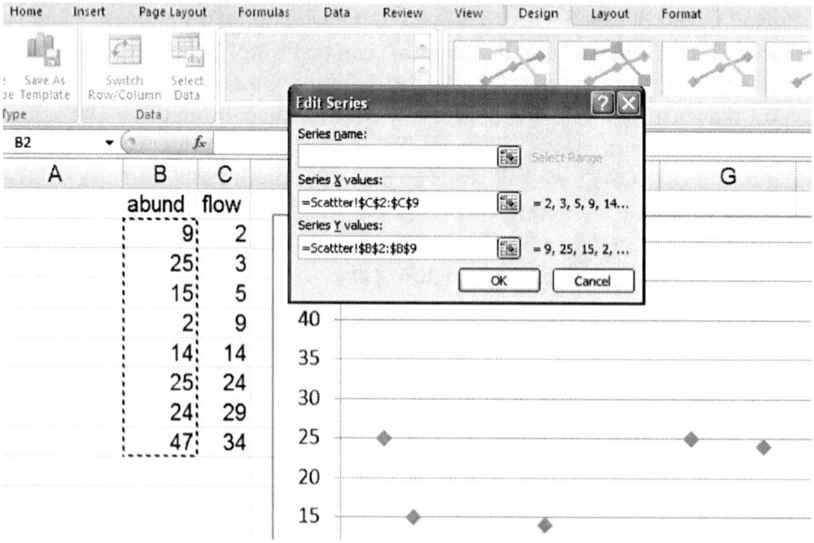

Figure 221. Once the data are selected, the plot is created but with a default layout

To select our data we simply highlight the appropriate range of cells. We can add extra data series but in this case we have only one. Now our graph is created and has a default layout. We will certainly need to alter this. Excel provides some built-in options and we can see these on the *Design* menu in the *Chart Tools* section on the ribbon (Figure 222).

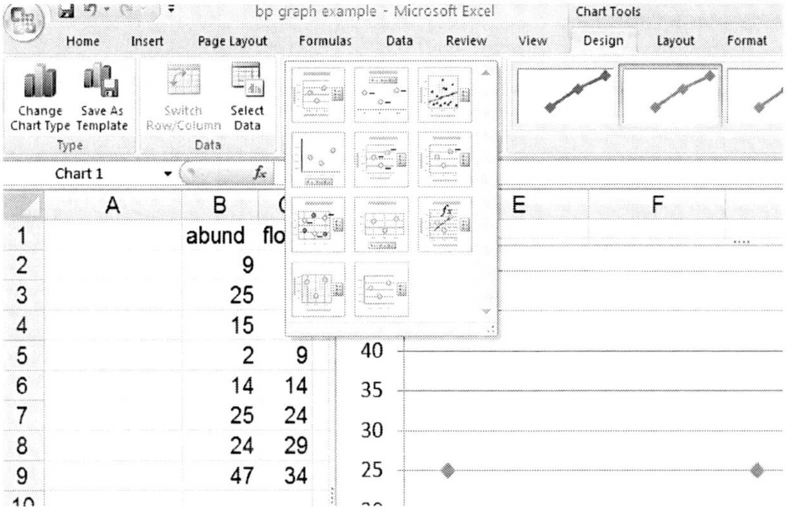

Figure 222. Excel 2007 provides some basic templates for scatter plots. None are suitable for a scientific scatter chart

With some graph types we may find a suitable default layout but in the case of our scatter plot none are available and we are better off adding and modifying the elements we require using the *Layout* menu (Figure 223).

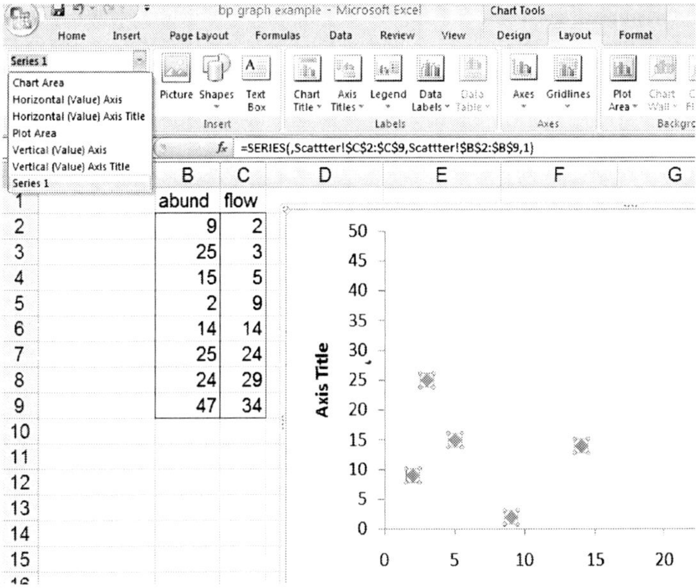

Figure 223. Various chart elements may be selected for editing from the ribbon menu in Excel 2007

Some elements can be simply removed by clicking them in the graph and using the delete key on the keyboard. We can also use the drop-down menu to select a particular element and use the *Format selected element* button to bring up a menu of formatting options. Alternatively many chart elements can be selected by right-clicking (Figure 224).

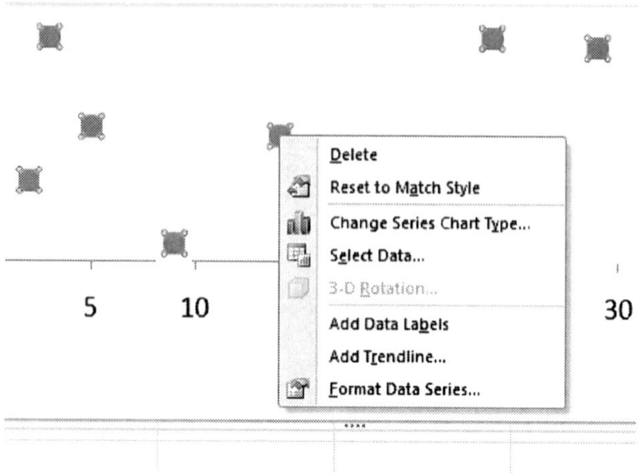

Figure 224. Right-clicking a graph element will usually bring up a context-specific menu. Here we format the data series and alter the type of points displayed

If we right-click a data point for example we can select to format the data series. This allows us to alter the plotting character and its colour for example. We can use the buttons in the menu to add axis titles. Once the blank title is in place we can select it and type our required text. If we wish to use special formatting, e.g. superscript, we can highlight the text we wish to make superscript and then select *Font* from the standard menu (Figure 225).

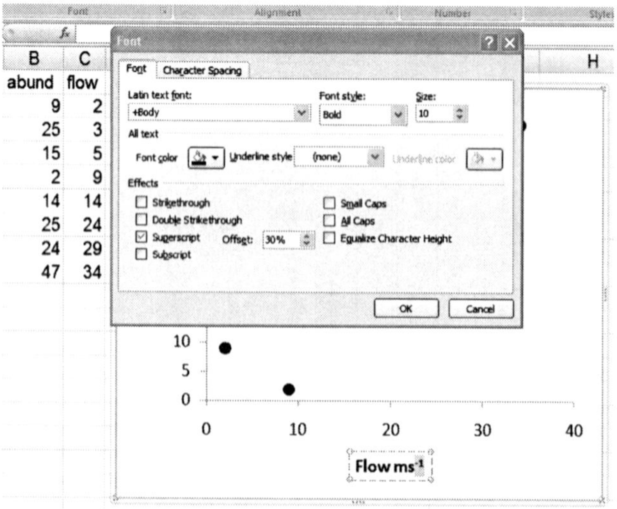

Figure 225. Axes titles can be formatted by highlighting the text and using the *Font* menu on the ribbon in Excel 2007

In our case we wish to add units to the x-axis title and a superscript is required. We can also alter other font characteristics like the font itself or the colour. The final scatter plot looks like Figure 226.

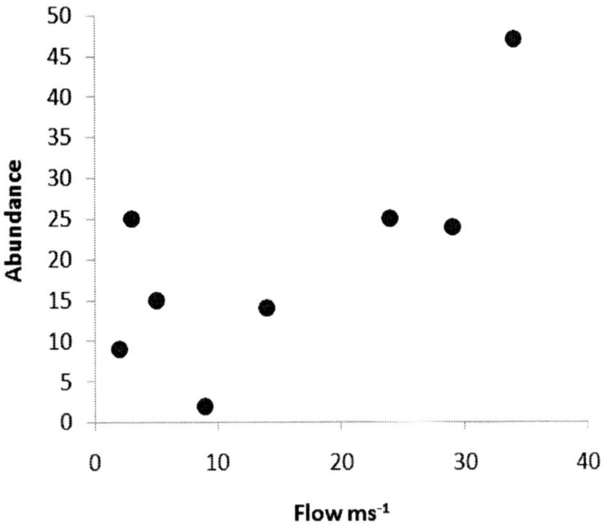

Figure 226. The completed scatter plot. Abundance of mayfly nymphs and stream flow in Guscott Stream, Devon

Here we have selected a plain circle as the plotting character. There are not many points and there is no overlapping so a filled character is acceptable. The *Layout* menu also allows us to add lines of best-fit to our plot (we did this in Section 11.2). Here we are not assuming a linear model and so a line is not appropriate.

Pie charts

The pie chart has fallen a bit out of favour in recent times but it can be useful for displaying proportional data. In Table 81 we can see data on the abundance of various bird species. Each species is recorded in a variety of habitats and the researcher wanted to know if particular bird species were associated with particular habitats.

Table 81. Abundance of common bird species at habitats in Sussex

	Garden	Hedgerow	Parkland	Pasture	Woodland
Blackbird	47	10	40	2	2
Chaffinch	19	3	5	0	2
Great tit	50	0	10	7	0
House sparrow	46	16	8	4	0
Robin	9	3	0	0	2
Song thrush	4	0	6	0	0

These data can be displayed using pie charts. We can display the information in two main ways: we can show a pie chart for each species or we can show the habitats. In other words we can select either columns or rows. We could select a row or a column from the data and then use the *Insert* menu to select a Pie chart. Alternatively, we can ensure that the active cell is not adjacent to any data and create a blank pie chart Figure 227.

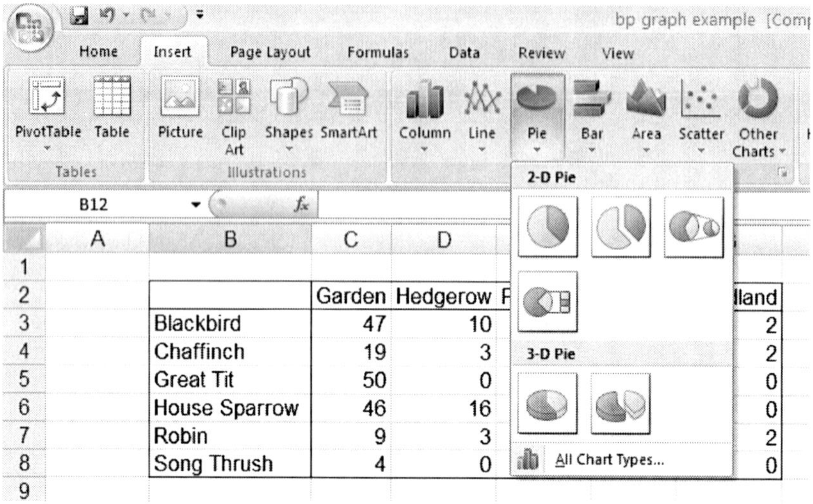

Figure 227. Options for creating pie charts in Excel 2007. The basic 2D chart is generally the most useful

Once we have created a blank chart, we can select the data we want using the button on the *Design* menu in the ribbon. This brings up a new menu box allowing us to select the cells that we wish to chart (Figure 228).

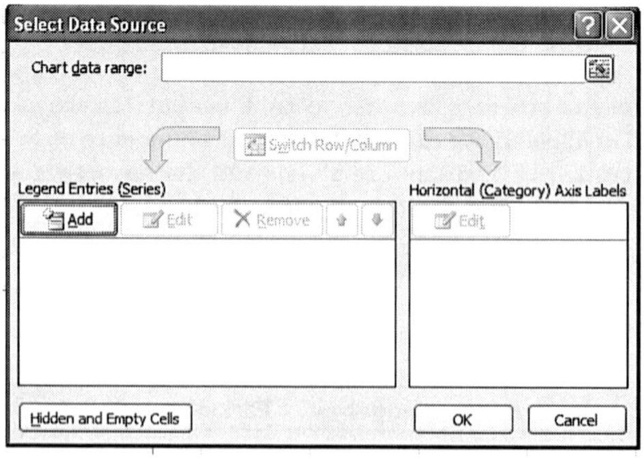

Figure 228. The window/menu for selecting data sources is common to all chart types

Once we have the window open, we can select the data we want. In our example we will select the first column (Figure 229).

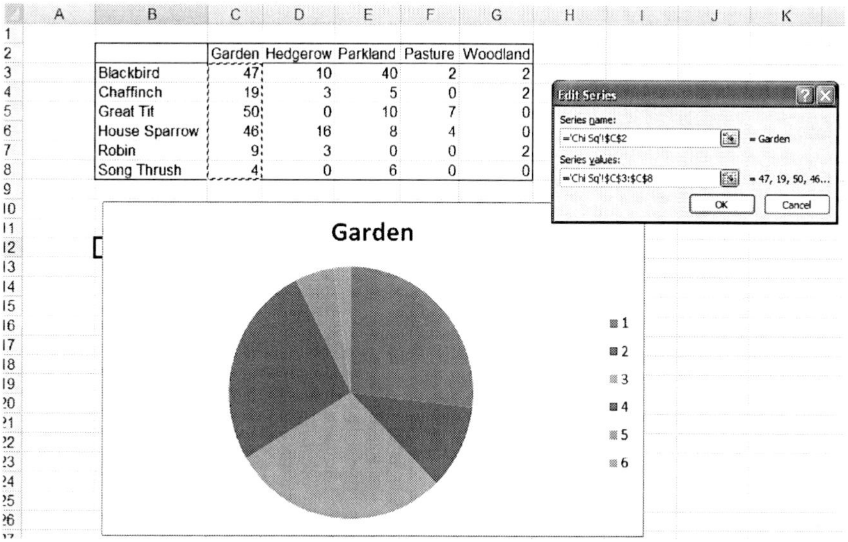

Figure 229. We can select rows or columns to display in a pie chart

The column we selected is for the garden habitat and we can select the header cell to act as the name for this data series. Excel helpfully adds the name to the graph as we go along. If we had selected a row to chart, we would choose the appropriate row label instead. We can see from Figure 229 that the slices of pie are not labelled. At the moment Excel does not know where the labels are. We must OK the series we have selected and then edit the category labels from the menu (Figure 228). Now we can select the labels for the slices of pie (Figure 230).

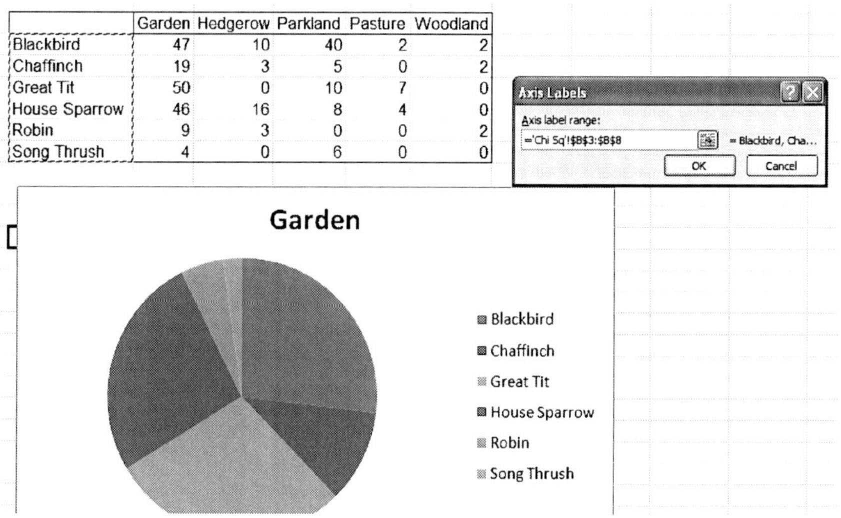

Figure 230. Selecting the labels for the slices of pie in a pie chart using the "edit category axis labels" facility

We simply highlight the range of cells where the labels are; in this case we choose the bird species names. Now our graph is tolerably complete; however, having quite a few segments can be confusing so we decide to label them. On the *Layout* menu on the ribbon, we select *Data Labels* (Figure 231).

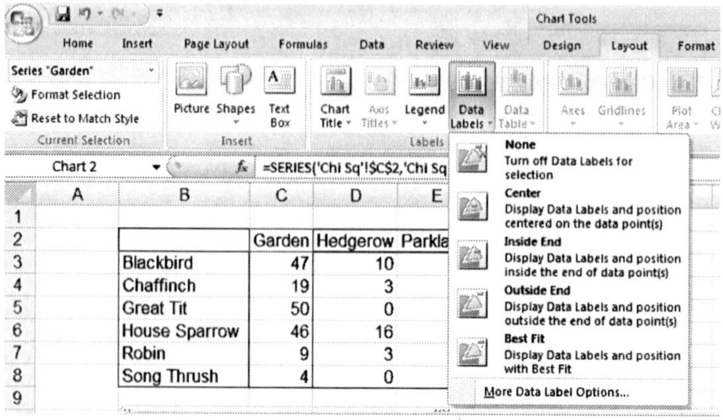

Figure 231. Adding data labels to pie charts can help readability, especially where there are many segments

We can see that there are some basic options but these tend to produce the values for the segments/slices and so it is best to use *More Data Label Options* (Figure 232).

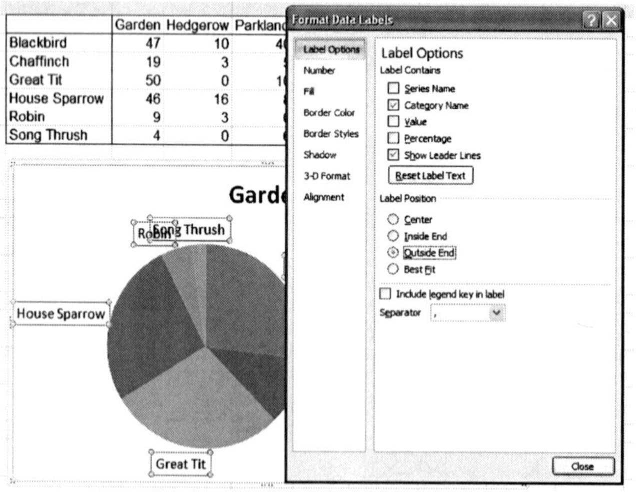

Figure 232. Selecting label options for a pie chart

In this case we wish to display the names of the bird species. We will also use leader lines; these only appear if the slices are very small and we move the label from its original position. In our case the top two labels are very close. Once we have created the labels, we can use the mouse to move them (Figure 233).

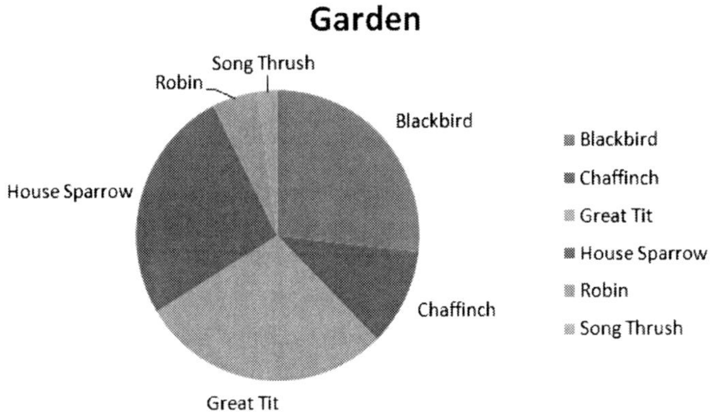

Figure 233. Completed pie chart showing proportions of bird species recorded in garden habitat

Once we move the labels from their original position, we see the leader lines. We have left the legend in place and because we have labelled the segments this is probably unnecessary and could be eliminated.

12.5 More about graphs in R

The basic graphical commands shown so far will undoubtedly allow you to present a wide variety of results in R; however, there are many additional commands that will allow you to tweak your graphs and produce exactly what you want.

12.5.1 Bar charts *barplot()* command

The bar chart is familiar to everyone and is a useful graphical tool that may be used in a variety of ways. The basic function is:

```
> barplot(data)
```

Before you can draw a graph, you need to get your data into an appropriate format. R has many ways of manipulating data but it is often easiest to assemble and manipulate your data in a spreadsheet (you can save in CSV format).

The first stage is to arrange your data in a CSV file. You may have your data arranged in columns or in rows. You may also have both row and column names. Don't forget that variable names in R can contain letters and numbers but the only punctuation allowed is a period.

The second stage is to read your data file into memory and give it a sensible name. When using barplots you may have both row and column names so don't forget to tell R that you are using row names if you are (using the *row.names = 1* option where 1 is the number of the row where the labels are, usually the first).

Barplot data as matrix or frame?

R stores data in a variety of formats. A simple list of values (could be numbers or text or factors) is called a vector. If you create a CSV file with several columns you may read this into R. The resulting object is a data frame. Finally we have a matrix. At first glance a matrix is like a data frame. The essential difference is that a data frame is treated as a series of columns whereas a matrix is treated as an entirety.

Bar plots are best drawn using matrix data and the sections that follow utilise data in that format. To get data from one format into another you can use the *as.matrix()* command so:

```
> datamatrix = as.matrix(dataframe)
```

This command will create a matrix (called *datamatrix*) from the data frame (called *dataframe*). Here is an example of a data frame:

```
      height abund
upper     12    21
lower      6    15
left      21     8
right     16     3
```

Here we have two columns (height and abund) and we have four rows. When the data were read into R, we used *row.names = 1* to set row names.

At first glance we cannot tell if these data are in a matrix or a frame. We can run a quick test to find out:

```
> names(ourdata)
```

Usually if we run the *names()* command we get the names of the data columns. If, however, we have a matrix our result is *NULL*.

Simple multi-category bar chart

Your data may consist of a simple row of means, e.g. here are some data on road deaths in Virginia. These data come with the basic distribution of R and are called VADeaths (which is a matrix). The means have been extracted below and assigned to the variable VADmeans (this is also a matrix).

```
Rural Male Rural Female   Urban Male Urban Female
     32.74        25.18        40.48        25.28
```

These data were extracted using the *colMeans()* function (there is also a *rowMeans()* function). We can see that there are four categories. To create a basic bar chart we simply call the *barplot()* function:

```
> barplot(VADmeans, main="Road Deaths in
Virginia",xlab="Categories", ylab="Mean Deaths")
```

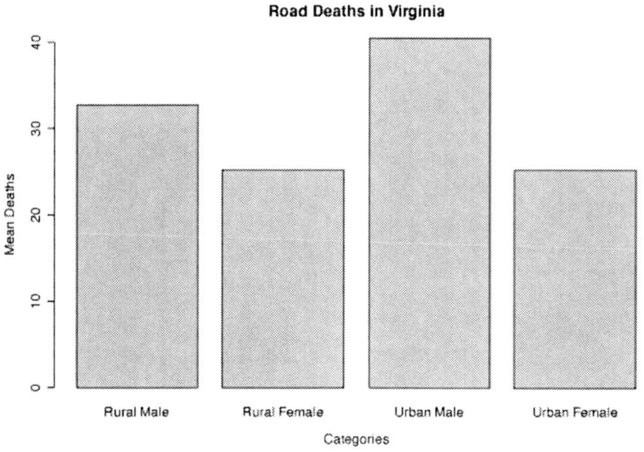

Figure 234. Basic bar chart in R

This produces a very basic plot; we have added a main title and labels for the *x* and *y* axes using fairly simple commands. When plotting a graph R opens a graphics window; if you select the window (by clicking on or in it) you can copy to the clipboard and paste into a variety of applications. Alternatively you can save the graph to a file.

Because our data were in matrix form, R has automatically added the names to the bars. If you are plotting a simple vector of numbers (or a single column of a data frame), you need to specify the names to be used:

```
> barplot(plotdata, names = row.names(plotdata))
> barplot(plotdata, names = c("a", "b", "c", "d"))
> barplot(plotdata, names = mynames)
```

In the first example we use the row names of the data frame itself. In the second example we create a short list of names explicitly (*a, b, c, d* in this case). In the third example we refer to another R object (imagine that we created a list of names earlier).

Stacked bar charts or not?

The VADeaths dataset consists of a matrix of values with both column and row labels, see the data in Table 82.

Table 82. Data on road casualties used for barplot

	Rural male	Rural female	Urban male	Urban female
50–54	11.7	8.7	15.4	8.4
55–59	18.1	11.7	24.3	13.6
60–64	26.9	20.3	37.0	19.3
65–69	41.0	30.9	54.6	35.1
70–74	66.0	54.3	71.1	50.0

If we attempt to produce a bar chart of these data we get something like Figure 235.

```
> barplot(VADeaths, legend= rownames(VADeaths))
```

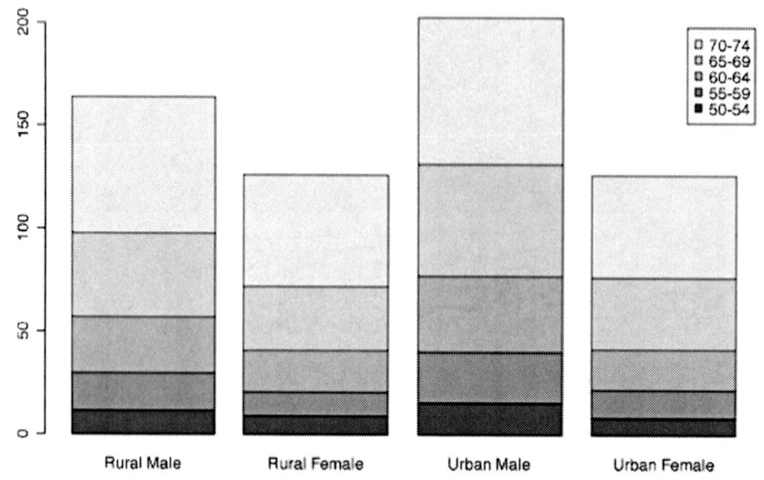

Figure 235. Stacked bar chart. Each bar is sub-divided by the data in the row categories

This time a legend was added using the legend command along with the row names of the dataset. We see that by default a stacked bar chart is produced. To unstack the bars and plot them alongside one another we use a new command:

```
barplot(VADeaths, legend= rownames(VADeaths), beside= TRUE)
```

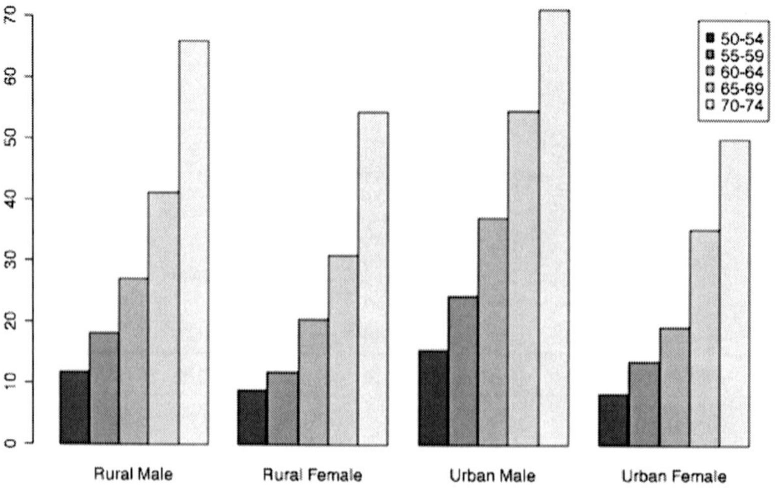

Figure 236. Bar chart with bars unstacked

This is fine but the colour scheme is kind of boring. Here is a new set of commands:

```
> barplot(VADeaths, beside = TRUE, col = c("lightblue",
"mistyrose", "lightcyan","lavender", "cornsilk"), legend
= rownames(VADeaths), ylim = c(0, 100))
> title(main = "Death Rates in Virginia", font.main = 4)
```

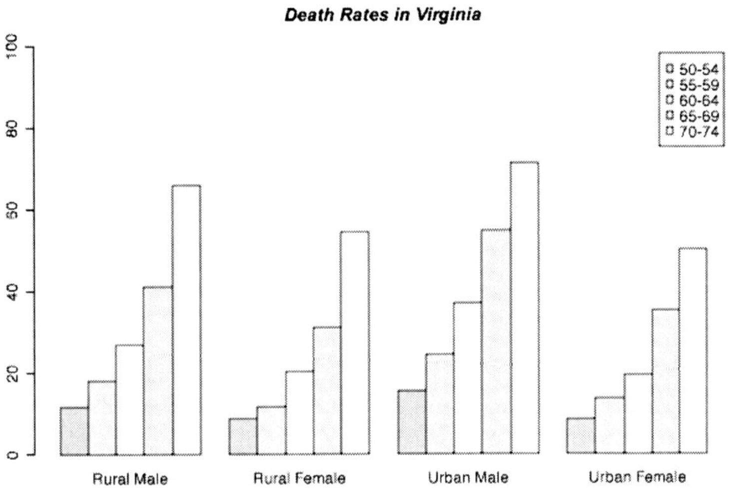

Figure 237. Unstacked bar chart with customised colours and a legend

This is a bit better. We have specified a list of colours to use for the bars. Note how the list is in the form *c(item1, item2, item3, item4)*. The command *ylim()* sets the limits of the *y*-axis; in this case a lower limit of 0 and an upper of 100. The command is in the form *ylim= c(lower, upper)* and note again the use of the *c(item1, item2)* format. The legend takes the names from the row names of the datafile. We set the *y*-axis limit to accommodate the legend box.

It is possible to specify the title of the graph as a separate command, which is what was done above. The command *title()* achieves this but it only works when a graphics window is already open. The command *font.main* sets the typeface, a value of 4 produces bold italic font.

Frequency bar plots

Sometimes you will have a single column of data that you wish to summarise. A common use of a bar chart is to produce a frequency plot showing the number of items in various ranges. Here is a vector of numbers:

```
75 67 70 75 65 71 67 67 76 68
```

These have been assigned to a variable called carb and we wish to make a frequency plot. Let's try:

```
> barplot(carb)
```

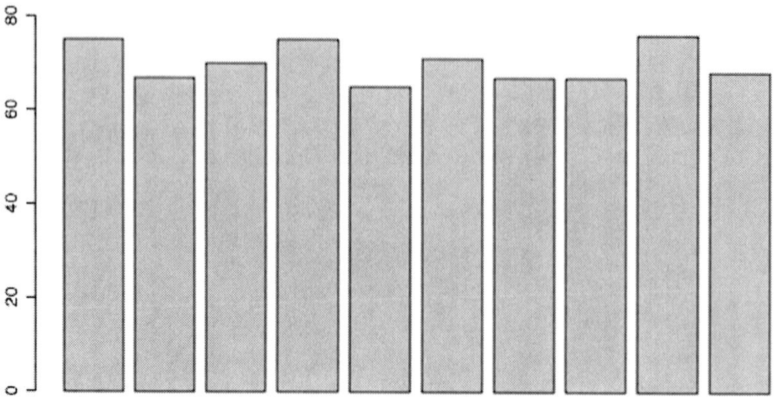

Figure 238. A simple bar plot. The intention was to produce a frequency plot but the basic *barplot()* command produces a bar for each datum

Our bar chart appears like Figure 238 but that wasn't really what we wanted at all. What has happened is that each item has been plotted as a separate entity. We need to tabulate the frequencies. Fortunately there is an easy way to do this using the *table()* function. Let's redraw the graph but using the following:

```
> barplot(table(carb))
```

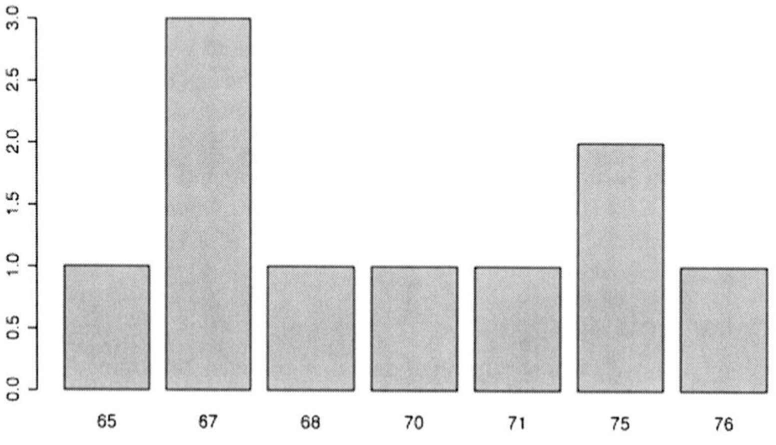

Figure 239. Frequency bar plot from *table()* data

This (Figure 239) is much better. Now we have the frequencies for the data arranged in several categories (sometimes called bins). This is a kind of substitute for a histogram that we met earlier (Section 4.2.1) but the x-axis is not a continuous range; each bar is a discrete interval. As with other graphs we can add titles to axes and to the main graph.

We can look at the *table()* function directly to see what it produces.

```
> table(carb)
```

This function produces:

```
carb
65 67 68 70 71 75 76
 1  3  1  1  1  2  1
```

We can see that the function has summarised the data into various numerical categories.

We may wish to show the frequencies as a proportion of the total rather than as raw data. To do this we simply divide each item by the total number of items in our dataset:

```
> barplot(table(carb)/length(carb))
```

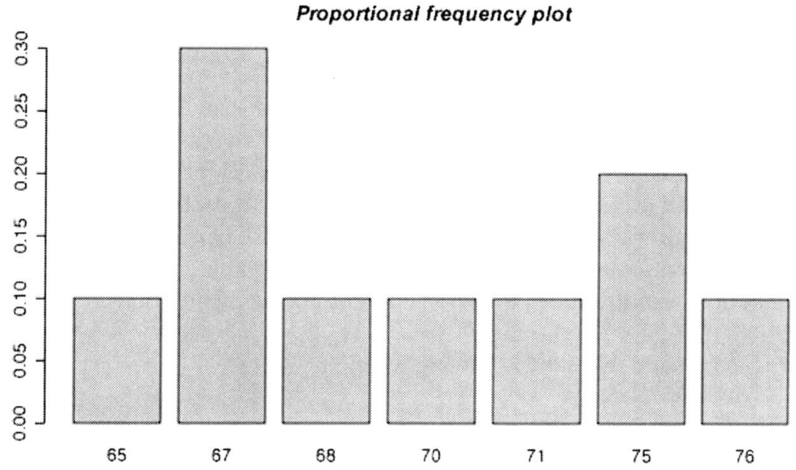

Figure 240. Frequency bar plot using proportions

This shows exactly the same pattern but now the total of all the bars add up to 1. Of course we also could produce a histogram (see Sections 4.2.1 and 12.5.2).

Horizontal bar plots

It is straightforward to rotate your plot so that the bars run horizontal rather than vertical (which is the default). To produce a horizontal plot you add *horizontal = TRUE* to the command, e.g.

```
> barplot(table(carb), horiz=T, col="lightgreen",
xlab="Frequency", ylab="Range")
> title(main="Horizontal Bar Plot", font.main= 4)
```

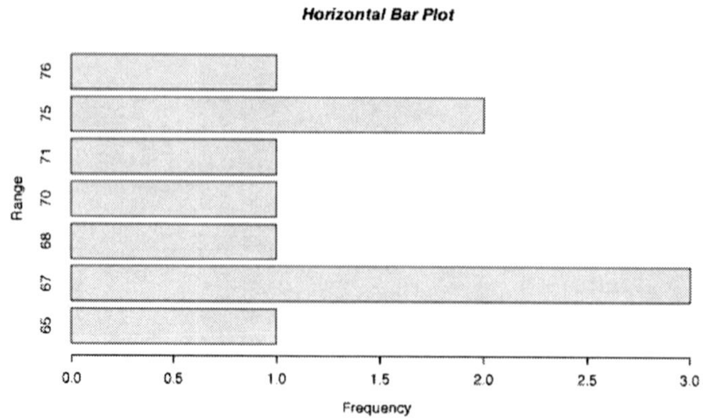

Figure 241. Horizontal bar plot

The bars now appear horizontally (Figure 241) but note that the *x*-axis is still the bottom of the graph and the *y*-axis is still the vertical axis. This time we have used the *title()* command to add the main title separately. The value of 4 sets the font to bold italic (try other values).

Adding error bars

In many cases the data we wish to present will be from samples. In other words we will have summary data in the form of average, variability and so on. The basic *barplot()* produces a single bar for each sample/item and we naturally select the average as our display item. We should add a measure of dispersion to our graphs; error bars showing the standard error would be quite acceptable. Previously we used error bars in Excel (Section 12.4.3) on our bar chart. We also used error bars in R when looking at logistic regression (Section 11.2.3). At that time we introduced the *segments()* command, albeit briefly.

The *segments()* command takes the following form:

```
> segments(x1, y1, x2, y2)
```

This draws a line from *x1, y1* to *x2, y2* on an existing graph. If we have values for means and standard error we can work out the co-ordinates to plot that will draw our error bars. We will return to this in detail with a worked example in Section 12.6.

12.5.2 Histograms, *hist()* command

The *barplot()* function can be used to create a frequency plot of sorts but it does not produce a continuous distribution along the *x*-axis. A true frequency distribution should have the bar categories (i.e. the *x*-axis) as continuous items. The frequency plot produced previously has discontinuous categories.

To create a frequency distribution chart we need a histogram, which has a continuous range along the *x*-axis. The command in R is:

```
> hist(variable)
```

Here is a vector of numbers saved as the variable test.data:

```
2.1 2.6 2.7 3.2 4.1 4.3 5.2 5.1 4.8 1.8 1.4 2.5 2.7 3.1
2.6 2.8
```

To create a histogram we type:

```
> hist(test.data)
```

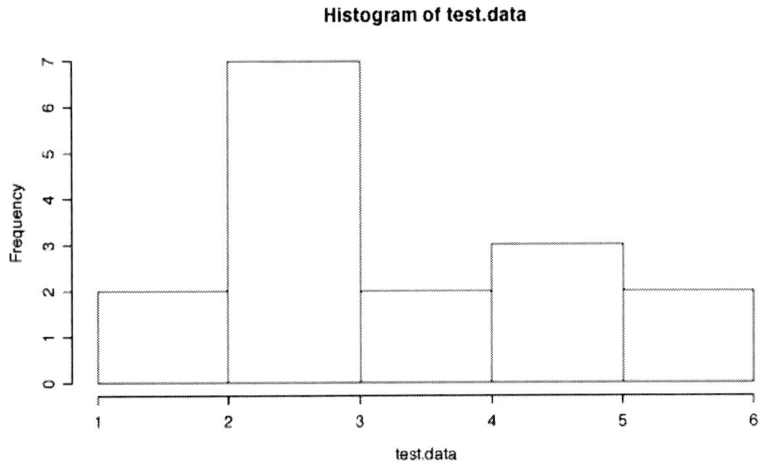

Figure 242. Histogram drawn using all defaults

To plot the probabilities (i.e. proportions) rather than the actual frequency we need to add the command *prob = TRUE* like so:

```
> hist(test.data, prob= TRUE)
```

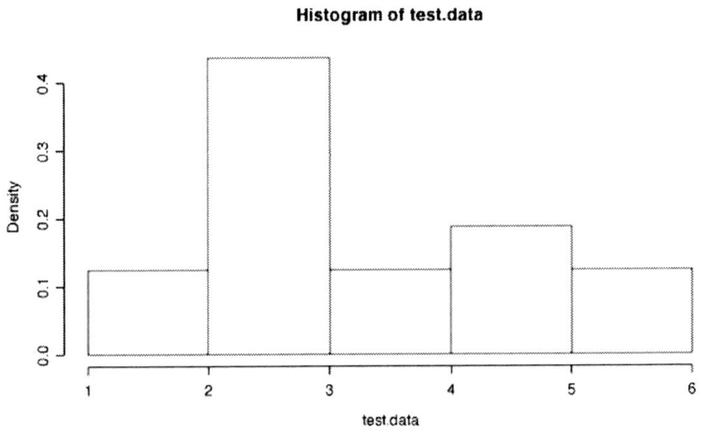

Figure 243. Histogram drawn using probabilities

This is useful but the plots are a bit basic and boring. We can change axis labels and the main title using the same commands as for the *barplot()* function. Here is a new plot with a few enhancements:

```
> hist(test.data, col="cornsilk", xlab="Data range",
ylab="Frequency of data", main="Histogram", font.main=4)
```

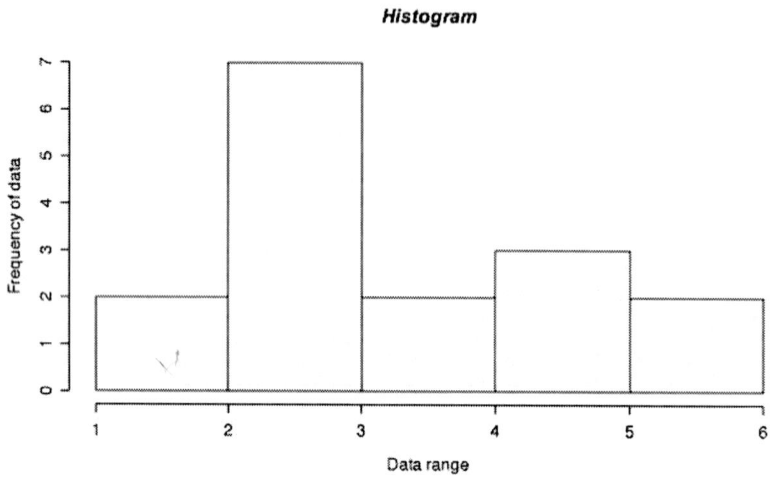

Figure 244. Histogram with coloured bars

These commands are largely self-explanatory. The 4 in the *font.main* command sets the font to italic (try some other values).

By default R works out where to insert the breaks between the bars. You can change the number of breaks by adding a simple command, e.g.

```
> hist(data.set,breaks=10) # 10 breaks, or just
hist(data.set, 10)
```

The # tells R that what follows is a comment, useful for creating your own library of commands.

Alternatively you can be more specific and set the breaks exactly:

```
> hist(data.set,breaks=c(0,1,2,3,4,5,10,20,max(data.set)))
```

Notice how the exact break points are specified in the *c(x1, x2, x3)* format. You can manipulate the axes by changing the limits, e.g. make the *x*-axis start at 0 and run to 6 by another simple command, e.g.:

```
> hist(test.data, 10, xlim=c(0,6), ylim=c(0,10))
```

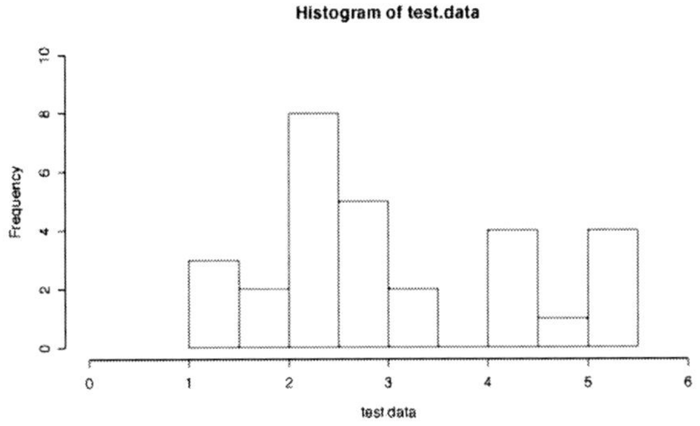

Figure 245. Histogram drawn with customised break points

This sets 10 break-points and sets the y-axis from 0–10 and the x-axis from 0–6. Notice how the commands are in the format *c(lower, upper)*. The *xlim()* and *ylim()* commands are useful if you wish to prepare several histograms and want them all to have the same scale for comparison.

12.5.3 Box–whisker plots, *boxplot()* command

Single sample plot

A box–whisker graph allows you to convey a lot of information on one simple plot. Traditionally they were used for data that were not normally distributed (i.e. non-parametric) but they are generally more useful. You can plot a single sample or create a more complex plot of categories within a dataset.

The basic function is *boxplot()*. Here is a vector of numbers saved as the variable *test.data*:

```
2.1 2.6 2.7 3.2 4.1 4.3 5.2 5.1 4.8 1.8 1.4 2.5 2.7 3.1
2.6 2.8
```

To create a box–whisker plot we type:

```
> boxplot(test.data)
```

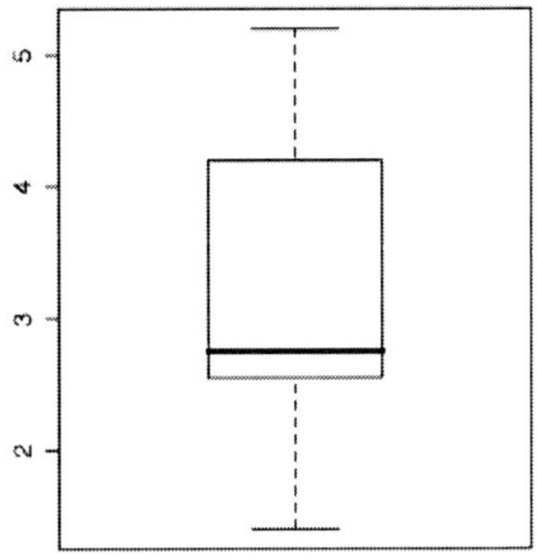

Figure 246. Box–whisker plot. The stripe shows the median, the box the inter quartiles and the whiskers extend to the extremes

Figure 246 is not the most exciting graph but we can jazz it up later. What we see is a box with a line through it. The line represents the median of the sample. The box itself shows the upper and lower quartiles. The whiskers show the range (i.e. the largest and smallest values). It is easy to see that this sample has a skewed distribution and is certainly non-parametric.

We can add axis labels, a main title and colour the box using simple commands. These commands are the same as for those used in producing bar plots and histograms. For example:

```
> boxplot(test.data, xlab="Single sample", ylab="Value
axis", main="Simple Box plot", col="lightblue")
```

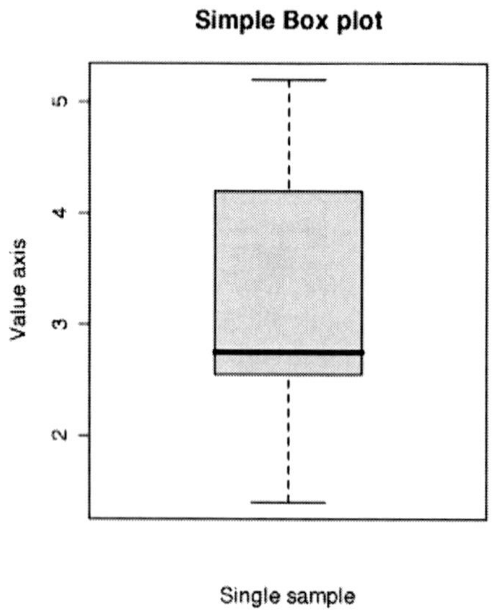

Figure 247. Box–whisker plot with coloured box

Let's make the data even more skewed and add an outlier:

```
2.1 2.6 2.7 3.2 4.1 4.3 5.2 5.1 4.8 1.8 1.4 2.5 2.7 3.1
2.6 2.8 12.0
```

We'll now redraw the graph. This time the main title will be added using a separate command:

```
> boxplot(test.data, xlab="Single sample", ylab="Value
axis", col="lightblue")
> title(main="Plot with outlier", font.main= 4)
```

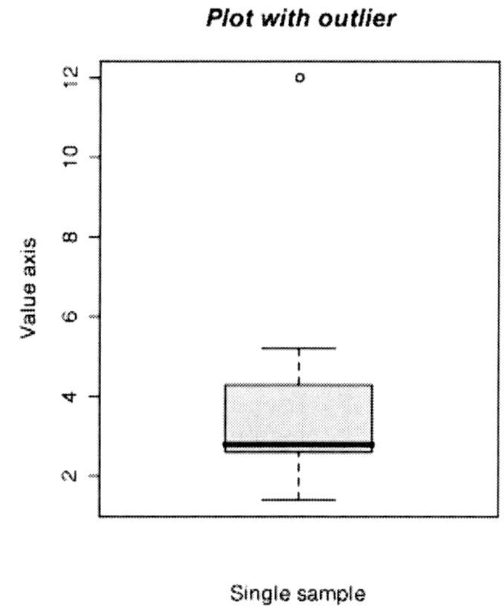

Figure 248. Box–whisker plot showing outlier

Now we see the outlier separately. R does not automatically show the full range of data (as was implied earlier). We can control the range shown using a simple command *range* = *n*. If we set *n* to 0 then the full range is shown. Otherwise the whiskers extend to *n* times the inter-quartile range. The default is set to *n* = 1.5.

```
> boxplot(test.data2, xlab="Single sample", ylab="Value
axis", col="lightblue", range=0)
> title(main="Plot with full-range", font.main= 4)
```

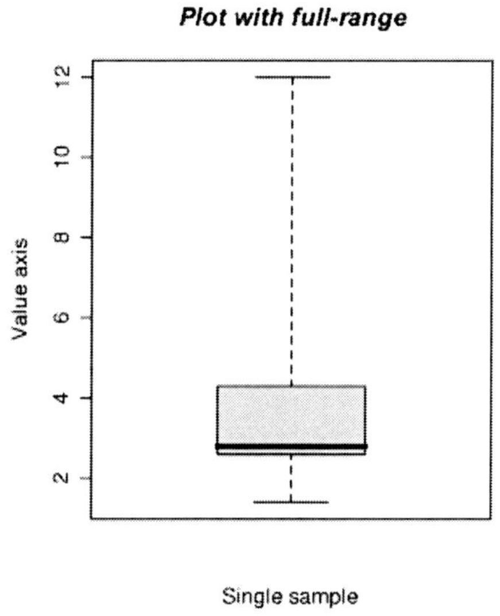

Figure 249. Box–whisker plot with whiskers extending to extremes

Plotting several samples on a box plot

So we can see how to represent a single sample but more often we wish to compare samples. For example, we may have raised broods of flies on various sugars. We measure the size of the individual flies and record the diet for each. Our data file would consist of two columns, one for growth and one for sugar as in Table 83.

Table 83. Data on effect of sugar diet on fly growth (only part of the data are shown)

Growth	Sugar
75	C
72	C
73	C
61	F
67	F
64	F
62	S
63	S

Here is shown only part of a larger dataset. We have one variable (growth) and several samples (i.e. the different sugars). To plot these we use the *boxplot()* command with slightly different syntax, e.g. *boxplot(y ~ x)*. This model syntax is used widely in R for setting up ANOVA and regression analyses for example (as we saw in Section 10.1).

To create a summary box plot we type:

```
> boxplot(growth ~ sugar, data=fly, xlab="Sugar type",
ylab="Growth", col="bisque", range=0)
> title(main="Growth against sugar type", font.main= 4)
```

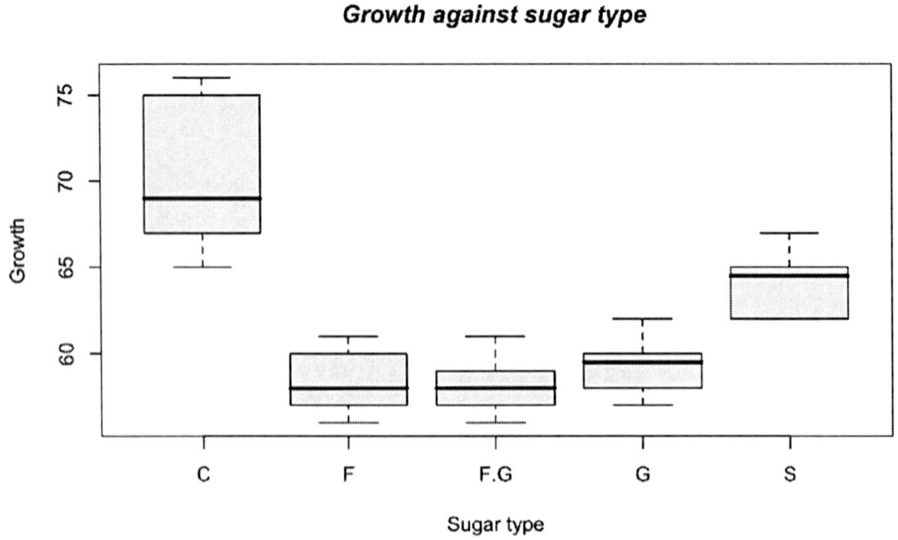

Figure 250. Box–whisker plot showing comparison between several samples of data

In essence our command says "take the growth data and split it according to sugar; look in the fly data object to find the variables". Now we can see that the different sugar treatments appear to produce differing growth in our subjects (Figure 250). Unlike our previous example, R has taken the names of the samples from the headings in the data frame.

Horizontal box plots

It is straightforward to rotate your plot so that the bars run horizontal rather than vertical (which is the default). To produce a horizontal plot, you add *horizontal = TRUE* to the command, e.g.

```
> boxplot(growth ~ sugar, data=fly, ylab="Sugar type",
xlab="Growth", col="mistyrose", range=0, horizontal=TRUE)
> title(main="Growth against sugar type - horizontal",
font.main= 4)
```

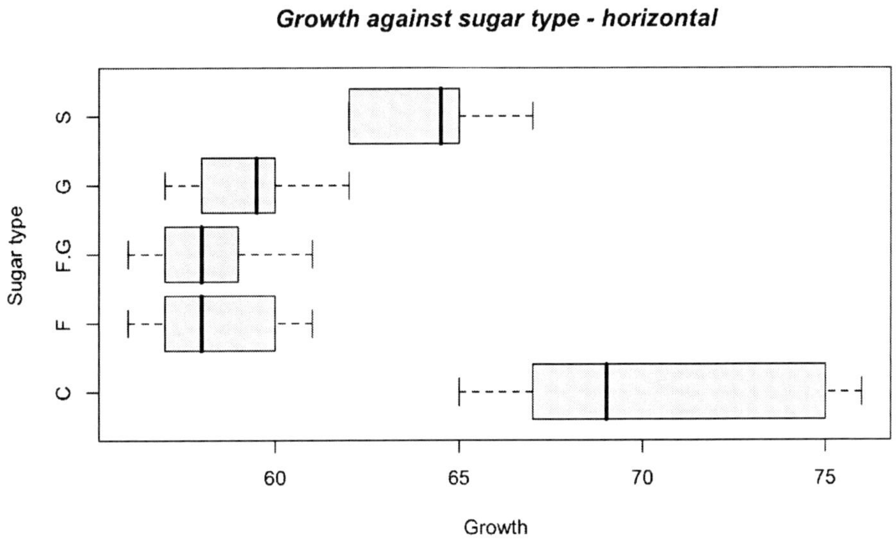

Figure 251. Box–whisker plot showing several samples drawn horizontally

Once again we have used the title command separately to add a main title. The 4 in the *font.main* command sets bold italic (try other values). Because the bars are presented horizontally we need to swap the titles; the x-axis is the horizontal one even though it is now representing the response variable.

12.5.4 Scatter plots, *plot()* command

A scatter plot is used when you have two variables to plot against one another. R has a basic command to perform this task. The command is *plot()*. As usual with R, there are many additional parameters that you can add to customise your plots.

The basic command is:

```
> plot(x, y)
```

where *x* is the name of your *x*-variable and *y* is the name of your *y*-variable. This is fine if you have two variables but if they are part of a bigger dataset then you have to remember to attach *(data.file)* your dataset. A more powerful command is:

```
> plot(y ~ x, data= your.data)
```

Note the use of the model syntax. This model syntax is used widely in R for setting up ANOVA and linear regression models for example (Section 10.1 and Chapter 11; see also its use in the box–whisker plot above).

R comes with a number of datasets built-in; these are used in the examples and can be useful. For example the dataset "cars" contains two variables, speed and dist.

To see a basic scatter plot, try the following:

```
> plot(dist ~ speed, data= cars)
```

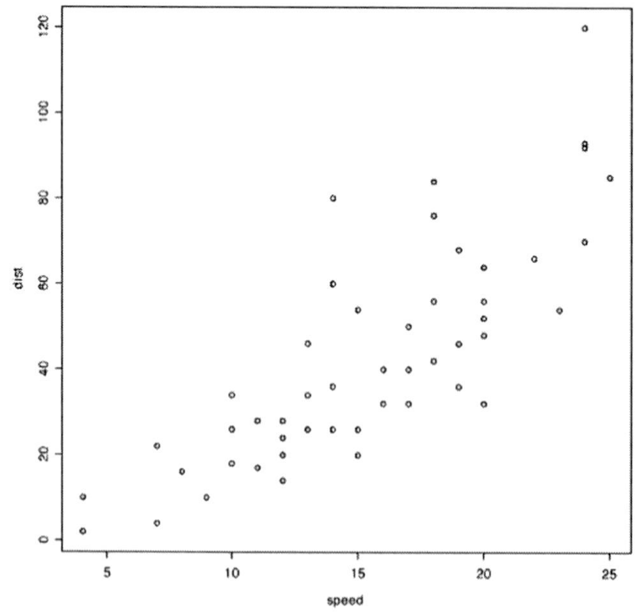

Figure 252. A basic scatter plot in R

This basic scatter takes the axis labels from the variables and uses open circles as the plotting symbol. As usual with R we have a wealth of additional commands at our disposal to beef up the display. A useful additional command is *abline()* to add a line of best fit. This is a command that adds something to an existing plot (like the *title()* command). For the above example we would type:

```
> abline(lm(dist ~ speed, data= cars))
```

The basic command uses *abline(a, b)* where *a* = slope and *b* = intercept. Here we use a linear model command (see Sections 8.4.2 and 11.1.2) to calculate the best-fit equation (try typing the *lm()* command separately, you get the intercept and slope).

If we combine this with a couple of extra lines we can produce a better-looking plot:

```
> plot(dist ~ speed, data= cars, xlab="Speed",
ylab="Distance", col= "blue")
> title(main="Scatter plot with best-fit line", font.main= 4)
> abline(lm(dist ~ speed, data= cars), col= "red")
```

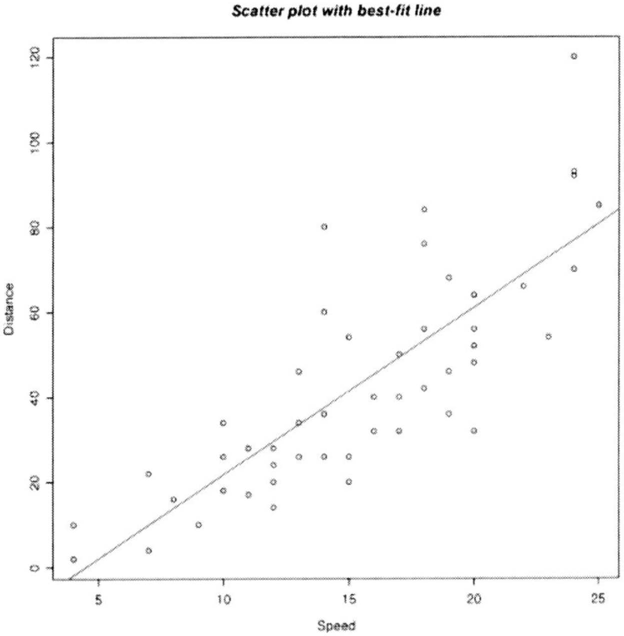

Figure 253. Scatter plot with trend line and custom axis labels

This illustrates several of the additional commands. We have set the axis labels and the colour of the plotting symbols. Next we added a main title and set the font to bold italic (try other values). Finally we set the best-fit line.

We can alter the plotting symbol using the command *pch* = *n*, where *n* is a simple number. We can also alter the range of the *x* and *y* axes using *xlim* = *c(lower, upper)* and *ylim* = *c(lower, upper)*. The size of the plotted points is manipulated using the *cex* = *n* instruction where *n* = magnification factor. Here are some commands that illustrate these parameters:

```
> plot(dist ~ speed, data= cars, pch= 19, xlim= c(0,25),
ylim= c(-20, 120), cex= 2)
> abline(lm(dist ~ speed, data= cars))
> title(main="Scatter plot with altered y-axis")
```

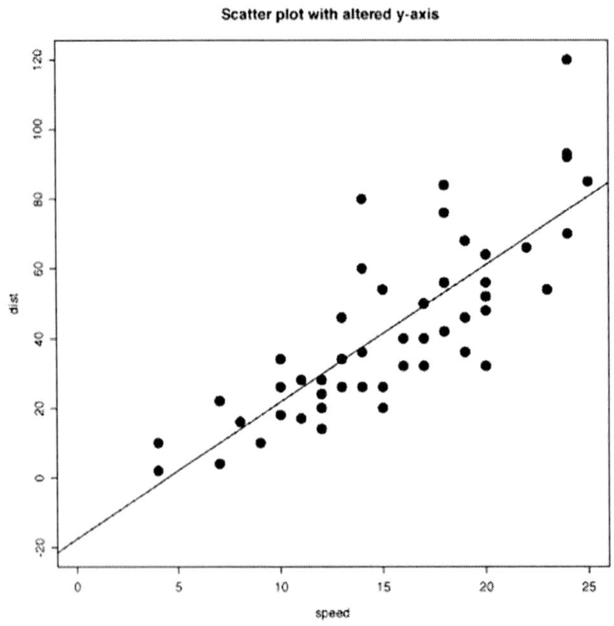

Figure 254. Scatter plot with trend line and custom plotting characters

In Figure 254 the plotting symbol is set to 19 (a solid circle) and expanded by a factor of 2. Both *x* and *y*-axes have been re-scaled (so we can see the line cross the *y*-axis clearly). The labels on the axes have not been specified and default to the name of the variable (which is taken from the dataset).

12.5.5 Stem and leaf plots, *stem()* command

A very basic yet useful plot is a stem and leaf plot. It is a quick way to represent the distribution of a single sample (we met this previously in Section 4.2). The basic command is:

```
> stem(variable)
```

Here is a vector of numbers saved as the variable test.data:

```
[1] 2.1 2.6 2.7 3.2 4.1 4.3 5.2 5.1 4.8 1.8 1.4 2.5 2.7 3.1 2.6 2.8
```

To see the stem plot of these data, type:

```
> stem(test.data)
  The decimal point is at the |

  1 | 48
  2 | 1566778
  3 | 12
  4 | 138
  5 | 12
```

R takes the values and re-arranges them in numerical order. The top row shows the smallest values. We can now see quite clearly that the data are not normally distributed. This is a useful command for moderately small samples as you can easily reconstruct the original data from the plot. Here we see that we have a smallest value of 1.4 then 1.8. The next row shows us values of 2.1, 2.5, 2.6, 2.6 and so on.

12.5.6 Pie charts, *pie()* command

Pie charts are not necessarily the most useful way of displaying data but they remain popular. We can produce pie charts easily in R using the basic command *pie()*.

To begin with, we need our data to be in the right format. This is one of the hard things about learning R; what sort of data/object is required for what. If you get your data into the wrong format it can be quite frustrating! The *pie()* command will work on a simple list of values (called a vector). The *pie()* command will also work on a matrix (we looked at an example in Section 12.4.1). It will not work on a data frame, which is the most common sort of data item that we use. This is not a great problem as we can get around this quite easily.

What we want to end up with is a dataset that looks something like Table 84.

Table 84. Data for pie chart. Each column will produce a separate slice of pie

First	Second	Third	Fourth	Fifth	Sixth
12	16	25	11	6	4

We can make our data in one of several ways. Here we'll focus on one way to keep things simple. To start with, organise your data into a CSV file. Make a file with multiple columns, give each column a title and a single value (to plot). This is the most usual way we are going to make and store our data. When we look at our data in R they look something like the following:

```
  First Second Third Fourth Fifth Sixth
1    12     16    25     11     6     4
```

We see an extra 1 at the beginning; this is telling us that there is only one row of data (it might also alert us to the fact that the item is a data frame). We need to have a matrix to make a pie chart and we could make our item into a matrix permanently using the *as.matrix()* command. Alternatively, we can convert it temporarily as we create the pie chart.

To produce a simple pie chart, type the following:

```
> pie(as.matrix(pie.data)[1,])
```

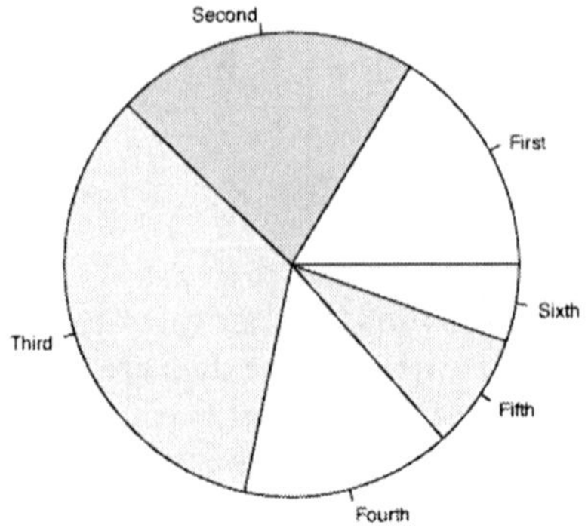

Figure 255. Pie chart drawn using R with all defaults

This is a basic chart; we can see that the names of the columns have been appended to each slice. We added the square brackets at the end [1,], telling R explicitly that we wanted the first row of data (and all columns). If we missed this out we would still get our chart but it would not have the slices labelled. We can add a title in the usual way using the *title()* command.

By default the slices are presented in anti-clockwise order; we can alter this by adding a simple command *clockwise = TRUE.*

The colours are set to pastel shades by default; to alter them you can add a list of colours to the command line in the form *col = c("col1", col2", col3").* Here is the finished article:

```
> pie(as.matrix(pie.data)[1,], clockwise=TRUE, col=
c("red", "orange", "yellow", "green", "blue", "purple"))
> title(main="Clockwise Pie Chart with custom colours",
font.main= 4)
```

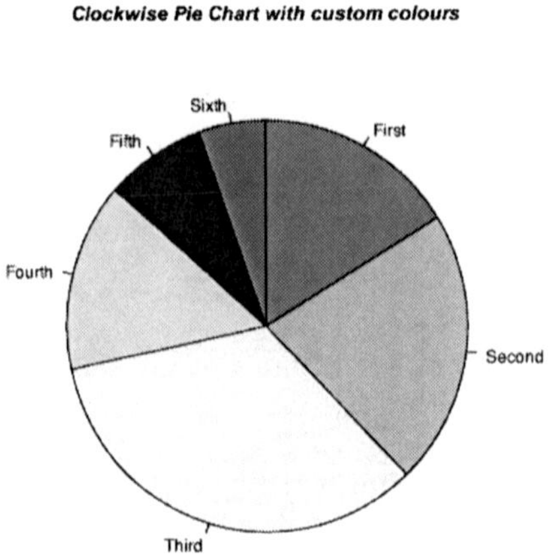

Figure 256. Pie chart drawn clockwise with custom colours

In Figure 256 we have clockwise slices with our own selection of colours. The title was set with a separate command and the font set to bold italic. The colours are a bit lurid but we have plenty to select from. We can see the colours available by typing one of the following:

```
> colours()
> colors()
```

Note that R is programmed to recognise British or US spelling.

If we have a more complex matrix we can select other rows or columns to form our pie chart. We use the square brackets to tell R what we require; the convention is [*rows*, *columns*]. So, if we require the first column we use [,1]. By leaving the rows part blank, all rows are selected by default.

12.5.7 Line plots

Previously we learnt about bar charts (Section 12.5.1), histograms (Section 12.5.2), box–whisker plots (Section 12.5.3) and scatter graphs (Section 12.5.4); however, there may be occasions when we wish to display data as a line, perhaps to show a time series. There is no specific line plot command in R so we must use other graph types and coerce the program to produce our line.

Plot types

If we produce a plot we generally get a series of points. The default symbol for the points is an open circle but we can alter it using the *pch* = *n* parameter (see the section on scatter plots). Actually the points are only one sort of plot type we can achieve in R (the default). We can use the parameter *type* = "*type*" to create other plots (Table 85).

Table 85. Summary of plot types that may be appended to the *plot()* command

Command	Plot type
type = "p"	Produces points.
type = "l"	Produces line segments.
type = "b"	Produces points joined by line segments.
type = "o"	Similar to "b" but the points are overlaid onto the line.
type = "n"	Produces a graph with nothing in it! This can be used to create a graph frame that you add lines to later.

So for example we may type: plot *(x, y, type* = *"b")* to produce a simple line plot with added points (in other words both lines and points).

Time series

Rather than a series of *x*, *y* data, you may have a single time series. Here is an example of data to illustrate:

```
> Vostok

   month temp
1  Jan   -32.0
2  Feb   -47.3
3  Mar   -57.2
4  Apr   -62.9
5  May   -61.0
6  Jun   -70.6
7  Jul   -65.5
8  Aug   -68.2
9  Sep   -63.2
10 Oct   -58.0
11 Nov   -42.0
12 Dec   -30.4
```

This shows the mean monthly temperatures for an Antarctic research station. The file was read using the standard *read.csv()* command and so contains two columns: month is a factor and temp is a numeric variable.

If we attempt to plot the whole variable, e.g. *plot(temp ~ month),* we get a horrid mess (try it and see). This is because the month is a factor and cannot be represented on an *x*, *y* scatter plot.

If we plot the temperature alone, we have the beginnings of something sensible:

```
> attach(vostok)
> plot(temp)
```

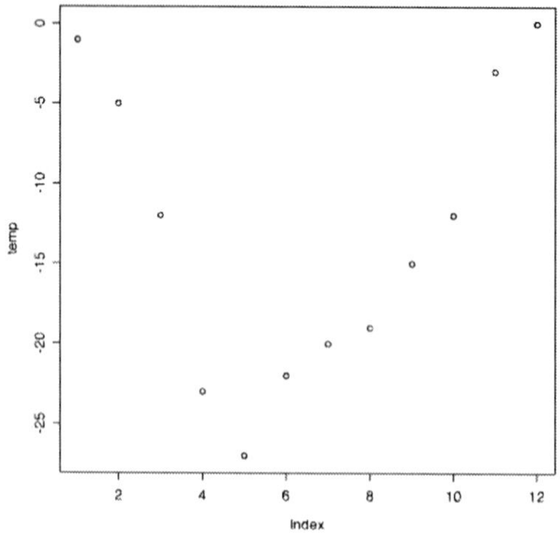

Figure 257. Scatter plot of time series. This first attempt produces a basic plot

Note that we used *attach()* here to allow R to get at the variables within the *vostok* data item. So far so good – there appear to be a series of points and they are in the correct order. We can easily join the dots to make a line plot by adding *(type = "b")* to the plot command (see

Table 85). Notice how R has used default labels for the axes: temp for the *y*-axis is taken from the values in the variable, but index is used for the *x*-axis because we have no reference (we only plotted a single variable).

What we need to do next is to alter the *x*-axis to reflect our month variable.

Custom axes

When we look at the time series plot produced above (Figure 257), we see that the *x*-axis needs a bit of work. Since the plot was made from a single variable (temp) there are no values for *x* and R substitutes a numeric index.

We need to scrap the current axes and start again with our own. It is simple to produce a plot with no axes, merely add *(axes = FALSE)* to the plot command like so:

```
> plot(temp, axes= FALSE)
```

R appends default labels to the axes so we need to remove those:

```
> plot(temp, axes= FALSE, xlab= "", ylab= "")
```

That does the job. We are going to add axis labels of course so could have specified them now but we use "" (double quotes) to illustrate how to produce blank ones (setting *xlab* = *FALSE* produces a label FALSE so we have to use "").

To add an axis, we use the *axis()* command. Axis 1 is the bottom of the plot (i.e. the *x*-axis); axis 2 is the left side of the plot (the *y*-axis). We can also specify the top (3) and the right side (4) if we wish. In its simplest form, axis(*n*) adds the axis specified with its default parameters. This won't do here because the default *x*-axis contains only index information. We need to tell R where to find the labels associated with the axis.

To generate an axis, we need to specify the length of it and the labels to be used. Here is what we need for our temperature example:

```
> axis(side = 1, at = 1:length(temp), labels = month)
```

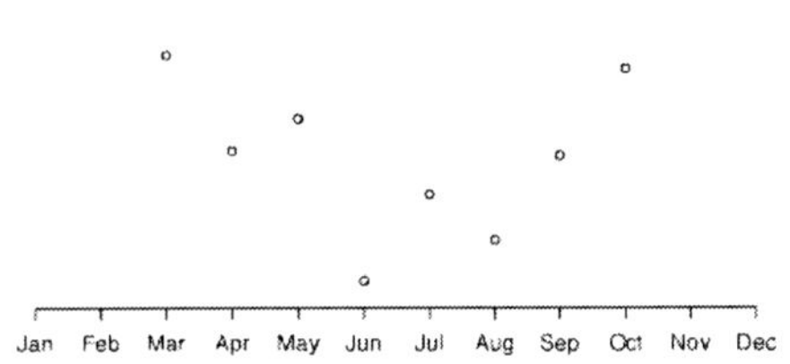

Figure 258. Time series as a scatter plot. The *x*-axis is customised to read from a text variable

The *side* = *1* part tells R that we want the bottom axis (the *x*-axis). The *at* = part works out how long the axis should be (the number of tick marks if you like). Finally we tell R to take the labels from the month data. Now we need to add in the *y*-axis and the axis labels. We could also add a title and perhaps the whole thing would look better if the dots were joined up to make a lineplot (which was after all the point of the exercise). Here is the whole series of commands from start to finish.

```
> vostok= read.csv(file.choose())
> attach(vostok)
> plot(temp, axes=FALSE, xlab="", ylab= "", type= "b")
> axis(1, at = 1:length(temp), labels = month)
> axis(2)
> title(main= "Time Series", font.main=4, xlab= "Month",
ylab= "Mean Temp C")
> box()
> detach(vostok)
```

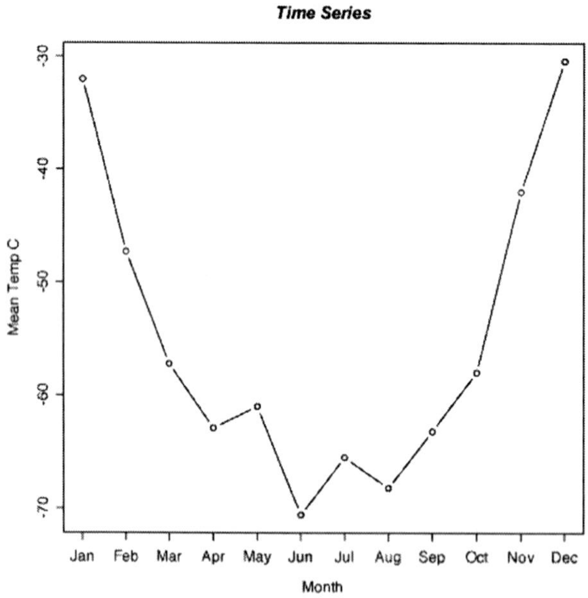

Figure 259. Completed time series plot with custom axes and points joined

The *box()* command merely adds a border around the plot. This looks a lot better. It is possible to alter the plot character and the colour of the lines using the *pch* = and *col* = commands that we've seen before.

12.6 Worked example graph data in R

We have seen some examples of producing graphs in the preceding sections. Here we will examine some data and use these as the basis for producing some graphs using R to highlight the differences to the approach used in Excel. We will use the same data that was used in Section 12.4.3, where we produced several graphs in Excel.

Bar chart

The first step is to get our data. Usually this will be in a spreadsheet and we need to ensure that we save a copy as a CSV file. We begin by importing the data to a named object.

```
> bf = read.csv(file.choose())
```

Now we can simply type *bf* to see the data. We might have data in one of two forms. In Section 12.4.3 we had the data as three separate columns, one for each sample. We might also have the data in recording format with two columns, one for the abundance data and one for the site name. The approach we take will depend on which layout we have. We will look at the multi-column data first. Here we see the data for our butterfly and habitat data:

```
> bf
   grass heath arable
1      3     6     19
2      4     7      3
3      3     8      8
4      5     8      8
5      6     9      9
6     12    11     11
7     21    12     12
8      4    11     11
9      5    NA      9
10     4    NA     NA
11     7    NA     NA
12     8    NA     NA
```

The samples do not contain the same number of observations (replicates) and R fills in the missing values with NA so that the final object is a rectangular data frame. What we want is to create a set of mean values and standard errors from these three samples, then we can produce our bar plot. There are a several ways we can do this but the flexible command *apply()* is what we will use.

The basic *apply()* command takes the following form:

```
> apply(data, row-col, FUN)
```

The *apply()* command applies a function to each column (or row) of a data frame. That is what we want and we apply the *mean()* command to each column as shown below. The 2 in the command tells R we want to apply the function to the columns; if we wanted to apply a function to the rows we replace the 2 with a 1:

```
> mn = apply(bf, 2, mean)
> mn
```

```
    grass      heath    arable
 6.833333         NA        NA
```

Something appears to have gone awry; we get NA for two of the samples. The problem is the NA values in the data. We can easily deal with them but we need to modify the command like so:

```
> mn = apply(bf, 2, mean, na.rm = TRUE)
> mn
    grass      heath    arable
 6.833333  9.000000 10.000000
```

We need to account for the missing values. We simply add *na.rm = TRUE* to the command and the means are reported correctly. We want to add error bars so we determine the standard deviation, which we will use to help calculate the standard error. We use a similar approach to above, like so:

```
> std = apply(bf, 2, sd, na.rm = TRUE)
```

To work out the standard error we need the number of values in each sample (recall that *SE = Std. Dev./Sqrt[n]*); however, if we try to do this using the *length()* command we get an error. Rather annoyingly, the *length()* command cannot deal with NA values (there are good reasons for this but it is still frustrating for us right now). We need a slightly different approach.

We have already determined the mean, if we could determine the sum then we can get the number of replicates as *mean = sum(x)/n*. We can do this in one go like so:

```
> n = apply(bf, 2, sum, na.rm = TRUE) / mn
> n
 grass  heath arable
    12      8      9
```

The final values we need are for standard error and we can use a simple formula:

```
> se = std / sqrt(n)
```

We have now created mean (*mn*) and standard error (*se*) and can construct our plot. Before we do that, we should consider how we would create these values if our data were in recording format, i.e. two columns, one for count and one for site:

```
> bf
    count   site
1       3  grass
2       4  grass
3       3  grass
4       5  grass
5       6  grass
6      12  grass
```

```
7       21  grass
8        4  grass
9        5  grass
10       4  grass
11       7  grass
12       8  grass
13       6  heath
14       7  heath
15       8  heath
16       8  heath
17       9  heath
18      11  heath
19      12  heath
20      11  heath
21      19 arable
22       3 arable
23       8 arable
24       8 arable
25       9 arable
26      11 arable
27      12 arable
28      11 arable
29       9 arable
```

The *apply()* command we used before will not work because the second column contains labels for the site names. What we must do is to *unstack()* the data to produce separate samples:

```
> unstack(bf, form = count ~ site)
$arable
[1] 19  3  8  8  9 11 12 11  9

$grass
 [1]  3  4  3  5  6 12 21  4  5  4  7  8

$heath
[1]  6  7  8  8  9 11 12 11
```

Here we tell R to use the count and site data explicitly using the *form* = part. Since we only have two columns, this is not strictly necessary; however, it is a good habit and helps you remember what data you are dealing with. Because the samples are unequal in size, R produces a list item rather than a matrix. The point is that lists are separate things of differing size whereas a matrix is a rectangular block of data. Recall from earlier where R padded out the data frame with NA.

We can use a modified form of the *apply()* command to get the summary values we need. Below we see the result of determining the mean:

```
> mn = sapply(unstack(bf, form = count ~ site), mean)
> mn
   arable     grass     heath
10.000000  6.833333  9.000000
```

Of course we do not need to type all these commands direct from the keyboard, once we have issued one command we can use the up arrow to recall the command and edit it. In this way we can quickly create mean, standard deviation and number of replicates. Because our data do not contain NA items we can use length as our function:

```
> sapply(unstack(bf, form = count ~ site), length)
arable  grass  heath
     9     12      8
```

In either form of data, we have now created mean and standard error items. In our example we called our summary values something short: *mn, std, n* and *se*. This saves a bit of typing! When we make our bar plot we will only need the *mn* and *se* items. The basic plot is created quite simply:

```
> bp = barplot(mn)
```

This time we give our plot a name. We will use this to help us make the error bars. Our plot is fine but the bars get a bit close to the top (R tries to fill the plot area most efficiently) and we need a bit of room for the error bars. If we type:

```
> mn + se
```

We will get a list showing us the largest values we should account for in our plot.

```
   arable     grass     heath
11.481366  7.589262 10.424001
```

We can now tell R how tall to make the *y*-axis. We can see that a *y*-axis that reaches up to about 12 would accommodate the upper error bars. If you decided to use standard deviation instead of standard error then you would need a bigger axis. We tell R the limits of the *y*-axis using the *ylim = ()* command:

```
> bp = barplot(mn, ylim = c(0, 12))
```

The resulting graph looks like Figure 260. At the moment there are no axis labels. We will add them later.

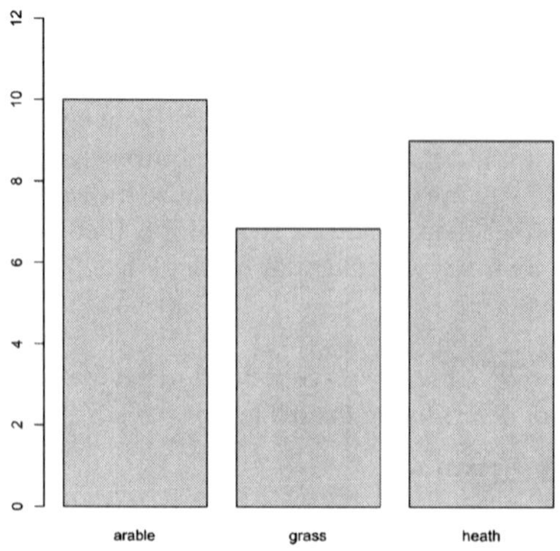

Figure 260. Basic bar plot of butterfly abundance data

We need to add error bars. We already calculated the standard error and saved the result to *se*. To draw the bars we use the *segments()* command, which draws short sections of straight line to an existing chart.

```
> segments(bp, mn + se, bp, mn - se)
```

The command requires the *x*, *y* co-ordinates of the start and end points of the line(s) you require. We called our plot *bp* for a reason: we wanted to preserve the information that R uses to generate the plot. We use it here as the *x* co-ordinate for our error bars. We will end up with a vertical line in the middle of each bar. The *y* co-ordinates are taken from the values of the bars. We want to go from a point above the bar (the distance above being determined by *se*) to a point below the bar (also determined by *se*). The result is a series of simple bars (Figure 261).

Once we have created our command, we can keep it in a simple text file. As long as you always call the bar plot *bp*, the mean value *mn* and the standard error *se*, you can simply copy and paste the command to create error bars on any bar chart.

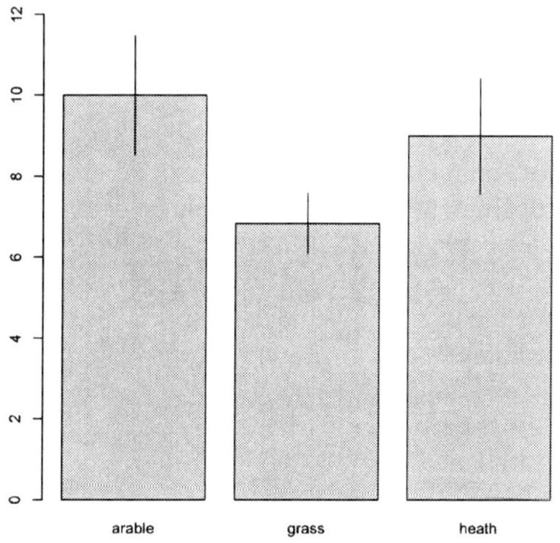

Figure 261. Simple error bars added to a bar plot using the *segments()* command

We should add titles next and it might also be nice to draw a line under the bars. We can do this with two simple commands:

```
> title(ylab = 'Abundance', xlab = 'Site')
> abline(h = 0)
```

The *abline()* command adds a horizontal line at 0 on the *y*-axis. The completed graph looks like Figure 262. Here we have added caps to the error bars:

```
> segments(bp -0.1, mn + se, bp + 0.1, mn + se)
> segments(bp -0.1, mn - se, bp + 0.1, mn - se)
```

The first command draws the top of the bars. The second command draws the cap on the bottom of the bars. See if you can work out how these commands work (remember that we need *x1*, *y1*, *x2*, *y2* as the co-ordinates in the command).

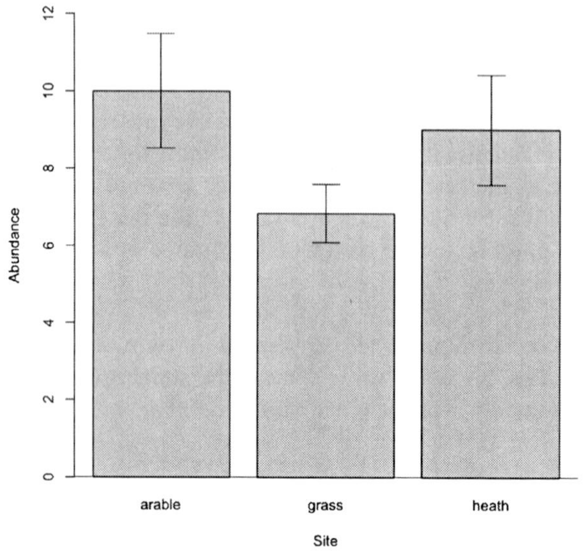

Figure 262. Final bar plot containing error bars and axis labels. Total butterfly abundance at three habitats in Shropshire. Bars show mean values, error bars show standard error (n = 9, 12, 8)

R uses default colours for the bars but we have seen previously how we can change the colour using the *col* = command (Section 12.5.1).

Creating bar plots with error bars is quite tricky but once you get the hang of them, they can be done with little problem. Generally, box–whisker plots are more useful and R is able to generate them with great ease, as we shall see next.

Box–whisker plot

The box–whisker plot is generally more useful than the bar chart. The former displays a greater range of information in a compact manner than the latter. Probably because of this, the routines for creating box–whisker plots in R are a lot more powerful and easier to use.

We will use the butterfly abundance data from previous examples. In the first case we have three separate columns (one for each sample) and in the second case we have two columns (one for count and one for site). Either way, we first start by reading the data from the spreadsheet (in CSV form) as we did previously when creating the bar chart.

When we have separate columns containing numerical values, we can use the following command:

```
> boxplot(bf, range = 0)
```

This produces something like Figure 263. We assume here that we called our data *bf*. The *range = 0* part tells R to extend the whiskers to the full range of the data. The default setting is 1.5 and this causes the whiskers to extend to 1.5 times the inter-quartile range. Any values outside of this are shown as single points, outliers.

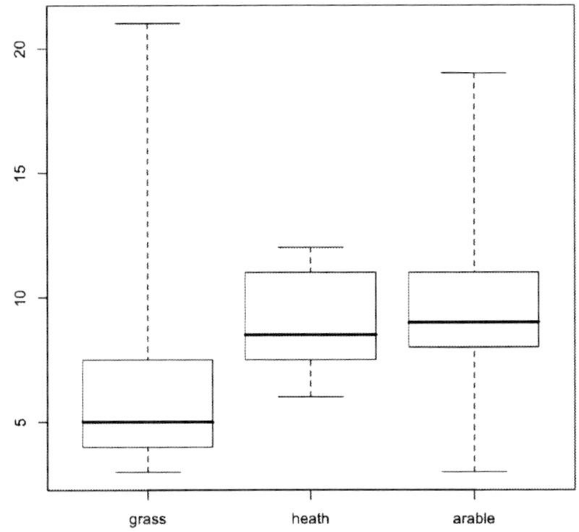

Figure 263. A simple box–whisker plot produced from data with three sample columns

When we have data in recording format, i.e. a separate column for each variable, we need to tell R what to plot:

```
> boxplot(count ~ site, data = bf, range = 0)
```

This produces a graph like Figure 264. Here we mention the variables explicitly; we also must tell R where to find these variables using the *data =* part.

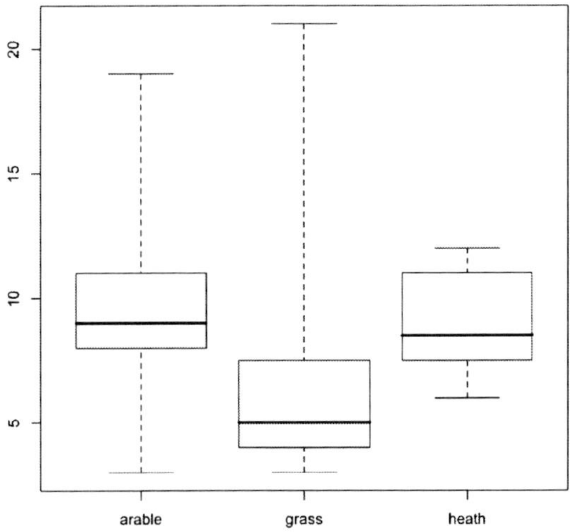

Figure 264. Simple box–whisker plot produced from data in recording format (i.e. with one column per variable)

Notice how the plots are identical apart from the order in which the samples are displayed. R has taken the sample names from the data. We can now finish the plot by adding axis titles:

```
> title(ylab = 'Abundance', xlab = 'Site')
```

We can choose to produce a box plot from any number of samples simply by typing their names into the command. For example:

```
> boxplot(bf$grass, bf$heath, bf$arable, range = 0)
```

This would produce a plot of each of the three samples. In this case the three samples are together in one data frame but they could be separate items; however, when we do this the samples are not labelled with the name – we get an index value (1–3 in this case). We can get over this by adding a *names* = part to the command:

```
> boxplot(bf$grass, bf$heath, bf$arable, range = 0, names
= c('grass', 'heath', 'arable'))
```

Notice how the list of names is in the form *c("name1", "name2", "name3")*. In the example above, the three samples were inside the *bf* data so we had to use *bf$name* to define each one. If we had used the *attach()* command or the objects were separate variables in the memory then we would not have needed to do that.

The *boxplot()* command is very flexible and powerful and is therefore the display method of choice for many situations.

Scatter plot

We use a scatter plot when we wish to show a relationship between two variables. In Section 12.4.3 we examined data on mayfly abundance in relation to stream flow; these data are shown again:

```
> mf
  abund flow
1     9    2
2    25    3
3    15    5
4     2    9
5    14   14
6    25   24
7    24   29
8    47   34
```

The first step is to read the data from a CSV file in the usual manner:

```
> mf = read.csv(file.choose())
```

Here we call the data *mf*. In general we will have our data in this form, one column for the response variable (the *abundance* in our case) and one for the predictor variable (the *flow* in our example). It is possible that we may have two separate items, which we wish to plot. We can deal easily with either case.

We can produce a scatter plot from our data by using:

```
> plot(abund ~ flow, data = mf)
```

Alternatively we can issue this command:

```
> plot(mf$flow, mf$abund)
```

In the first instance we select the response and predictor variables from within the *mf* data item. Unless the *mf* item has been attached using *attach(mf)*, then we need to tell R where the variables are to be found.

In the second case we name the *x* and *y* variables individually. We use $ to read the columns inside the *mf* data frame. If the data were attached or otherwise in the memory, then we could type the names without the first part:

```
> plot(flow, abund)
```

The axes will be labelled with either the column names or the names we typed, according to which option we selected. If we use the *y ~ x* approach, then the column names will appear. If we use *x, y,* then we get whatever we typed. We can create our own axis labels using *xlab* and *ylab* commands:

```
> plot(mf$flow, mf$abund, xlab = 'Flow', ylab = 'Mayfly
Abundance')
```

This produces a scatter plot that looks like Figure 265.

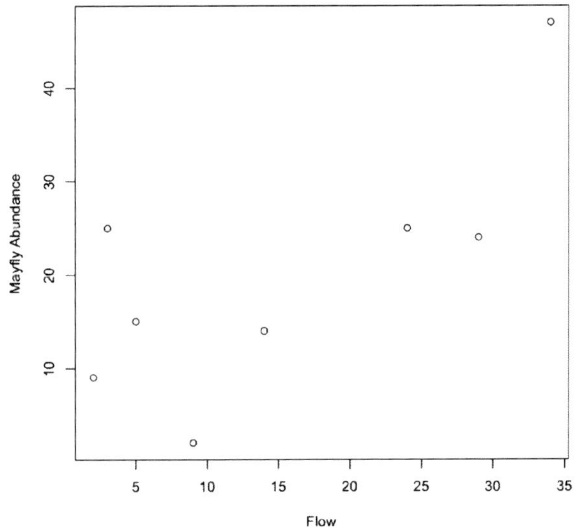

Figure 265. Simple scatter plot. Created using default plotting points

We may wish to alter various parts of the graph. For example, the plotting points are a trifle small. We can use additional commands to alter the plot character as well as its size and color.

```
> plot(mf$flow, mf$abund, xlab = 'Flow', ylab = 'Mayfly Abundance', pch = 16, cex = 2, col = 'blue')
```

Here we change the character to a solid circle (*pch* = *16*), make it twice normal size (*cex* = 2) and coloured blue (*col* = *"blue"*).

We often want to add units to axis titles and they might involve superscripts or subscripts. We can do this in R but it required a tiny bit of juggling. We need to use the *expression()* command. This is a special command that allows us to create labels with odd formatting.

We make superscripts using the ^ character and subscripts by enclosing in square brackets []. To create a label for our *x*-axis, we would use the following instead of regular text in the *xlab* = part of the command:

```
> expression("Flow ms"^"-1")
```

This would make the –1 appear as superscript to read *Flow* ms^{-1}.

To create a subscript in a label, we might use something like the following in the *ylab* = command:

```
> expression("Mayfly Abundance"["per net"])
```

This would make per net appear as subscript to read *Mayfly* $Abundance_{\text{per net}}$.

We can edit our plotting command to produce the final result:

```
> plot(mf$flow, mf$abund, xlab = expression('Flow ms'^'-1'), ylab = expression('Mayfly Abundance'[' per net']), pch = 16, cex = 2, col = 'blue')
```

This produces a graph that looks like Figure 266.

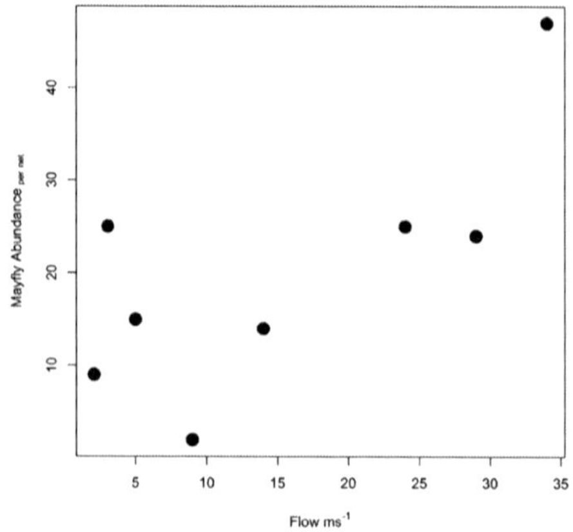

Figure 266. Scatter plot using custom plotting characters and colours and with axis titles including special formatting

We might consider adding a line of best-fit to our plot. In this case we should not because we are not looking at a linear relationship (i.e. we did not use Pearson correlation, we used Spearman's rank). The command to add a line is *abline()*, which we met in Section 12.5.4.

Pie chart

Pie charts are used to display proportional data. The human eye/brain is not too good at interpreting angular information and pie charts have fallen a bit out of favour recently. Most times you can find an alternative method to produce a display. Here we will use the bird/habitat data that we looked at previously (Table 81 in Section 12.4.3).

The first thing we need to do is to get our data into R. We generally want to get our data into a CSV file so that we can use the *read.csv()* command; however, we want to use information arranged by rows as well as columns. In our set of data we have two sorts of category: columns that represent habitats and rows that represent bird species. We would like to be able to present data by row and to use the name of the species as a label. It is easier to manage if we set the first column (bird species names) to become the row names for our data. We can set this easily by adding a new parameter to our basic command like so:

```
> bird = read.csv(file.choose(), row.names = 1)
```

This allows us to find and select the file from our computer and also tells R to take the first column and use it as row names rather than as data. Once we have imported the data, we see something like this:

```
> bird
              Garden Hedgerow Parkland Pasture Woodland
Blackbird         47       10       40       2        2
Chaffinch         19        3        5       0        2
Great Tit         50        0       10       7        0
House Sparrow     46       16        8       4        0
Robin              9        3        0       0        2
Song Thrush        4        0        6       0        0
```

We can tell that the row names are set explicitly as we see the names of the bird species; usually we would see plain numbers representing each row. To make a pie chart we need either a data matrix or a single vector of numbers. The data we actually have is a data frame. The simplest way to get around this is to use the *as.matrix()* command.

We can check our data and see what we have using the *str()* command (think of it as short for structure). In general, if we have a data frame we will see the words *data.frame* along with some summary information regarding each column. If we have a matrix then we will see information regarding the main body of the matrix and additional information about the row and column data:

```
> str(bird)
'data.frame': 6 obs. of  5 variables:
 $ Garden   : int  47 19 50 46 9 4
```

```
 $ Hedgerow: int  10 3 0 16 3 0
 $ Parkland: int  40 5 10 8 0 6
 $ Pasture : int  2 0 7 4 0 0
 $ Woodland: int  2 2 0 0 2 0
```

```
> str(pd)
 int [1:6, 1:5] 47 19 50 46 9 4 10 3 0 16 ...
 - attr(*, "dimnames")=List of 2
  ..$ : chr [1:6] "Blackbird" "Chaffinch " "Great Tit" "House Sparrow " ...
  ..$ : chr [1:5] "Garden" "Hedgerow" "Parkland" "Pasture" ...
```

It is usually pretty clear; in the example above we see the same information presented in two forms. The first line tells us clearly that we have a data frame with six rows and five columns. The data were converted and saved to a new object (*pd*) using the *as.matrix()* command.

```
> pd = as.matrix(bird)
```

We might decide to create a new data item/object by this method but we can simply incorporate the *as.matrix()* command inside the command to draw the pie chart:

```
> pie(as.matrix(bird)[1,])
> pie(as.matrix(bird)[,1])
```

In the first case we present the first row and in the second case we select the first column to draw. The data are converted to a matrix (on-the-fly in this instance) and the segments/slices are labelled appropriately. The second option produces a pie chart that looks like Figure 267.

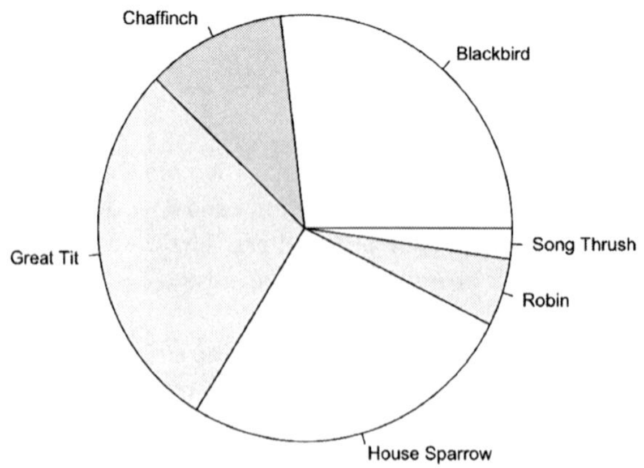

Figure 267. Pie chart produced from matrix data. Bird species visits to garden habitat

We need to decide which row or column we want to present and replace the number in the square brackets with the appropriate value. We might also use the name of the row or column header:

```
> pie(as.matrix(bird)['Robin',])
> pie(as.matrix(bird)[,'Parkland'])
```

Notice how the names must be in quotes (single or double is fine as long as you are consistent).

12.7 Graphs: a summary

In the preceding sections we have seen a variety of graphs and discussed the various methods to produce them using Excel and R. Whenever you have data, you should strive to find the best way to represent and summarise the data graphically. Your graphs are potentially the main way in which you will report your results (or at least they may be the most likely to be read) and it is therefore important that your graphs convey as much information as possible but in the clearest manner. In Table 86 we see a summary of the different types of graph and their potential uses.

Table 86. Summary of graph types and their uses

Type of graph	Uses
Bar chart	Data in categories. Data in each category can be further split into several bars or as a stacked chart. When used to summarise sample data, error bars can be added.
Box–whisker plot	Sample data summary. Conveys a lot of information in a compact space. This is probably the most useful general plot type.
Scatter plot	Correlations. Showing the relationship between two variables.
Pie chart	Proportional data. The human eye is not too good at working out angular information so the pie chart might be better replaced with a bar chart.
Histogram	Data distributions. The classic method of displaying data distribution. Used in conjunction with a density plot (see below) can be helpful.
Density plot	Data distributions. Can be useful as an alternative to the histogram or plotted over the top of an existing histogram. Also useful to compare two distributions on the same plot.
Stem and leaf plot	Data distributions. This type of plot allows you to see the distribution of the data easily and also the original data.

Whatever type of graph you produce, it is a good idea to step back and review your work once it is complete. Check that all the elements of the graph play a role and that the graph is not too cluttered. An often-overlooked (but important) element of the graph is the caption. This provides an opportunity to add information that allows the reader to gain a more complete picture of the situation. Ideally the reader should be able to look at your graph, read its caption and not really need to look at the text at all!

12.8 Writing papers

The aim of a paper is to disseminate your results as widely as possible. It is all about communication and the "who, what, where, when" theme still applies. You need to say who you are, what you did, where you did it and what the subject was.

12.8.1 Elements of a scientific paper

Before you attempt to write a paper you should have read plenty of papers on related subjects. This will familiarise you with the layout and format of a scientific paper. If you are sending to a particular journal, check their format carefully because they will have slightly different rules to others even though the basic format remains the same. Table 87 describes the basic elements of a scientific paper.

Table 87. The elements of a scientific paper

Element	Purpose
Title	The title is the first thing that is generally seen so it needs to be clear and succinct and yet make the reader want to dive in.
Contact details	This is not just advertising; other researchers may want to contact you to discuss your work and perhaps to collaborate in further research. Include your institution, address and email.
Abstract	This is a complete précis of the work, including stats results. If you have a lot of results you can include only the most important ones. This needs to be brief but remember that this may be the only thing that is read. Don't include citations.
Keywords	Usually the journal offers you the chance to add a few keywords. If your paper ends up being added to a database, these may be the search items. In addition, the reader can see what you think are the key elements. You can add the species name for example.
Introduction	This is where you set the background to the subject area. It is also the place where you show off how much background reading you have done. Lead the reader into the subject in general and hit them with your specific aims at the end. In general, detailed hypotheses are not given but in student works they might be expected.
Methods	This is where you describe what you did and where. You should include site descriptions and possibly even species details (sometimes they are better in the introduction). The point is that a reader should be able to repeat your work following the instructions you provide. Avoid equipment lists and bullet points.

Element	Purpose
Results	This section details what you found. Raw data are not generally included although species lists could go in as long as they are not too long. The point here is to summarise: present averages, dispersion and how big the samples are. Give stats results too. Say everything in words as well as referring to graphs (or tables).
Discussion	This is where you say what you think it all means and to summarise the findings. Try not to bring in too much that is new (the bulk of your citations go in the introduction).
Acknowledgements	This is where you say thank you. Avoid over sentimentality but do put in people like landowners and anyone who put up money.
References	Everything you cite in the text goes in the reference list. Check out the formatting requirements of the journal; a good start is to use the standard Harvard layout.
Appendix	This is for stuff that you really don't want anyone to look at but which really dedicated obsessives might want to see. Long, complicated species lists and tables of raw data might be things you'd include for example.

12.8.2 Writing style

The idea behind writing a paper is to get across the main elements of your work as succinctly and clearly as possible. Different branches of science have subtly different styles but there are many elements in common. It is most common to use a passive style in reporting what was done. In other words we say, "the vegetation was sampled" rather than "I sampled the vegetation". This leads to a rather impersonal style and this is exactly the point. A research paper should be about the work that was carried out, not about how great you are. If you turn out a good paper, people will realise how great you are!

The passive style forces you to think carefully about how you phrase your sentences and leads to clearer work. Try to use short words and avoid jargon wherever possible, unless of course the word you want is specific to your field and is the best available. Your aim is to aid understanding, so use the shortest and simplest way to get across each point and avoid ambiguity.

12.9 Plagiarism

Plagiarism is a form of stealing. Essentially it involves you setting forth someone's work and passing it off as if it were your own. Of course you need to use previous knowledge in your work but you need to acknowledge where your knowledge/information came from.

There are three types of knowledge:

- Common knowledge. This is stuff that everyone knows that does not need to be referenced, e.g. "the fox is a mammal", "France is in Europe". These are generally understood by everyone to be true and self-evident. They are not likely to be central to your hypothesis.

- Knowledge and opinions that belong to someone else. For example, "IUCN consider this species to be seriously endangered". You should not state, "This species is seriously endangered" without referring to the original authors. The conclusion that this species is endangered came from someone else's work, so you refer to them. This sort of fact tends to be in your introductory material.
- Knowledge and opinions that you have developed yourself. These are generally in your results and conclusions. They stem from your research so there is no need to reference them. Of course if you are referring back to some previous work that you prepared, naturally you should refer to this as if it were someone else's.

The key to avoiding plagiarism is to know how and when to cite references.

12.10 References

References come in two parts. There is a bit in the text that essentially says "look at this for information" and a list at the end that gives the original sources. The first part is called the citation and you generally quote the source as (author, date) in brackets. The second part is the references themselves. If you do not include citations in the text, the list of sources is more properly called a bibliography (essentially a list of the material you used in preparing the work).

There are two main ways to indicate that you wish to refer to some other work: you can give the citation in brackets like so (Gardener, 2010) or (Figure 23), or you can use numerical form. (number citations as superscripts[1] or [1] are the two most common). When we use citations as (name, date), our reference list is sorted in alphabetical order. When we use a number, we sort the references by the order in which they appear (i.e. numerically). Generally, the (name, date) approach is preferred. Once you are familiar with a subject, you tend to recognise names of authors and a name means something whereas a number is meaningless without the corresponding reference. The number style tends to be adopted in more "popular" works and in places where you do not wish to interrupt the narrative. For most scientific publishing, the (name, date) style is most appropriate.

There are two main ways to use a citation. We usually refer to an area of study and give examples of the research, e.g. "Studies have shown that this plant is common in this type of habitat (Crawley, 1999)". On the other hand you may refer to the individual directly, e.g. "Crawley (1999) showed that this plant is common in this habitat". In the second form, we do not need to repeat the author in the brackets. As a general rule it is good to get your citations at the end of sentences so the more impersonal first example would be preferred. Where we have more than one author, our approach depends on how many there are. If there are only two, we give both, e.g. "(Gardener & Gillman, 2002)". Where there are more than two, we present only the first author and add *et al.* to the name, e.g. "(Chamberlain *et al.*, 2010)". The *et al.* part is Latin for "and others" (it is short for *et alia* so the *al.* bit has a full stop) and is usually in italics. Some journals prefer you to give up to three authors names so you should check before you write too much. Where we have two different references by the same author(s) and in the same year, we append letters, e.g. "(Gardener & Gillman, 2001a, 2001b)".

Where you need to refer to a specific part of a larger work, perhaps a figure or table, you can append this to the citation, e.g. "(Smith, 2008, figure 2.1 p. 87)". This allows you to refer to another part of the work later but still only include a single reference at the end.

The reference list itself is an important part of the report. This is where you demonstrate where all your sources came from. There are many different formats of setting out reference lists and different journals require different layouts; however, the underlying set-up is the same and is based on the Harvard system. A reference needs to contain a variety of information:

- Author name(s). We give all the names along with the initials (note that the initials do not appear in the citation).
- Year of publication. Only the year is given although there are some exceptions so it is a good idea to check the rules for the publication you are aiming at.
- Title of the work. For most published works, this is not a problem but if you are referring to a website then it can be tricky. If you cannot find an official title, you may have to make one up ("homepage" is a good standby).
- The place it was published. This is a broad category as the exact information depends on the type of publication. For a journal article we give: journal, issue and pages. For a book we give the publisher and the city of publication. For a website we give the URL.

When we refer to a website and therefore present a URL, we add an additional piece of information, [date accessed]. This is because websites can be changed.

All references need an author, date, title and the place published. You can see that this is not so different from the basic tenets of biological recording: who, what, where, when. Before you write too much, you should check out the style required.

12.11 Poster presentations

If you give a talk you may be able to speak to fifty people in a room. At a large conference there are usually many talks being given simultaneously and there may be people who were unable to get to your talk. The poster session allows you to make a presentation, which is left on display for hours, sometimes days. It can potentially reach hundreds of people at the meeting because it is hanging around for so long. Generally, there is a set session where you stand by your poster and present it to anyone who expresses an interest; otherwise it stands alone.

Your poster needs to have impact; it may be glimpsed from some distance away so make text large. Do not put too much detail. You are attempting to summarise your work in a single page, albeit a big one. You need to reduce the information to the absolute minimum to get your message over. All the major elements need to be present: title, introduction, methods, results, discussion etc., which can be a challenge.

There is no set layout for a poster; you may have the background information at the top and follow down the page or you may use a more radical layout – as long as the reader can follow what you have done that is okay. Many posters are created from a single sheet of paper (A1 or A0 for example) but you can make perfectly good posters from separate sheets of A3. Each conference will have a slightly different area available.

- Use big fonts. Can you read the headings from 2–3 metres away?
- Reduce clutter, use bullet points and keep text short and snappy.

It can be useful to have a supply of A4- or A5-sized posters as handouts that people can take away (you can pin a container under the poster). Make sure that you can read it at this size without a magnifying glass!

12.12 Giving a talk (PowerPoint)

PowerPoint is a great tool for presenting your findings to an audience in a talk setting; however, it can also be something of a distraction so here are a few key things to remember about using PowerPoint:

- Make text large.
- Avoid long sentences.
- Use bullet points as a memory aid for yourself and don't type up your talk onto the screen.
- Make graphs and images large and clear.
- Avoid fancy animation effects, they merely distract from your main points.
- Try to bring in items as bullet points one at a time rather than all at once; the audience will start to read your items before you get to them and will therefore not be listening.
- Set out your talk/slides exactly like a paper, i.e. title, introduction, aims, methods, results, discussion, final conclusion(s), acknowledgements.
- Have a blank slide at the end.

The acknowledgements slide is very useful as it signals the end of your talk and you can end with a statement like, "Finally I would like to thank the following people, thank you for listening". This clearly signals the end and people can applaud (or throw things as appropriate) and then you can be ready to accept questions.

13. Summary

By the time you reach this point, you will have realised that there is more to *Data Analysis* than simply looking at averages. The process should begin before you even have any data to analyse. Start by planning what it is you want to do and decide upon the best way to collect the data you need. The elements of planning include research about the species you are looking at, as well as methods of sampling and how you are going to record the data once you have it. You should have a good idea of what analytical approach you are going to use before you get anywhere near using a calculator.

A pilot study may well be a useful stage in the planning process and the experience may help you to refine your original ideas and hypotheses, and enable you to collect more meaningful data. This refinement is part of the scientific method, continuous evaluation and re-evaluation. At each stage you should evaluate your work and ensure that you have done the best job before proceeding to the next stage.

Writing down the data is a stage in itself and should not be taken lightly. The way you record your data can aid your analysis later so it is important to get this right. If you record your data appropriately then it may be used easily with little further processing. If, however, you record your data in an inappropriate manner, you may require a lot of extra time and effort in re-arranging your data.

Once the data are collected, the main analytical processes can begin. This will usually start with summary data, for example averages. You may produce graphical summaries to help you visualise the data and confirm the approach for the next stage. These graphical summaries will often include checking the distribution of the data, for example using histograms. Once you have affirmed your analytical approach, you will move on to undertaking the actual analyses. There are two stages in this: one is doing the mathematical part and the other is preparing summaries of those results (graphs and/or tables). A final (but often overlooked) part of the analysis is to make some sense of results in the context of the biological system in which they were collected. You will need to do this as part of the presentation of the work.

The presentation and interpretation of the work is the final part of the process and closes the loop of the scientific process that began with the planning at the beginning. The presentation of your work may be quite short and informal, perhaps a simple group meeting involving a few PowerPoint slides. At the other end of the scale, your presentation may be a lot more complex and involve a dissertation running to many pages. Whatever the type of presentation, there are elements in common to them all. The structure for all presentations should include: an introduction, methods, results and some interpretation. Within that structure you have a lot of freedom; some presentations will focus more on interpretation and others will focus more on actual results.

The stages summarised here form the basis for the scientific method. You get an idea, work out how to tackle the problem, collect appropriate data to test your idea, and then you carry out some analysis and report on your results. Your original idea may be supported or not supported. Science thus moves on in this looping process, each statement is treated sceptically until unequivocal empirical evidence is produced. All good scientists, from the great and lofty to the humble student, use the same process to advance (so there is hope for us all).

Glossary

Alternative hypothesis	Often written as H_1. This is what you expect to find; however, it is not easy to prove mathematically so the concept that is usually tested is the null hypothesis (H_0).
Analysis of variance (ANOVA)	Analysis of variance. A statistical process that allows comparison of more than two samples. Data should be normally distributed.
Association	A form of linking together data. When data are in categories, the links between categories are called associations. See Chi-squared.
Average	A measure of the central tendency of a sample of numerical data. See Mean, Median, Mode.
Bar chart	A graphical method of displaying values for various categories. See also Histogram.
Box–whisker plot	A graphical method of displaying numerical sample data. The plot shows five quartile values.
Analysis ToolPak	An add-in for Excel that permits a range of statistical analyses to be conducted.
Bibliography	A list of sources that were used in the preparation of a document. They are not referred to explicitly in the text of the document (i.e. they are not cited), compare to References.
Braun-Blanquet	Josias Braun-Blanquet, a Swiss botanist who developed methods of examining plant communities. The Braun-Blanquet scale is a simplified abundance scale (see also Domin scale). There are several variants but the main scale runs thus: + = < 1%, 1 = 1–5%, 2 = 5–25%, 3 = 25–50%, 4 = 51–75%, 5 = >75%.
Chi-squared	A statistical method for examining the association between categorical factors, developed by Pearson.
Citation	A reference to a source of information within the text of a document. The citation itself is brief (name, date) and the full source is listed in the reference section.
Coefficient	Usually refers to multiple regression, each factor has a coefficient that is analogous to the slope in a simple straight-line relationship. See Multiple Regression.
Confidence interval	A measure of the variability of a numerical sample. Expressed as a value and a percentage (or proportion), the percentage of the data that lie within a certain distance from the mean, e.g. $CI_{0.95} = 1.5$ indicates that 95% of the data lie within 1.5 units of the mean for the sample.

Constancy	A term used in the NVC system. Five quadrats are used in the NVC system and constancy refers to how many of the five a species occurs in (so it is a measure of frequency). Usually written as a Roman numeral I–V.
Contingency table	A table of observed values for observations in various categories. Used in the Chi-squared test for association, e.g. categories could be habitat and invertebrate order.
Correlation	A link between two variables, e.g. stream speed and mayfly abundance. If one value (e.g. speed) increases and the other (e.g. abundance) decreases the correlation is negative. If both factors change in the same direction then the correlation is positive. See Spearman's rank and Pearson's product moment.
Critical value	A value regarded as the cut-off point for a statistical test. In some tests when this value is exceeded the result is regarded as significant. For other tests the result is regarded as significant if the calculated value is less than the critical value.
DAFOR scale	A relative abundance scale (therefore ordinal) D = dominant, A = abundant, F = frequent, O = occasional, R = rare. Can be applied to any organism and is defined by the user for convenience. Often the letters are converted to numerical values, e.g. D = 5, R = 1.
Degrees of freedom	Related to the sample size of data. Usually the degree of freedom is the sample size –1 but there are variations according to the statistical test being applied.
Dependent variable	In a correlation or regression this is the variable that is thought to be affected by others. See Response variable.
Density plot	A method of displaying the distribution of a numerical sample as a continuous line. An alternative approach to a histogram (see also Tally plot and Stem–leaf).
Dispersion	A term used in relation to the spread of data in a sample, e.g. standard deviation, range.
Domin scale	An abundance scale used for determining plant percentage cover. Named after a Czech botanist the scale goes from 0–10 like so: 1 = <4% with 1–3 individuals, 2 = <4% with 4–10 individuals, 3 = <4% with more than 10 individuals, 4 = 4–10%, 5 = 11–25%, 6 = 26–33%, 7 = 34–50%, 8 = 51–75%, 9 = 76–90%, 10 = 9–100%.
Error bars	A way of illustrating graphically a measure of the spread of the data. For example in a bar chart the bars may represent the mean values whilst error bars could represent standard deviation, standard error or confidence interval.
Goodness of fit	A type of statistical analysis that compares the observed frequencies in a number of categories to be compared to theoretical frequencies. Similar in approach to the Chi-squared test.
Histogram	A graph to show frequency distribution of a sample. Like a bar chart except that the bars should touch, i.e. they show a continuous range of values split into convenient categories (or bins).
Hypothesis	Something that you are trying to test. Usually this is a single thing that might be proven. See Null hypothesis and Alternative hypothesis.

Independent variable	In a correlation or regression this is the variable that controls the level of the dependent variable; however, we should beware of cause and effect. See Predictor variable.
Intercept	A straight-line relationship between two numerical variables can be represented by the equation $y = mx + c$. The intercept is where the straight line crosses the y-axis (c in the equation). See Multiple regression.
Kruskal–Wallis	A statistical approach that allows comparison of more than two samples when the data are not normally distributed.
Logistic regression	A way of examining relationships between numerical variables where the dependent factor is binary, i.e. is either 0 or 1 (presence or absence) or has two alternatives.
Mean	A measure of the central tendency of a numerical sample. Calculated as the sum of the values in the sample divided by the number of observations. Used for normally distributed data. See Parametric.
Median	A measure of the central tendency of a numerical sample. Calculated as the middle value when they are ranked in ascending numerical order. Used for non-normally distributed data.
Mode	A measure of the central tendency of a numerical sample. Calculated as the most frequent value. Usually only used for very large samples and rarely used in statistical testing.
Multiple regression	A statistical method that examines the link between a single dependent variable and several independent variables. The relationship between the dependent and independent variables is assumed to be a straight line of the form $y = m_1x_1 + m_2x_2 + c$.
Non-parametric	When a sample of data is normally distributed (i.e. is skewed) the data are described as non-parametric.
Null hypothesis	Often written as H_0. This is a hypothesis that you disprove mathematically in order to support your alternative hypothesis (H_1), which is what you expect to find. Think of it as the dull hypothesis, e.g. there is no difference; there is no correlation; there is no association.
NVC	National Vegetation Classification. This is a standardised method of surveying for plant communities.
p-value	A measure of the likelihood of a result happening by random chance. In statistics, a value of $p = 0.05$ is taken as significant, i.e. there is only 5% chance that the result could have occurred randomly.
Parametric	Normal distribution. Describes the frequency distribution of a sample where the data are symmetrical about the middle and form a bell-shaped curve.
Pearson's product moment	A statistical method for determining the relationship (correlation) between two factors that are normally distributed. The relationship is assumed to be a straight line in the form of $y = mx + c$.
Pie chart	A graphical method of displaying data as a proportion. Each slice of the pie represents the proportion towards the total. Pie charts can always be represented as a bar chart instead.
Pivot table	A summary table generated by Excel. Pivot tables are a useful way to rearrange and summarise data, especially when the data is in the form of biological records.

Post-hoc test	Literally "after this". A method of comparing samples pair by pair after a multi-sample test is applied. See Tukey HSD test.
Predictor variable	A predictor variable is one that has some effect on another variable (directly or indirectly). In a scatter plot this would be represented on the *x*-axis. Effectively we are saying that levels of this variable help us to predict levels in the response variable.
Quartile	A value half-way between two extremes. The main quartiles lie 1 and 3 quarters of the way along a series of values laid out in size order (i.e. ranked), so are half-way between the middle and one end.
R program	A statistical programming language used for many types of analysis. If is free and open source.
Range	The difference between the maximum and minimum values in a sample of numerical data. Equates to the 0th and 4th quartiles. See Box–whisker plot.
References	A list of sources used in the preparation of a document. All sources listed are cited in the text, usually as (name, date), compare to Bibliography.
Residual	A measure of how far away from the line of best-fit a datum is.
Response variable	This is the variable that is of interest to the researcher. It is the variable that is affected by changes in the predictor variable(s). In a scatter plot, this would be represented on the *y*-axis.
Running mean	A cumulative average (median might also be used). The mean value is calculated each time a new observation is made. This may be used to help determine when enough data has been collected.
Scatter plot	A graphical method of displaying the relationship between two numerical values. See Correlation.
Skewed	Usually refers to not normally distributed data (i.e. non-parametric), where the frequency distribution has its highest point skewed from the middle (in normal distribution the highest point would be in the middle of the distribution).
Slope	A straight-line relationship between two numerical variables can be represented by the equation $y = mx + c$. The slope is a measure of how steep the straight line is (m in the equation). See Multiple regression.
Spearman's rank	A statistical method for determining the relationship (correlation) between two factors. These factors do not need to be normally distributed and the relationship does not have to be linear (although it should not be a U or inverted U-shape).
Standard deviation	A measure of the spread of data from the middle (mean) in normally distributed samples.
Standard error	A measure of the spread of data from the middle in normally distributed data. It is calculated as the standard deviation divided by the square root of the sample size.
Stem and leaf	A type of frequency distribution graph. The values in a sample are rewritten so that they appear to form a tally plot. The advantage over the tally plot is that the original data values can be reconstructed from the graph.

t-test	Usually called the Student's *t*-test after the pen name of the original author. The *t*-test is a statistical method for determining the difference between two samples of normally distributed data.
Tally plot	A simple form of frequency distribution where simple tally marks are placed against categorical bins representing size class.
Tukey HSD test	A form of post-hoc test. Tukey's honest significant different test is carried out after analysis of variance to look at pairwise comparisons.
U-test	The *U*-test is a statistical test for examining the difference between two samples when the data are not normally distributed. It is often called the Mann–Whitney *U*-test although Wilcoxon if often attributed as the author.
Variance	A measure of the spread of data (see Dispersion). Variance is the standard deviation squared.
z-test	A version of the *t*-test for determining the statistical difference between two samples of normally distributed data. The *z*-test is essentially a *t*-test with large sample sizes (>25).

Index

Lightning Source UK Ltd.
Milton Keynes UK
UKOW012258100112

185102UK00002B/5/P

9 781907 807121